Gas Kinetics

M. F. R. Mulcahy

Commonwealth Scientific and Industrial
Research Organization Australia
Division of Mineral Chemistry

A HALSTED PRESS BOOK

John Wiley & Sons
New York

Published in the U.S.A. by Halsted Press,
a Division of John Wiley & Sons, Inc., New York

First published in Great Britain 1973

Library of Congress Cataloging in Publication Data

Mulcahy, M. F. R.
Gas Kinetics

"A Halsted Press book."
Includes bibliographical references.
1. Chemical reaction, Rate of. 2. Gases.
I. Title.
QD502.M84 533'.2 73–9382
ISBN 0–470–62361–6

Contents

Preface

Gas Kinetics is the name given to the body of facts, theories and methods of investigation relating to the kinetics of chemical reactions occurring in the gas-phase. This book aims to provide an account of the subject suitable for advanced students. At the same time, it aspires to offer to anyone interested a compact survey of the field and a selective guide to recent literature. Its general scope will be evident from the table of contents. The approach is primarily factual and experimental. As far as space allows, the kinetic phenomena are presented as the context for theoretical discussion rather than vice versa. Some stress is laid on the diversity and power of modern experimental methods. In Chapters 2, 3, and 4, the various techniques currently used to determine the rates of reactions and obtain insight into their mechanisms are described, and examples of their application are presented from recent investigations. In this way, I hope to keep the reader in sight of the laboratory; and to underline the fact, so easily obscured in a text-book, that the original 'data' which it is the business of theories to generalize and explain are not simply graphs and tables with indistinct antecedents but precious facts of Nature won, often enough, only by adventurous enterprise.

This is not to imply, of course, that understanding the facts is less important than discovering them. Basic theoretical concepts are presented in Chapter 1 and their discussion is resumed, in a somewhat more descriptive way, in Chapter 5. Thus, Chapter 1 provides the frame of reference for Chapters 2, 3 and 4, and these in turn contain, among other matters, much of the experimental background to Chapter 5.

The influence of heat- and mass-transfer on reaction rates is first considered in Chapter 2, and various circumstances in which either or both effects are important are noted in Chapters 4, 6 and 7. An appreciation of the properties of branching chains described in Chapter 6 is a necessary preliminary to the brief account of the kinetics of combustion in Chapter 7.

A few words should be said about the references to the literature of which a good many will be found at the end of each chapter. None of these is essential to a first reading. For the most part, their purpose is to direct the more specialized reader to the original or further evidence, or to a more detailed or more advanced treatment of the matter in hand. They have been chosen, as far as possible, with an eye to their clarity of presentation. There is no attempt to 'cover' the literature on any topic; nor should the reference cited be regarded as necessarily indicating the original discovery or treatment. Similar considerations apply to the selective bibliographies which are also appended to each chapter.

Finally, it is a pleasure to record a debt of gratitude to my colleagues in research and other members of the CSIRO Division of Mineral Chemistry who in ways too various to mention have facilitated the writing of the book; nor should I fail to acknowledge the contribution of my wife, Jeanne Mulcahy, and of my family, whose practical interest made it possible.

M. F. R. MULCAHY

1 Basic phenomena and theoretical background

Chemical kinetics is the study of the rate and mechanism of chemical change. Its function is first to determine the rates at which chemical reactions take place in different circumstances and then to arrive at an understanding of the established facts. The latter endeavour necessarily entails the study of reaction mechanisms. At the most macroscopic level this involves elucidating the intermediate steps by which many reactions proceed to their final products; and at the most microscopic level it includes considering the events that occur during individual molecular collisions.

Gas kinetics is concerned with reactions occurring in the gas phase. For a reaction to take place the molecules of the reactants must come into contact, and whatever status gas kinetics has as a separate discipline derives from the fact that the nature of molecular encounters is best understood when they occur in the gas phase. In addition, the energy states of the reacting species are unperturbed by solvent effects and other complicating influences. Thus, in seeking to understand rate phenomena, gas kinetics is able to draw upon a well-established corpus of fact and theory derived from molecular dynamics and spectroscopy. This is much less the case when a liquid or solid phase is involved. For these reasons, the study of gaseous reactions is more likely to be able to distinguish the factors that relate most specifically to the nature of chemical change. On the other hand, an important limitation needs to be recognized. By far the greatest number of compounds sufficiently volatile to be studied with reasonable convenience in the gas phase are covalent compounds; and hitherto gas kinetics has been predominantly—though not altogether entirely—concerned with reactions of covalent bonds. Nevertheless it will be found in the following pages that this lack of generality is well compensated by the depth of insight into the nature of chemical reactivity that has been achieved. And there is every reason to expect, at least on scientific grounds, that this will be broadened as well as deepened still further in the near future.

1–1 Effects of concentration and temperature on reaction rate. Terminology and units

With rare exceptions, the rate of a chemical reaction increases with increasing concentrations of the reactants and with increasing temperature. For a non-reversible reaction at constant temperature:

$$aA + bB + cC + \rightarrow qQ + rR + sS + \qquad \text{1–I}$$

we find, in general,

$$-\frac{d[A]}{dt} = -\frac{a}{b}\frac{d[B]}{dt} = \cdots = \frac{a}{q}\frac{d[Q]}{dt} = \frac{a}{r}\frac{d[R]}{dt} = \cdots = f([A],[B],\ldots) \qquad \text{1–1}$$

where the symbols [A], [B], etc., denote gram-molecular concentrations. Unless otherwise indicated, our units will be mole cm^{-3}. Frequently, though by no means always, the function $f([A], [B]\ldots)$ takes a simple form and a few such cases will be mentioned briefly. The rates of decomposition and isomerization reactions

$$A \rightarrow qQ + rR + \qquad \text{1–II}$$

$$A \rightarrow Q \qquad \text{1–III}$$

often conform to the relation:

$$-\frac{d[A]}{dt} = k[A] \qquad \text{1–2}$$

When this is the case, the reaction is said to be of the *first order*; k is the first-order *rate constant* for the reaction and has the units $seconds^{-1}$. Another such reaction, or the same reaction in different circumstances, may follow the rate law:

$$-\frac{d[A]}{dt} = k[A]^2 \qquad \text{1–3}$$

in which case it is said to be of the *second order*. The rate constant, which, of course, is not the same as that in equation 1–2, has the units cm^3 $mole^{-1}$ s^{-1}. Again, the rate of reaction

$$aA + bB \rightarrow qQ + rR \qquad \text{1–IV}$$

may be described by the relation

$$-\frac{d[A]}{dt} = \frac{a}{q}\frac{d[Q]}{dt} = k[A][B] \qquad \text{1–4}$$

This reaction is said to be first order in A, as also in B, and to have an 'overall order' of two. The units of k are again cm^3 $mole^{-1}$ s^{-1}. *Third-order* reactions are defined analogously; for example, if the rate law for reaction 1–IV were to be

$$-\frac{d[A]}{dt} = -\frac{a}{b}\frac{d[B]}{dt} = k[A][B]^2 \qquad \text{1–5}$$

the reaction would be first order in A, second order in B and would have an overall order of 3. Our units for k are cm^6 $mole^{-2}$ s^{-1}. For reactions of order other than 1 the units of k depend on those used for concentration. Concentration units in common use alternative to the above are mole litre^{-1} and molecule cm^{-3}. When we have occasion to use the last-named units, they will usually be denoted by $[n_i]$ for the species i.

The order of a reaction is not necessarily related to its stoichiometry. The stoichiometry must always be taken into account, however, when defining the rate constant. Thus k in equation 1–4 refers to the disappearance of A. Definition in terms of the appearance of Q would yield a constant q/a times larger. Neglect of this kind of ambiguity in statements of 'the rate constant of the reaction' is not uncommon.

Sometimes the concentrations of all reactants but one remain effectively constant during the reaction. This may be because they are replaced as soon as they are consumed or because their initial concentrations are much greater than the concentration of the reactant that changes. When the latter disappears according to a first-order law, its behaviour may be discussed conveniently in terms of a *pseudo first-order rate constant* k_ψ. For example, if B in reaction 1–IV is in large excess throughout the reaction, equation 1–4 or 1–5 can be replaced by

$$-\frac{d[A]}{dt} \simeq k_\psi[A] \qquad 1\text{–}6$$

Naturally k_ψ is constant only for the particular value of [B]; determination of its values at different constant concentrations of B would reveal the nature of its dependence on [B]; that is, the order of the reaction in B.

It is often convenient to discuss the kinetics of gas–phase reactions in terms of partial pressures, for example to replace equation 1–4 by

$$-\frac{dp_A}{dt} = k'_p p_A p_B \qquad 1\text{–}7$$

It is worth noting that, except for first-order reactions, the relation between a rate constant defined in this way and the corresponding one referring to concentrations depends on the temperature; thus in the case cited, $k = RTk'_p$ for ideal gases.

The influence of temperature on reaction rate is almost always profound. It is best related to the rate constant which, in general, varies with absolute temperature according to equation 1–8

$$k = A\exp(-C/T) \qquad 1\text{–}8$$

where A and C are independent of temperature or very nearly so. A is called the *pre-exponential* or *frequency factor* and has the same units as k. Equation 1–8, like equations 1–1 to 1–7, is purely an empirical statement of the facts. For theoretical reasons, it is customary to multiply C by the gas constant R. Thus we have the *Arrhenius equation*:

$$k = A\exp(-E_a/RT) \qquad 1\text{–}9$$

E_a is known as the *activation energy* or more precisely as the 'Arrhenius', 'experimental', 'apparent', or 'overall' activation energy according to circumstances. It has the dimensions of energy, mole^{-1}. Our units are kcal $mole^{-1}$, R having the value 1.987×10^{-3} kcal $mole^{-1}$ deg^{-1}. Other common units are cal $mole^{-1}$, and, among physicists, electron-volt $molecule^{-1}$. The SI unit is joule $mole^{-1}$.§ The value of E_a is usually determined graphically from an *Arrhenius plot*; that is, by plotting $\log_{10} k$ against $1/T$, the slope of the straight line so obtained being equal to $-E_a/2.303R$; A is found by substituting the value of E_a in equation 1–9 at one or more temperatures. A and E_a are often referred to as the *Arrhenius parameters*; their theoretical significance decreases from a maximum for a reaction known to occur in a single molecular event to almost zero for a reaction of undetermined mechanism.

Differentiation of equation 1–9 with respect to T yields a relation of some practical significance:

$$\frac{1}{k}\frac{dk}{dT} = \frac{E_a}{R} \cdot \frac{1}{T^2} \tag{1–10}$$

That is, the *relative* temperature coefficient of the rate constant of a given reaction decreases with the square of the mean temperature at which it is measured. A reaction which increases in rate by 20 per cent per degree rise in temperature at 800 K does so by only 1 per cent per degree at 1300 K (assuming, of course, that the reaction mechanism remains unchanged over the interval). This means that, all else being equal, kinetic studies require more precise control of temperature the lower the temperature at which they are conducted.

A reaction in which the chemical change is completed in a single molecular event is called an *elementary reaction*; for example

$$CH_3NC \rightarrow CH_3CN \tag{1–V}$$

Most reactions are more complex than this and proceed by way of a *reaction mechanism* consisting of a series of elementary reactions. In the gas phase, these usually involve free atoms or radicals. The decomposition of ozone, for example, occurs in two elementary steps:

$$O_3 + M \rightarrow O_2 + O + M \tag{1–VI}$$

$$O + O_3 \rightarrow 2O_2 \tag{1–VII}$$

(M represents any molecular species which happens to be present, including O_3; its role in this type of reaction will be explained later). When elementary reactions are involved which regenerate active free radicals as fast as they consume them, the overall reaction is a *chain reaction*. For example the synthesis of hydrogen bromide takes place by the following chain mechanism:

$$Br_2 + M \rightleftharpoons 2Br + M \tag{1–VIII}$$

$$Br + H_2 \rightarrow HBr + H \tag{1–IX}$$

$$H + Br_2 \rightarrow HBr + Br \tag{1–X}$$

§ 1 kcal = 4.184×10^3 joule; 1 electron-volt $molecule^{-1}$ = 23.062 kcal $mole^{-1}$.

The characteristics of chain reactions are considered in Chapter 3.

Elementary reactions are classified by their *molecularity*, that is according to the number of species taking part. Reaction 1–V, for example, is *unimolecular*; reactions 1–VII and 1–IX are *bimolecular*; and the reverse of reaction 1–VI or 1–VIII is *termolecular*. This is in accordance with current usage which equates molecularity with reaction order (for elementary reactions). The terminology is somewhat confused, however, by the fact that the order may be non-integral and may change with the conditions. Reaction 1–V, for example, becomes second order at low pressures and is then better symbolized as

$$CH_3NC+M \rightarrow CH_3CN+M \qquad \text{1–XI}$$

This situation is sometimes described as 'bimolecular' and sometimes as 'unimolecular in the second order region'. It is worth noting that the terms 'unimolecular', etc., are applicable only to elementary reactions, whereas 'first order', etc., apply generally.

1–2 Energetics of molecular collisions. Potential energy diagrams and surfaces

For a reaction to take place between two molecular species

$$A+B \rightarrow C+D$$

some at least of the constituent atoms of A and B must be brought into contact. Theoretical discussion therefore begins plausibly with the simple postulate that the transformation of each pair of A and B molecules to C and D molecules takes place in the course of a single molecular collision. Three stages of reaction can then be distinguished: the initial impact, the redistribution of atoms between the colliding molecules, and the separation of the products as independent species. It is, of course, implicit in the problem that the three stages must merge one into the other. Nevertheless the distinctions are useful. The first clue towards assessing the influence of each stage on the reaction rate comes from the fact that the frequency of molecular collisions deduced by kinetic theory is almost always several orders of magnitude greater than the reaction rate. Therefore collisions, though necessary for reaction, are not sufficient. (In conformity with this, the collision frequency normally is much less dependent on temperature than the reaction rate.) The separation of the product molecules following a reactive collision can best be regarded as analogous to a collision between C and D in reverse and hence as occurring after the new molecules are essentially formed. This focuses attention on the influence of the 'chemical' events of the second stage, which are rate-determining. The question then arises as to what distinguishes a reactive from a non-reactive collision, and to illustrate this we need to review some basic facts relating to the energetics of molecular collisions and intramolecular atomic displacements.

At this point it is instructive to consider a very simple but concrete example, namely the exchange reaction between hydrogen atoms and hydrogen molecules

$$H + H_2 \rightarrow H_2 + H \qquad 1\text{–XII}$$

This reaction is the archetype of a particular kind of reaction, namely bimolecular atomic exchange, but it has sufficient generality for our present purpose. The kinetics have been investigated experimentally by determining the rate at which para-hydrogen molecules are converted to the ortho form as the result of the exchange.[1] In an unreactive collision, the mutual potential energy between the H atom and the H_2 molecule varies with the distance, r, between their centres according to the curve

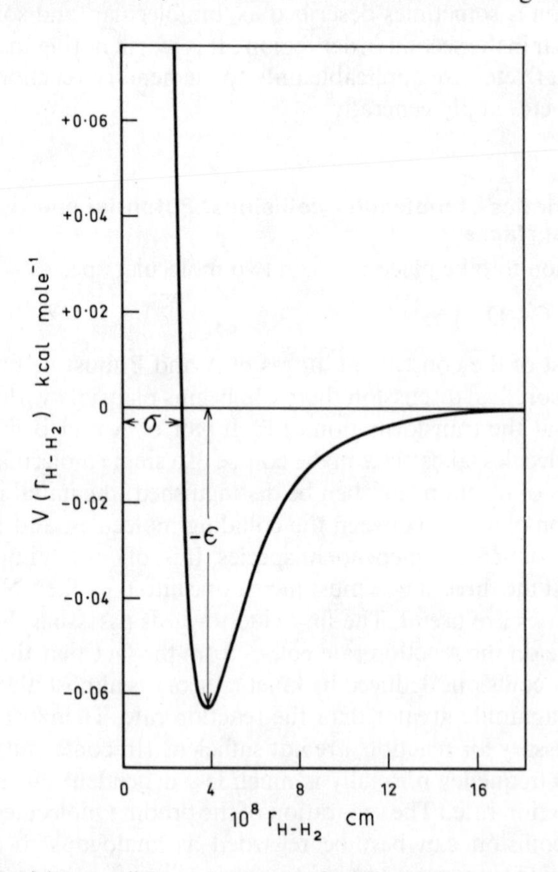

Fig. 1–1 Potential energy diagram showing the mutual potential energy of a hydrogen atom and a hydrogen molecule in an unreactive collision as a function of the distance between centres r_{H-H_2}.

shown in Fig. 1–1. This type of curve is characteristic of collisions between non-polar molecules and, in general, its parameters can be calculated by standard procedures from the viscosity or compressibility of the gas.[2] The potential energy is approximated sufficiently well by the Lennard–Jones function:

$$V(r) = 4\varepsilon\left[\left(\frac{\sigma}{r}\right)^{12} - \left(\frac{\sigma}{r}\right)^{6}\right] \qquad 1\text{–}11$$

where ε, σ, and the energy zero are defined as in Fig. 1–1. Since, in the absence of any extraneous influence, the sum of the potential and kinetic energies remains constant, particles initially at rest at infinite separation will approach to the distance σ and then recede (elastically) to their original positions. If either particle initially possesses a component of velocity along the line of centres, the minimum distance of separation will be greater or less than σ depending on the direction of the velocity component. The steepness of the curve at short distances shows, however, that a relatively large amount of kinetic energy is required to bring the particles significantly closer than σ. For this reason σ can be regarded as the *collision diameter*. In the general case its value is not very different from that calculated from the viscosity by elementary kinetic theory. The shortest distance effected by an ordinary collision, that is, one with an initial kinetic energy of about $\bar{k}T$, can be estimated from equation 1–11. In the present instance at 300 K (at which temperature the reaction rate is appreciable), it is about 0.9σ. This permits the hydrogen atom to approach no closer than about 2.3 Å from either atom of the H_2 molecule: a distance which is still large compared with that achieved by reaction (0.75 Å). This accounts for the fact that most collisions are unreactive.

The energy required to bring about the relative motion of atoms at *intra*molecular distances which necessarily occurs in the course of reaction must be of a similar magnitude to that required to cause similar displacements in stable molecules. For the two atoms in the H_2 molecule, the latter energy is given by the familiar potential energy curve illustrated in Fig. 1–2 (where the energy zero is taken at the minimum). This curve is derived from experimental spectroscopy, but the potential energy of the system of three atoms at comparable mutual distances must be calculated theoretically. The various methods, which to a greater or less degree draw upon knowledge of the properties of the reactants such as that contained in Fig. 1–2, are considered in Chapter 5. Reaction is conceived as a continuous progress between the configurations shown below:

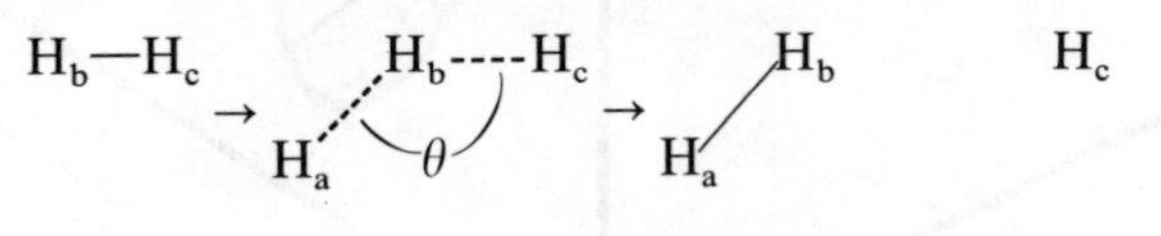

and the potential energy is calculated as a function of the three independent variables: the distances r_{ab} between H_a and H_b and r_{bc} between H_b and H_c, and the angle of approach θ. The result is a *potential energy hypersurface* which is the four-dimensional equivalent of Fig. 1–2. If one variable, for example θ, is kept constant, the variation of the energy with respect to

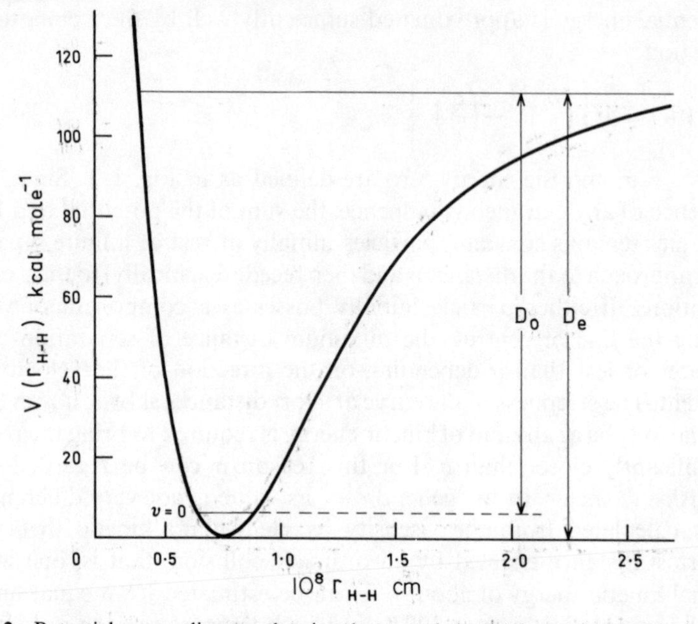

Fig. 1–2 Potential energy diagram showing the mutual potential energy of two hydrogen atoms in a hydrogen molecule as a function of the internuclear distance r_{H-H}.

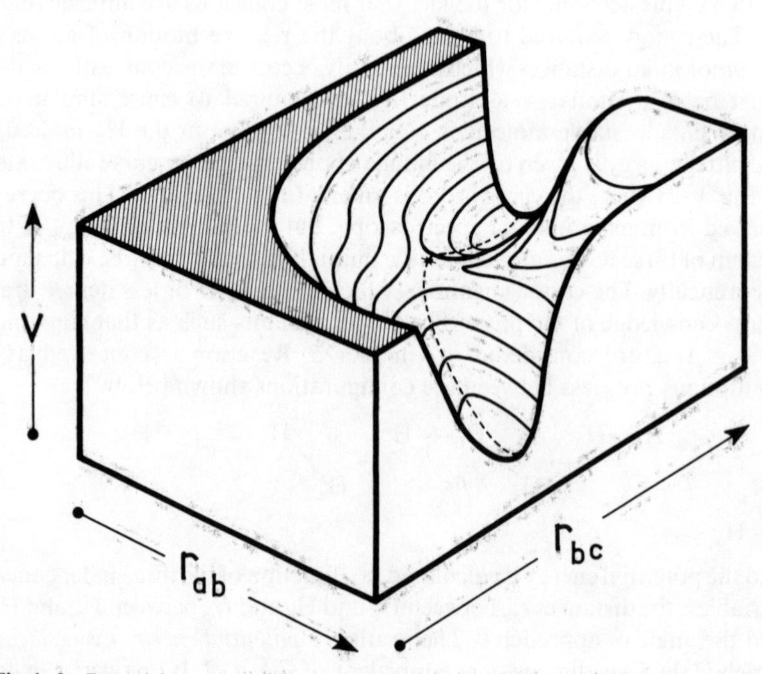

Fig. 1–3 Potential energy surface showing the mutual potential energy of three atoms a, b, c as a function of the internuclear distances r_{ab} and r_{bc} at constant angle abc (schematic).

the other two can be represented graphically by a three-dimensional *potential energy surface* such as is illustrated diagramatically in Fig. 1–3. When, either before or after reaction, the H atom is at infinite distance from the molecule, only one independent variable remains (r_{ab} or r_{bc}) and the hypersurface reduces to Fig. 1–2. Between these extremes, the calculations show primarily that, as H_a approaches H_b and H_c recedes, the potential energy passes through a maximum. They also show that the height of the maximum is least when the approach of H_a to H_b—H_c is collinear ($\theta = 180°$). This means that reaction is most likely to occur in this configuration; and the most relevant potential energy surface is that which represents the energy as a function of r_{ab} and r_{bc} with constant θ at 180°. The dotted line in Fig. 1–3 shows the most probable *reaction path* or *trajectory* up the sloping floor of the 'valley' with the 'walls' formed by the r_{bc}–energy relation to a 'saddle-point' maximum at * followed by descent into the valley formed by the r_{ab}–energy relation.

The alternative method of representing a potential energy surface is by a contour map. Figure 1–4 is an example showing the saddle-point region of the present reaction. The height of the maximum is about 9 kcal mole^{-1}. Leaving aside the relatively minor effect of quantum-mechanical tunnelling, this means that reaction can occur only if the colliding particles together possess initially not less than the equivalent

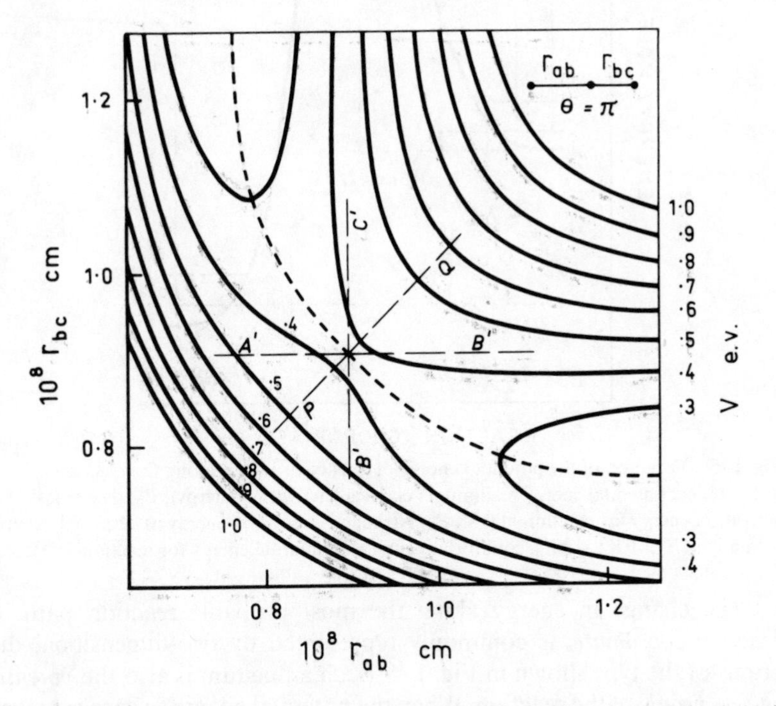

Fig. 1–4 Potential energy contour map of the 'saddle-point' region showing the potential energy of three hydrogen atoms H_a, H_b, H_c as a function of the internuclear distances r_{ab}, r_{bc} at constant angle abc = 180° (based on calculations by Porter and Karplus[(3)]).

of 9 kcal kinetic energy. If they do not, the system fails to reach the saddle point and returns to its original state.

Calculation of potential energy surfaces has seldom been attempted for systems of more than four atoms or groups which can be treated as single atoms (CH_3 for example). Nevertheless, on theoretical grounds it is safe to generalize the result quoted for the $H—H_2$ system: for reaction to occur, the molecules must be supplied with at least a critical energy. This normally is much greater than the kinetic energy brought into the great majority of collisions, but it is less than the energy required to break the appropriate chemical bond(s) in isolation, unless, of course, bond rupture itself happens to be the reaction under consideration e.g. $H_2 \rightarrow H+H$. Anticipating the discussion in the next section, we shall refer to the critical energy as the activation energy (though we shall see later that this identification is not exact).

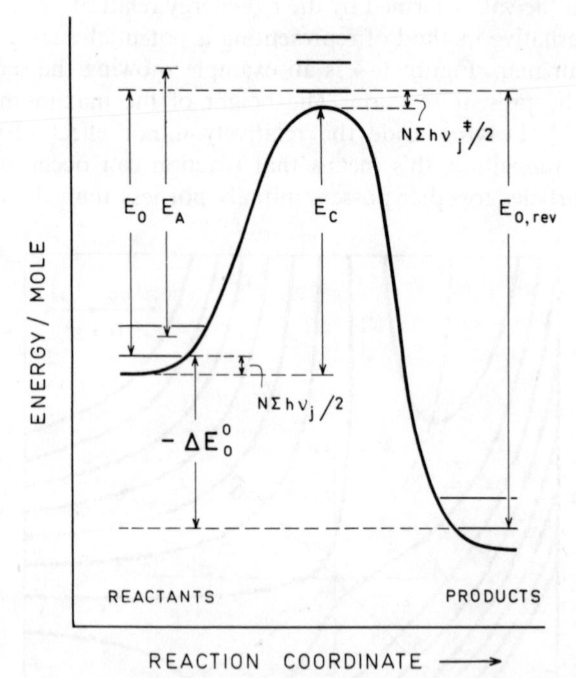

Fig. 1–5 Variation of the potential energy of a reactive system along the reaction coordinate. E_c, potential energy maximum ('classical' activation energy); E_0, theoretical activation energy for reaction at 0 K; E_A, Arrhenius activation energy as observed experimentally at $T > 0$ K; $-\Delta E_0^0$, equilibrium change in internal energy for reaction at 0 K.

The change in energy along the most probable reaction path, or *reaction coordinate*, is commonly represented by two-dimensional diagrams of the type shown in Fig. 1–5. Such a diagram is also the *potential energy profile* of the reaction. When the potential energy surface is known, the abscissa can be given a precise meaning, but often it is merely used to represent, in a general way, the stage reached by simultaneous changes in interatomic distances and angles accompanying reaction which are more or less imperfectly known. Figure 1–5, to which we shall return later, refers to an exothermic reaction or to an endothermic reaction if the direction of the reaction coordinate is reversed. It will be appreciated that the difference between the activation energies for the same reaction proceeding in opposite directions is equal to the heat of reaction (or internal energy change) and that the activation energy for an endothermic reaction can never be less than the heat of reaction.

To incorporate the existence of a critical energy into a theory of reaction rate we may proceed either by way of the kinetic theory of gases or by statistical thermodynamics. The first way is called *collision theory* and the second *activated-complex* or *transition state*, or (vaingloriously) *absolute rate theory*. The two approaches are complementary, each having its particular insights and obscurities.

1–3 Elementary collision theory of reaction rate

1–3–1 Bimolecular reactions

In its simplest and still very prevalent form, collision theory postulates that the critical energy required for reaction is provided by the kinetic energy brought to molecular collisions by the relative motion of the colliding partners. The total number of collisions ω_{AB} between molecules of two kinds, A and B, per unit time per unit volume is derived from Maxwell's distribution law of molecular velocities in textbooks of kinetic theory. If the molecules are regarded as 'hard spheres' (elastic billiard balls with no mutual attraction) having diameters σ_A and σ_B and masses m_A and m_B, the formula is

$$\omega_{AB} = [n_A][n_B]\sigma_{AB}^2\left(\frac{8\pi \bar{k}T}{m_{AB}^*}\right)^{1/2} = [n_A][n_B]z_{AB} \qquad 1\text{–}12$$

where $\sigma_{AB} = \frac{1}{2}(\sigma_A+\sigma_B)$, $m_{AB}^* = m_A m_B/(m_A+m_B)$, and z_{AB} is the *collision frequency* for unit molecular concentrations. If A is identical with B, the right-hand side must be divided by 2. It is assumed that only those collisions can lead to reaction in which the kinetic energy of relative motion along the line of centres is greater than a critical energy ε_c which, for the moment, is identified with the height of the saddle point on the potential energy surface. The fraction of such collisions, that is, the chance that the colliding partners have ε_c distributed in this way between them, is $\exp(-\varepsilon_c/\bar{k}T)$. Although ε_c is necessary for reaction, it may not be sufficient: the molecules may need to collide in the correct place or with a particular mutual orientation. A dimensionless *steric factor* P, necessarily less than unity, is therefore added to complete the relation:

$$-\frac{d[n_A]}{dt} = \{Pz_{AB}\exp(-\varepsilon_c/\bar{k}T)\}[n_A][n_B] \qquad 1\text{–}13$$

Expressed in molar units equation 1–13 becomes

$$-\frac{d[\mathrm{A}]}{dt} = \{PZ_{\mathrm{AB}} \exp(-E_c/RT)\}[\mathrm{A}][\mathrm{B}] \qquad 1\text{–}14$$

where $Z_{\mathrm{AB}} = Nz_{\mathrm{AB}}$, N being Avogadro's number. Z_{AB} is often also rather loosely alluded to as the 'collision frequency'. Its value is not very sensitive to the nature of A and B and, for the size of molecules normally encountered in the gas phase, is in the region of $(3 \pm 2) \times 10^{14}$ cm^3 mole^{-1} s^{-1} between 300 and 1000 K. The influence of van der Waals' attraction on the collision frequency can be allowed for[4] on the basis of equation 1–11 by multiplying Z_{AB} by the factor $2.7\,(\varepsilon/\bar{k}T)^{1/3}$. Typically this increases Z_{AB} by a factor of 2 to 3; but to date the correction has not often been applied.

Equation 1–14 reproduces the experimental rate law for a second-order reaction (equation 1–4) if the term in the curly brackets is identified with the rate constant:

$$k = PZ_{\mathrm{AB}} \exp(-E_c/RT) \qquad 1\text{–}15$$

The temperature dependence of k is of the correct form (equation 1–9) except for the square root dependence of Z_{AB} on T. But the latter is almost always small compared with the influence of the exponential term, and the curvature it would contribute to a plot of $\log k$ against $1/T$ is seldom if ever outside the experimental error. Thus E_c is roughly identified with the activation energy E_a. The value of A derived from an Arrhenius plot of the experimental data should therefore be of the order of 10^{14} cm^3 mole^{-1} s^{-1} or less. This, in fact, is generally the case for genuine bimolecular reactions; values of about 10^{14} corresponding to $P \approx 1$ are not uncommon though it is more usual to find P between 1 and about 10^{-4}. Table 1–1 gives the Arrhenius parameters for some bimolecular reactions. Cases where P is much greater than 1 are rare (for reactions between uncharged species). However, this does not include the class of second-order unimolecular reactions considered in the next section for which P undoubtedly can be greater than unity and for which a special explanation is available.

In summary, collision theory provides an acceptable though imprecise description of bimolecular reactions for which, in a general way, it predicts the observed kinetic behaviour. Nothing so far has been said, however, about elementary reactions of the type

$$\mathrm{A} \rightarrow \text{Products} \qquad 1\text{–XIII}$$

which more often than not are found to be first order, not second order as the above considerations would predict:

$$\mathrm{A} + \mathrm{A} \rightarrow \text{Products} + \mathrm{A} \qquad 1\text{–XIV}$$

Furthermore the existence of termolecular reactions has yet to be explained. In fact, these two classes of reaction are closely related, but we shall consider the former, that is, unimolecular reactions first.

Table 1–1 Arrhenius parameters for some bimolecular elementary reactions

Reaction	$\log_{10} A$ ($cm^3\ mole^{-1}\ s^{-1}$)	E_a ($kcal\ mole^{-1}$)
$H + H_2 \rightarrow H_2 + H$	14.7	8.7
$H + O_2 \rightarrow OH + O$	14.2	16.2
$N + O_2 \rightarrow NO + O$	13.2	7.9
$O + N_2 \rightarrow NO + O$	14.1	75.4
$H + C_2H_6 \rightarrow H_2 + C_2H_5$	14.1	9.7
$F + CCl_4 \rightarrow FCl + CCl_3$	13.0	10.2
$Cl + CH_4 \rightarrow HCl + CH_3$	12.8	3.8
$I + C_2H_5I \rightarrow I_2 + C_2H_5$	14.0	17.1
$CH_3 + CH_4 \rightarrow CH_4 + CH_3$	11.8	14.7
$CH_3 + NH_3 \rightarrow CH_4 + NH_2$	10.8	9.8
$CF_3 + CH_4 \rightarrow CF_3H + CH_3$	12.0	11.3
$C_6H_5 + H_2 \rightarrow C_6H_6 + H$	10.7	6.5
$NO + O_3 \rightarrow NO_2 + O_2$	11.8	2.4
$SO + O_2 \rightarrow SO_2 + O$	11.5	6.5
$CH_3 + CH_3 \rightarrow C_2H_6$	13.4	~0
$CH_3 + SO_2 \rightarrow CH_3SO_2$	10.8	1.5
$C_2H_5 + C_2H_2 \rightarrow C_4H_7$	11.0	7.0
$HI + C_3H_6 \rightarrow C_3H_7I$	10.9	23.4

1–3–2 Unimolecular reactions

A bimolecular reaction involving an exchange of atoms must occur during a collision or not at all. But a molecule undergoing unimolecular reaction, that is, a reaction within itself, may rebound from an energetic collision containing the necessary activation energy but yet take a short time to react. This time-lag, during which the energy received by the molecule in a haphazard way becomes concentrated in the appropriate bond or locus of reaction, is responsible for the first-order kinetics. We shall see shortly that this follows from the fact that the molecule may not have reacted before it undergoes another collision. Since it already has energy well above average, such a collision will very probably deprive it of much of its excess energy and so render it incapable of reaction. Applying this concept to reaction 1–XIII and denoting the critically energized molecule by A*, we arrive at what is known as the Lindemann–Hinshelwood mechanism:

$$A + A \rightarrow A^* + A \qquad \text{(a)}$$

$$A^* + A \rightarrow A + A \qquad \text{(b)}$$

$$A^* \rightarrow \text{Products} \qquad \text{(c)}$$

If reaction (c) did not occur, 'reaction' (a) would be exactly balanced by 'reaction' (b) and the concentration of A* would be stationary at the value corresponding to the Boltzmann equilibrium distribution of energies. That is,

$$[A^*]_{eq} = (k_a/k_b)[A] = K[A] \qquad 1\text{–}16$$

An analogous situation is assumed to hold in the presence of reaction (c), the stationary concentration of A* now being determined by the balance between all three reactions:

$$\frac{d[A^*]}{dt} = 0 = k_a[A]^2 - (k_b[A] + k_c)[A^*] \tag{1–17}$$

whence

$$[A^*]_{ss} = k_a[A]^2/(k_b[A] + k_c) \tag{1–18}$$

But

$$k_b = \lambda_{AA*}Z_{AA*} \approx \lambda_{AA*}Z_{AA} \tag{1–19}$$

where λ_{AA*} is the (high) probability that A* is de-energized on colliding with A. Hence, the reaction rate is given by

$$-\frac{d[A]}{dt} = d\frac{[\text{Products}]}{dt} = k_c[A^*]_{ss} = k_c K[A]\left\{\frac{[A]}{[A] + k_c/(\lambda_{AA*}Z_{AA})}\right\} \tag{1–20}$$

This equation admits of two limiting cases. If $\lambda_{AA*}Z_{AA}$ is sufficiently large compared with k_c, it becomes, in the limit,

$$-\left(\frac{d[A]}{dt}\right)_\infty = k_c K[A] \tag{1–21}$$

Thus we have an explanation of the first-order rate law. It is implicit in the theory, however, that the kinetics will remain first order only so long as $[A] \gg k_c/(\lambda_{AA*}/Z_{AA})$. If a *unimolecular rate constant* for the reaction is defined in terms of experimental quantities:

$$k_U = -\frac{1}{[A]}\frac{d[A]}{dt} \tag{1–22}$$

it follows from equations 1–20 and 1–21 that

$$k_U = k_c K\left\{\frac{[A]}{[A] + k_c/(\lambda_{AA*}Z_{AA})}\right\} = (k_U)_\infty\left\{\frac{[A]}{[A] + k_c/(\lambda_{AA*}Z_{AA})}\right\} \tag{1–23}$$

$(k_U)_\infty$ being the value of k_U at infinite [A]. Since the second term in the denominator is independent of [A], the observed value of k_U should decrease with decreasing [A], that is with decreasing pressure (p_A). In the limit, at zero [A] or p_A, k_U becomes directly proportional to [A] or p_A:

$$(k_U)_0 = K\lambda_{AA*}Z_{AA}[A] = k_a[A] \tag{1–24}$$

or

$$-\left(\frac{d[A]}{dt}\right)_0 = (k_U)_0[A] = k_a[A]^2 \tag{1–25}$$

Hence the reaction should become second order at low pressures. These effects will be discussed in a little more detail in Chapter 5. Here we simply note that they are reproduced by experiment. Figure 1–6 shows experimental values of $k_U/(k_U)_\infty$ for two unimolecular reactions graphed against

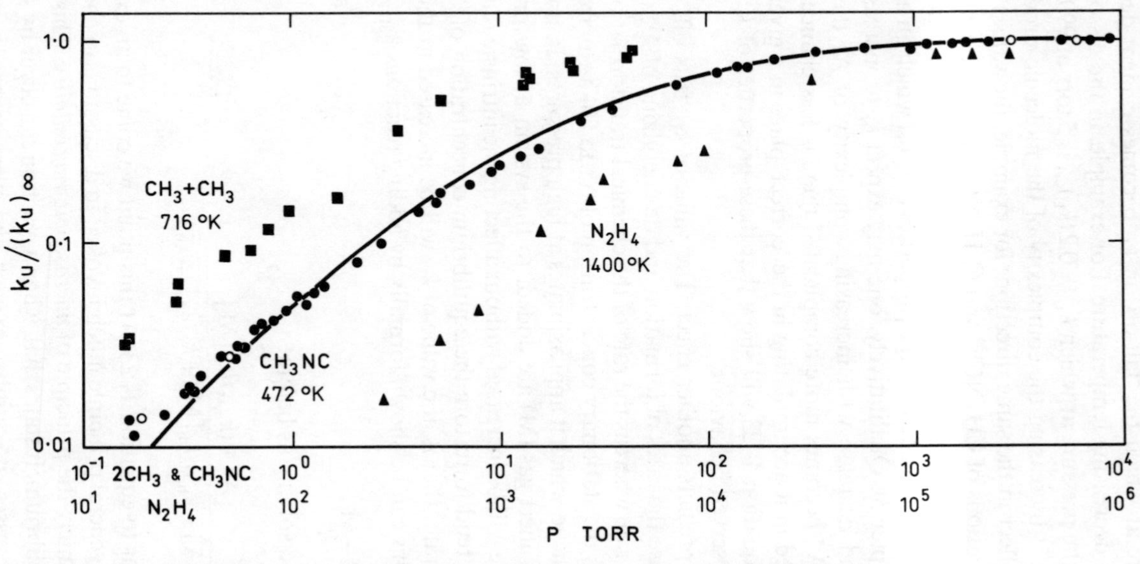

Fig. 1–6 Dependence of the unimolecular rate constants on pressure for the isomerization of CH_3NC[5] and the decomposition of N_2H_4[6]; and dependence of the bimolecular rate constant on pressure for the recombination of CH_3 radicals.[7] The line drawn through the experimental points for CH_3NC was calculated by Schneider and Rabinovitch[5] using RRKM theory. (See Chapter 5, Section 5–2.)

pressure on logarithmic scales. The values of $(k_U)_\infty$ are obtained by extrapolation as discussed in Chapter 5. It is evident, that at 472 K the isomerization of methyl isocyanide is effectively first order above 10^3 torr and, as shown by the 45° slope of the graph, second order below 0.5 torr.

The pressure at which the 'fall-off' in k_U becomes evident decreases somewhat with decreasing temperature. For example, in the dissociation of ethane $p_{0.2}$, the pressure at which $k_U = 0.2\,(k_U)_\infty$, is 5 torr at 999 K and 3 torr at 913 K.[8] Increasing the complexity of the molecule produces a more striking effect in the same direction; for example, the values of $p_{0.2}$ for the isomerizations of CH_3NC[5] and C_2H_5NC[9] at 533 K are 12 and 0.1 torr respectively.§ Equation 1–23 refers these effects to the ratio $k_c/\lambda Z$ or since λZ is roughly constant, to the properties of k_c (of which an account is given in Chapter 5). Qualitatively, one may expect k_c to increase with temperature and decrease with increasing complexity of A; the latter because when A* becomes more complicated there is less chance of the energy contained in it accumulating in the 'correct' place in a given time. Inspection of equation 1–23 will show that these properties of k_c are in line with the observed behaviour.

The theory predicts another effect. The value of k_U 'falls off' at low pressures because the rates of formation and de-energizing of A* depend on the collision rate whereas the rate of the chemical transformation does not. This causes the stationary concentration of A* to fall away from the equilibrium value to which it approximates at high pressures. If, however, molecules of an inert gas (M) are added, to the system at low pressure, the collision rate will be increased without affecting the intrinsic reactivity of A*. This will tend to restore the equilibrium concentration of A* and so increase the rate; that is, the value of k_U will be increased at the same value of [A]. This can be shown formally by adding reactions a_M, b_M,

$$M + A \rightleftharpoons A^* + M \qquad (a_M,\ b_M)$$

to the above and solving as before:

$$k_U = k_c K \left\{ \frac{[A] + (k_{a,M}/k_a)[M]}{[A] + k_{b,M}/k_b[M] + k_c/\lambda_{AA*} Z_{AA}} \right\} \qquad 1\text{–}26$$

This is analogous to equation 1–23. At this point we come to an example of an important general principle much invoked in discussions of chemical mechanisms, namely the principle of *microscopic reversibility*. This refers to chemical equilibrium. It states that, when a reaction can occur by several independent processes, at equilibrium each and every process is in equilibrium with its contrary.[10] In the present circumstances, in the absence of reaction (c), the equilibrium number of energized molecules relative to the total is completely determined by the Boltzmann distribution law. Hence, for ideal gases, it is independent of the presence or otherwise of

§See also Table 1–3, p. 31.

a foreign gas. Otherwise expressed, we have

$$\frac{[A^*]_{eq}}{[A]} = \frac{k_a\{[A]+(k_{a,M}/k_a)[M]\}}{k_b\{[A]+(k_{b,M}/k_b)[M]\}} = \frac{k_a}{k_b} \qquad 1\text{–}27$$

whence it follows that

$$k_{a,M}/k_a = k_{b,M}/k_b \quad \text{and} \quad k_a/k_b = k_{a,M}/k_{b,M} = K \qquad 1\text{–}28$$

in accordance with the principle just enunciated. Since the individual values of k_a, etc., do not depend on the extent of reaction, these relations, though derived from the special case of equilibrium, are valid generally. Applying this result to equation 1–26, we see that addition of sufficient M will increase the value of k_U up to but not beyond the same limit $(k_U)_\infty$ achieved by increasing [A]. This again agrees with experiment.

Different molecules are not equally efficient, however, in exchanging energy on collision. Indeed, a study of the increments in reaction rate brought about by the addition of different inert molecules enables their relative efficiencies to be measured. On the basis of the foregoing relations it is easily shown that, if the value of k_U at some low pressure of pure A is increased to the same extent on the one hand by increasing p_A to $(p_A+\Delta p_A)$ or, on the other, by adding a partial pressure of inert gas p_M, then

$$\frac{\Delta p_A}{p_M} = \frac{k_{b,M}}{k_b} = \frac{\lambda_{MA*}Z_{MA*}}{\lambda_{AA*}Z_{AA*}} \qquad 1\text{–}29$$

If, as previously, Z_{AA*}, Z_{MA*} are taken as equal to Z_{AA}, Z_{MA}, which can be calculated from transport properties, the relative collision efficiencies of A and M for de-energizing A* (or energizing A) can be determined. Alternatively and preferably, if the second-order region is accessible, the relative efficiencies can be obtained from the limiting values of $(k_U)_0/(p_A+p_M)$ in the presence and absence of M (see equation 1–25 and equation 1–32 below). Table 1–2 lists typical values of $\lambda_{MA*}/\lambda_{AA*}$. They vary by less than a factor of 20 from the simplest to fairly complicated M molecules and, in general, increase with the boiling point of M. The latter correlation suggests that van der Waals' attraction influences the rate of energy exchange.[11,12] Eventually the effect must 'saturate' when $\lambda_{MA*} = 1$. In the isomerization of CH_3NC induced by collisions with *n*-hydrocarbons, this happens when the carbon chain is longer than four or five atoms.[15] The reactant molecule itself is usually found to be abnormally efficient (as is exemplified in Table 1–2) and λ_{AA*} is generally assumed to be unity.

Equation 1–25 shows that at the second-order limit the rate of reaction is equal to the rate of collisional energization which, in this case, is synonymous with the rate of activation. It might therefore be expected that the rate constant k_a would be given by the bimolecular equation 1–15 and the value of A in the Arrhenius expression for k_a would have an upper limit of about 10^{14} cm^3 mole^{-1} s^{-1}. In fact, considerably greater values of A are commonly observed. It is important to note that equation 1–15 is

Table 1–2 Relative collision efficiencies for transfer of energy in unimolecular reactions ($\lambda_{MA^*}/\lambda_{AA^*}$) A = reactant molecule; M = inert molecule

M	b.p. (degrees K)	*Isomerizations*		*Decompositions*	
		CH_3NC(13)§	*cyclo*-C_3H_6(14)‖	C_2H_6(12)‖	NO_2Cl(11)§
He	4	0.14	0.05	0.17	0.15
H_2	20	0.15	0.12	—	0.15
Ne	27	0.16	—	0.18	0.22
N_2	77	0.24	0.07	0.25	0.34
Ar	87	0.17	0.07	0.26	0.30
Kr	121	0.18	—	0.31	0.36
C_2H_6	190	0.48	—	*1.0*	—
CO_2	195	0.60	—	0.33	0.49
SF_6	210	—	—	—	0.49
cyclo-C_3H_6	240	—	*1.0*	—	—
NO_2Cl	258	—	—	—	*1.0*
H_2O	273	—	0.74	0.88	—
CH_3NC	332	*1.0*	—	—	—
Reaction temp. degrees K		554	763	873	476

§ Determined at second-order limit.
‖ Determined in 'fall-off' region.

based on the particular assumption that reaction becomes possible when the *relative kinetic energy* of the colliding particles along their line of centres is greater than E_c. This corresponds to an accumulation of the energy in two translational degrees of freedom, one for each molecule. Formulae of similar form would be obtained if two degrees of freedom of other kinds were chosen; for example, two molecular rotations or a single vibration. The plausibility of ascribing the critical energy E_c to the energy of head-on collision rests on the all-or-nothing character of the collisions in a reaction involving an exchange of atoms. In the present case, however, the energy remaining in the molecule after collision plays the essential role. The condition for reaction, therefore, is that the molecule should contain not less than E_c in its *internal* degrees of freedom. The probability that it does so increases with the number of degrees of freedom available. If the system behaves classically, the probability at equilibrium is given to a good approximation by

$$\frac{[A^*]_{eq}}{[A]} = \frac{1}{(s-1)!}\left(\frac{E_c}{RT}\right)^{s-1} \exp(-E_c/RT) \qquad 1\text{–}30^{(16)}$$

where $2s$ is the number of specified internal degrees of freedom and E_c must be $\gg sRT$, as is usually the case. It is reasonable not to consider rotational energy of the molecule as a whole as contributing to internal configurational changes; in which case, equation 1–30 refers to an assembly of s active classical oscillators. For a molecule of N atoms, s is not greater than $3N-6$ or, for a linear molecule, $3N-5$. Combining equations 1–16, 1–24, and 1–30 leads to an expression for k_a:

$$k_a = \frac{\lambda_{AA*} Z_{AA}}{(s-1)!} \left(\frac{E_c}{RT} \right)^{s-1} \exp(-E_c/RT) \tag{1–31}$$

This equation cannot have much quantitative significance because classical behaviour is assumed for the molecular vibrations. Nevertheless it successfully predicts that the value of A should have a *lower* limit at about 10^{14} cm^3 mole^{-1} s^{-1} and should increase with increasing values of s; that is, with increasing complexity of the reactant molecule. Some experimental values of A are $10^{15.1}$ and $10^{15.7}$ for the decompositions of O_3 and N_2H_4 respectively, and 10^{19} for the isomerization of CH_3NC.[17]

When the reactant is highly diluted with inert gas the limiting low-pressure rate law becomes

$$-\frac{d[\mathrm{A}]}{dt} = k_{a,M}[\mathrm{A}][\mathrm{M}] \tag{1–32}$$

as is exemplified in Fig. 1–6 by the plot for the decomposition of N_2H_4 in an excess of argon. Experimental determination of the ratio $k_{a,M}/k_a$ provides a direct measure of the relative activating–deactivating efficiencies of A and M. This fact is independent of the validity of equation 1–31, but compatible with it since all terms except λ and Z are independent of the nature of the activating molecule.

1–3–3 Termolecular reactions

The great majority of this small but important class of reactions are exothermic reactions of the type

$$\mathrm{A+B+M \rightarrow AB+M} \tag{1–XV}$$

Either A or B or both are free atoms or radicals and M may be identical with A or B. The reaction rate is commonly found to decrease gently with increasing temperature. Examples are

$$\mathrm{I+I+M \rightarrow I_2+M} \tag{1–XVI}$$

$$\mathrm{O+O_2+M \rightarrow O_3+M} \tag{1–XVII}$$

$$\mathrm{OH+OH+M \rightarrow HOOH+M} \tag{1–XVIII}$$

Such reactions are bimolecular decompositions in reverse. The *third body* M must be present at some stage when the reactive species are in contact in order to remove part of the heat of reaction from the collision complex as kinetic energy; otherwise the complex would fall apart again. Nevertheless, even in the absence of M, the complex has a finite lifetime and this enables the kinetics to be explained by a variation of the Lindemann–Hinshelwood mechanism:

$$\mathrm{A+B \underset{c}{\overset{a}{\rightleftharpoons}} AB^*} \tag{a, c}$$

$$\mathrm{AB^*+M \rightarrow AB+M} \tag{b}$$

Reactions (b) and (c) are the alternative fates of the highly energized complex AB*, the symbols being chosen to conform with the unimolecular mechanism discussed in Section 1–3–2. Analogous treatment yields the relations

$$\frac{d[\mathrm{AB}]}{dt} = k_{\mathrm{b}}[\mathrm{AB}^*]_{\mathrm{ss}}[\mathrm{M}] = \left\{\frac{k_{\mathrm{a}}k_{\mathrm{b}}}{k_{\mathrm{b}}[\mathrm{M}]+k_{\mathrm{c}}}\right\}[\mathrm{A}][\mathrm{B}][\mathrm{M}] \qquad 1\text{–}33$$

When $k_{\mathrm{c}} \gg k_{\mathrm{b}}[\mathrm{M}]$, that is, when the average lifetime of AB* is short compared with the time between collisions, equation 1–33 approximates to the limiting relation at zero [M]:

$$\left(\frac{d[\mathrm{AB}]}{dt}\right)_0 = (k_{\mathrm{a}}k_{\mathrm{b}}/k_{\mathrm{c}})[\mathrm{A}][\mathrm{B}][\mathrm{M}] \qquad 1\text{–}34$$

This accounts for the third-order kinetics. It also allows a rough theoretical estimate to be made of the termolecular rate constant (k_{t}) for a reaction between atoms. Since the activation energy for reaction (a) in such cases is small or zero, both k_{a} and k_{b} will be about equal to the collision frequencies. The lifetime of AB* can plausibly be set at about the period of a molecular vibration, namely 10^{-13} s. Thus,

$$k_{\mathrm{t}} = k_{\mathrm{a}}k_{\mathrm{b}}/k_{\mathrm{c}} \approx 10^{14} \times 10^{14}/10^{13} \approx 10^{15}\,\mathrm{cm^6\,mole^{-2}\,s^{-1}}$$

In fact, the values of k_{t} for the recombinations of O, H, N, and halogen atoms are all within an order of magnitude of 10^{15} at 300 K when M is a simple molecule.[18]

Equation 1–33 implies that k_{t} decreases with increasing total pressure. (The *rate*, of course, *in*creases, but non-linearly.) Eventually, when $k_{\mathrm{b}}[\mathrm{M}] \gg k_{\mathrm{c}}$, [M] disappears from the rate expression and the reaction becomes second order:

$$\left(\frac{d[\mathrm{AB}]}{dt}\right)_\infty = k_{\mathrm{a}}[\mathrm{A}][\mathrm{B}] \qquad 1\text{–}35$$

In some respects it is more revealing to consider this effect in reverse by recasting equation 1–33 into bimolecular form:

$$\frac{1}{[\mathrm{A}][\mathrm{B}]}\left(\frac{d[\mathrm{AB}]}{dt}\right) = k_{\mathrm{B}} = (k_{\mathrm{B}})_\infty\left\{\frac{[\mathrm{M}]}{[\mathrm{M}]+k_{\mathrm{c}}/(\lambda_{\mathrm{MAB*}}Z_{\mathrm{MAB}})}\right\} \qquad 1\text{–}36$$

$(k_{\mathrm{B}})_\infty$, the experimental second-order constant at high pressure, is identified with k_{a}. This equation is formally identical with equation 1–23 for the reverse unimolecular reaction and shows the same 'fall-off' in the value of k_{B} with decreasing pressure p_{M}. (It will be recalled that p_{M} includes p_{A} and p_{B}.) Figure 1–6 shows experimental values of $k_{\mathrm{B}}/(k_{\mathrm{B}})_\infty$ for the reaction

$$CH_3 + CH_3 + M \rightarrow C_2H_6 + M \qquad 1\text{–XIX}$$

As is the case with unimolecular reactions, increasing the reaction temperature displaces the fall-off to higher pressures.[19] Likewise, there is the same effect of molecular complexity.[18,20] Combinations giving rise to products containing up to four atoms are invariably termolecular at, say, 500 K and 500 torr, whereas combinations of alkyl radicals, additions of polyatomic radicals to alkenes and the like, which give more complicated products, are equally invariably bimolecular. As to the influence of the nature of M on k_t, equation 1–34 leads to the relation

$$k_t = \lambda_{M,AB*} Z_{M,AB*} K' \approx \lambda_{M,AB*} Z_{M,AB} K' \qquad 1\text{–}37$$

where K' is the 'equilibrium constant' for the formation of AB* and is independent of M. Relative values of k_t with different M's therefore reflect directly the relative efficiencies for de-energizing AB*. As would be expected, the experimental values follow the same constitutive trend as that shown at the low-pressure limit of unimolecular reactions (Table 1–2). Termolecular reactions are considered further in Section 5–3.

1–3–4 Applicability of equilibrium statistical formulae

Collision theory bases its formulae for reaction rates on expressions derived for gas molecules in the condition of statistical equilibrium characterized by the Boltzmann distribution of energy states. The very occurrence of reaction implies that this is an approximation, and the important question arises as to what degree the approximation is justified. If the great majority of molecules which receive sufficient energy to render them capable of reaction are returned to their original condition without actually reacting, the reaction constitutes a small perturbation of the equilibrium state, and the equilibrium formulae are valid to the extent that this is so. Such is the situation with unimolecular reactions at their high-pressure limits and with termolecular reactions when genuinely third order. This is evident from the presence of the 'equilibrium constants' K and K' in the rate expressions 1–21 and 1–37. A similar situation occurs with bimolecular reactions when the steric factor (P in equation 1–15) is very small.

In general, however, bimolecular reactions are less easily disposed of. *A priori*, the approximation is likely to be least accurate for bimolecular reactions which are otherwise described as unimolecular reactions at the second-order limit. Kinetically, these owe their existence to departure from the equilibrium *concentration* of the critically energized molecules, and, as Fig. 1–6 shows, the *uni*molecular rate constant may be vastly less than its equilibrium value. At the low-pressure limit, however, the reaction rate becomes identical with the *rate of production* of the energized molecules; and the question is to what degree it is legitimate to calculate this rate (k_a) on the assumption that it is the same as it would be if the equilibrium concentration were maintained; in other words, if there were no reaction. This is what is assumed by equation 1–31, but we should note that we are not primarily concerned with the accuracy of this particular equilibrium expression which is at best an approximate one. The problem concerns the applicability of the *correct* equilibrium formula—whatever form this may take for the particular reaction—to the calculation of the activation rate in the non-equilibrium situation. If, as previously, we assume that at higher pressures the potentially reactive molecules are de-energized by one or a very few collisions with 'ordinary' molecules, it follows from microscopic reversibility than in the equilibrium situation they are produced by equally few collisions between 'ordinary' molecules. For a normal reaction the latter form the great bulk of the molecules present and it is reasonable to suppose that their behaviour is not greatly influenced by a deficiency of molecules in the rare states of high energy. Hence in the reacting gas at low pressure 'collisions between the ordinary molecules will have the same effect as in the equilibrium gas—for they do not know the "doom" that awaits them if collision renders them reactive'.§ Consequently, the rate of collisional activation in the reacting gas will be much the same as in the equilibrium gas.

This statement can be tested experimentally by considering kinetically a reaction system of the type

$$AB+M \underset{k_{-D}}{\overset{k_D}{\rightleftharpoons}} A+B+M \qquad \text{(D, −D)}$$

The equilibrium constant is given by the relation

$$K_D = (k_D)_{eq}/(k_{-D})_{eq} \qquad 1\text{–}38$$

where $(k_D)_{eq}$, $(k_{-D})_{eq}$ are the rate constants of the opposing reactions in the state of chemical equilibrium, which necessarily includes the Boltzmann equilibrium for all species present. K_D can be determined without recourse to kinetic measurements. The rate constants, k_D, k_{-D} are ordinarily determined well away from chemical equilibrium. In these circumstances, as discussed above, reaction (−D) occurs within a Boltzmann distribution of the molecules whereas reaction (D) does not. Since therefore $k_{-D} = (k_{-D})_{eq}$ to a good approximation, a test of the relation

$$K_D = k_D/k_{-D} \qquad 1\text{–}39$$

affords an estimate of the same approximation for k_D; that is, of the error incurred in calculating k_D by means of equilibrium statistics.

For the dissociation–recombination of diatomic molecules at short reaction times in shock waves at high temperatures (a very severe test) equation 1–39 has been found to hold to within an order of magnitude, which is about the accuracy imposed by most of the experiments and the extrapolation required to convert the two measurements to a common

§ The quotation, slightly altered, is from reference 21. The assumption that the reactive molecules are produced exclusively by collisions between molecules with energies well below the critical energy is known as the *strong collision assumption*.

temperature.[22,23] More precise measurements are available for the systems

$$NOCl+M \rightleftharpoons NO+Cl+M \qquad \text{1–XX, –XX}$$

and

$$O_3+M \rightleftharpoons O_2+O+M \qquad \text{1–XXI, –XXI}$$

The values of k_{20} obtained from k_{-20} via equation 1–39 vary from close agreement to about half the directly determined value depending on the nature of M.[24] And the indirect values of k_{21} for $M = O_3$ determined over a range of temperatures agree within ± 25 per cent with most of the directly determined values.[23]

A conservative estimate based on the available experimental evidence indicates that deviations from statistical equilibrium are not likely to affect the reaction rate by more than a factor of 2 or 3 even in reactions which are fast by ordinary standards. The equilibrium assumption must, of course, fail eventually, if the reaction becomes fast enough to consume the activated molecules faster than they can be replaced by collisions. Theoretical investigations, however, indicate that for genuine bimolecular reactions the error incurred by the assumption is less than 10 per cent if E_c/RT is greater than 10 and it becomes smaller at higher values of E_c/RT.[25] *De facto*, the relation

$$k/k_{rev} = K \qquad \text{1–40}$$

is used a good deal to obtain otherwise inaccessible kinetics parameters, as we shall see more specifically in Section 1–5–1.

1–4 Activated-complex theory of reaction rate

A potential energy surface constructed for an elementary reaction contains much more information relevant to understanding the reaction rate than is utilized by the simple collision theory so far discussed. The theory now to be outlined makes full use of this information and combines it with considerations drawn from statistical thermodynamics. When the spatial coordinates of the reacting system correspond to the saddle point on the potential energy surface (Fig. 1–3) the system is said to be in the *transition state*, mid-way, as it were, between reactants and products. The corresponding complex of atoms, which necessarily contains the activation energy, is known as the *activated complex*. Its properties are those of an ordinary stable molecule in all but one respect. A small displacement of an atom or group of atoms from its equilibrium position in a stable molecule causes an increase in potential energy and consequently generates a restoring force. This corresponds to positive ('upwards') curvature of the potential energy surface in the direction of the displaced coordinate, as shown for example, by the two-dimensional 'surface' in Fig. 1–2. An activated complex behaves in the same way when its coordinates are displaced in any direction except along the reaction path. In this direction in the region of the saddle point the potential energy surface is flat or slightly negatively curved and the restoring force is zero or negative (see Fig. 1–3). This is the unique property of the activated complex. It has the consequence that when the vibration which follows on an internal displacement is resolved in terms of the $3N-6$ normal vibrational modes of the complex, the frequency of one mode is found to have an imaginary value. This mode is identified with the decomposition of the complex to the reaction products.

1–4–1 Derivation of the rate expression

Since the decomposition of the activated complex gives rise to the reaction products, the reaction rate is obtained by multiplying the concentration of activated complexes by the rate constant for their decomposition. Using as example the reaction

$$A+BC \rightarrow ABC^{\ddagger} \rightarrow AB+C \qquad \text{1–XXII}$$

(A, B, and C being atoms), the rate is given by

$$k_r[n_A][n_{BC}] = \frac{d[n_{AB}]}{dt} = k_{\ddagger}[n_{ABC^{\ddagger}}] \qquad \text{1–41}$$

k_r being the rate constant in units of molecular concentration and ‡ the standard symbol for an activated complex. This expression can be developed in two ways which attach somewhat different meanings to $k_{\ddagger}$ and $[n_{ABC^{\ddagger}}]$ but lead to the same result. Our procedure is a simplified version of that presented by Le Roy, Ridley, and Quickert.[26] The alternative (and original) method can be found in the classic paper by Eyring and, in a slightly different version, in the book by Glasstone, Laidler, and Eyring (see Bibliography). Both procedures make the approximation already familiar from collision theory that the reaction does not disturb the Boltzmann distribution. This being so:

$$[n_{ABC^{\ddagger}}] = K_{\ddagger}[n_A][n_{BC}] \qquad \text{1–42}$$

and

$$k_r = k_{\ddagger}K_{\ddagger} \qquad \text{1–43}$$

The equilibrium constant $K_{\ddagger}$ is given by the statistical expression:[27]

$$K_{\ddagger} = \frac{Q'_{ABC}}{Q_A Q'_{BC}} \exp(-\varepsilon_c/\bar{k}T) \qquad \text{1–44}$$

Q'_i is the molecular partition function *per unit volume* of species i and includes the zero-point energy level of i; ε_c, as previously, is the difference in potential energy between the complex and the reactants, not the difference between the lowest quantum levels (ε_0) in terms of which equation 1–44 is more often expressed. The difference will be appreciated from Fig. 1–5. Thus ε_c is the height of the saddle point of the potential energy surface (Fig. 1–3).

In accordance with the usual approximation, Q'_i is factorized into the separate partition functions for the electronic, translational, rotational, and vibrational degrees of freedom:

$$Q'_i = (Q_e Q_t Q_r Q'_v)_i \qquad 1\text{–}45$$

When, as will be assumed, the temperature is too low for higher electronic states to be excited, $Q_e = g$, the statistical weight of the ground state, for example 1 for H_2, 2 for H, 3 for O_2. The standard expressions for Q_t, etc., are:[27]

$$Q_t = (2\pi m\bar{k}T)^{3/2}/h^3 \quad \text{(per unit volume)} \qquad 1\text{–}46$$

$$\left.\begin{aligned} Q_r &= \pi^{1/2}(8\pi^2\bar{k}T)^{3/2}(I_1 I_2 I_3)^{1/2}/\sigma h^3 \quad &\text{(non-linear molecule)} \\ Q_r &= 8\pi^2\bar{k}TI_1/\sigma h^2 \quad &\text{(linear molecule)} \end{aligned}\right\} \qquad 1\text{–}47$$

$$Q'_v = \prod^{3N-a} \exp(-hv_j/2\bar{k}T)\{1-\exp(-hv_j/\bar{k}T)\}^{-1} \quad \text{(harmonic oscillator)} \qquad 1\text{–}48$$

$$= \exp(-\varepsilon_z/\bar{k}T) \prod^{3N-a} \{1-\exp(-hv_j/\bar{k}T)\}^{-1} \qquad 1\text{–}49$$

$$\equiv Q_v \exp(-\varepsilon_z/\bar{k}T) \qquad 1\text{–}50$$

where I_1, I_2, I_3 are the principal moments of inertia and σ the number of indistinguishable orientations of the molecule (e.g. $\sigma = 1$, 2, 3, 6, and 12 for HCl, H_2, NH_3, CH_3, and CH_4 respectively); a is 5 for a linear and 6 for a non-linear molecule, v_j are the normal frequencies, Q_v is the partition function reckoned from the lowest vibrational levels and ε_z is the total zero-point energy. Each of equations 1–46 to 1–50 can be further factorized, if required, into separate partition function for each degree of freedom, for example

$$Q_r = q_{r,1} q_{r,2} q_{r,3} \quad \text{(non-linear molecule)} \qquad 1\text{–}51$$

Equations 1–45 to 1–50 are applicable both to stable molecules and activated complexes, but in proceeding from 1–48 to 1–50 for an activated complex, the term in 1–48, which corresponds to the imaginary frequency of vibration along the reaction coordinate, is factorized out:

$$Q'_{\ddagger,v} = q_v^* Q_v^{\ddagger} \exp(-\varepsilon_z^{\ddagger}/\bar{k}T) \qquad 1\text{–}52$$

where

$$q_v^* = \exp(-hv^*/2\bar{k}T)\{1-\exp(-hv^*/\bar{k}T)\}^{-1} \qquad 1\text{–}53$$

The frequency v^* is usually assumed to be sufficiently low for q_v^* to reduce to the partition function for a classical oscillator, namely,

$$(q_v^*)_{cl} = \bar{k}T/hv^* \qquad 1\text{–}54$$

Hence, returning to reaction 1–XXII and equation 1–44 and combining equations 1–45 to 1–54 yields:

$$K_{\ddagger} = \frac{\bar{k}T}{hv^*} \cdot \frac{Q_{ABC}^{\ddagger}}{Q_A Q_{BC}} \cdot \exp\{-(\varepsilon_c + \Delta\varepsilon_z)/\bar{k}T\} \qquad 1\text{–}55$$

where $Q^{\ddagger}_{ABC} = Q_{ABC\ddagger}/q^*_v$ and $\Delta\varepsilon_z$ is the difference between the total zero-point energies of complex and reactants. By identifying υ^* with $k_{\ddagger}$ (the frequency of decomposition of the complex) in equation 1–43, an expression is obtained for k_r:

$$k_r = \frac{\bar{k}T}{h} \cdot \frac{Q^{\ddagger}_{ABC}}{Q_A Q_{BC}} \exp(-\varepsilon_0/\bar{k}T) \qquad 1\text{–}56$$

where

$$\varepsilon_0 = \varepsilon_c + \Delta\varepsilon_z \qquad 1\text{–}57$$

as is evident from Fig. 1–5. The general form of this equation is not peculiar to the type of reaction chosen for illustration; in particular, the term $\bar{k}T/h$ follows from general properties of the activated complex in the reaction coordinate and, provided equation 1–54 and the identification just mentioned are applicable, it always appears in the rate expression whatever the nature of the reaction.

There are certain difficulties, peculiar to equation 1–56, which are associated with the symmetry numbers σ of the rotational partition functions. A convenient and, it can be shown,[28] valid procedure to overcome these is to omit the symmetry numbers altogether from equation 1–56 and multiply the right-hand side by a statistical factor $l^{\ddagger}$. This is the number of equivalent ways by which the activated complex can be formed from the reactants or, alternatively, the number of equivalent routes the reaction can take (to form the same products). The following are some examples:

$$H + HCl \rightarrow H \cdots H \cdots Cl^{\ddagger} \qquad l^{\ddagger} = 1$$

$$H + H_2 \rightarrow H \cdots H \cdots H^{\ddagger} \qquad l^{\ddagger} = 2$$

$$H + D_2 \rightarrow H \cdots D \cdots D^{\ddagger} \qquad l^{\ddagger} = 2$$

$$H + NH_3 \rightarrow H \cdots H \cdots NH_2^{\ddagger} \qquad l^{\ddagger} = 3$$

The equivalence of $l^{\ddagger}$ for the second and third reactions is noteworthy. Substitution of the rotational symmetry numbers in equations 1–47 and 1–56 in the ordinary way would lead to the result that the rate of the $H + D_2$ reaction is greater on this account than that of the $H + H_2$ reaction. Similarly, it would obscure the fact that the $H + H_2$ reaction has two chances to occur where the $H + HCl$ reaction has one. These results are at variance with common sense. Further discussion of this matter will be found in reference 28.

A further comment needs to be made concerning the derivation of equation 1–56. There is no reason to believe that the activated complex, though fleeting, is not a real entity. But the reader will doubtless have noticed that, since υ^* (and q^*_v) is an imaginary quantity, equation 1–43 implies that $K_{\ddagger}$ and $[n_{ABC\ddagger}]$ are also imaginary. The difficulty can be avoided[26] or at least made more palatable by considering q^*_v (which is dimensionless) to be part of the term represented by $k_{\ddagger}$ rather than $K_{\ddagger}$;

k_r then becomes the product of a real rate constant $(q_v^* \upsilon^*)$ and a pseudo (but real) equilibrium constant:

$$k_r = (q_v^* \upsilon^*)K^\ddagger \approx \frac{\bar{k}T}{h} K^\ddagger \qquad 1\text{–}58$$

The transmission coefficient

In principle, the expression for k_r given by equation 1–56 is incomplete. The use of $(q_v^*)_{cl}$ in its derivation tacitly assumes the potential energy profile along the reaction path to be flat at the maximum over a much greater distance than the average de Broglie wavelength associated with the relative motion of the separating particles. The approximation is probably commensurate with other sources of error for many reactions above 300 K; but for reactions of light particles, notably hydrogen atoms, this may not be the case and the complete partition function q_v^* should be used. The latter can be expressed in terms of $(q_v^*)_{cl}$:

$$q_v^*/(q_v^*)_{cl} = \Gamma^* = h\upsilon^*\{2\bar{k}T \sinh(h\upsilon^*/2\bar{k}T)\}^{-1} \qquad 1\text{–}59$$

Γ^* therefore should be inserted in equation 1–56. When υ^*/T is very small, $\Gamma^* \to 1$, but the next degree of approximation gives a higher value:

$$\Gamma^* = 1 + (h\,|\,\upsilon^*\,|\,/\bar{k}T)^2/24 \qquad 1\text{–}60$$

From another point of view, the factor Γ^* can be regarded as a 'correction' for quantum-mechanical tunnelling through the energy barrier and it has commonly been derived on this basis (see, for example reference 29). It has been estimated that for the exchange reaction between H and H_2, the effect of Γ^* is to increase the value of k_r by a factor of about 6.5 at 300 K and about 1.4 at 1000 K; for the reaction between D and D_2 the corresponding factors are 2.9 and 1.1.[30]

Formally, Γ^* is often included in a dimensionless *transmission coefficient* κ by which $\bar{k}T/h$ should be multiplied. The effect of Γ^* is to increase the reaction rate above that given by the 'standard' expression (1–56), but other influences contributing to κ may produce the opposite effect. Effects of this kind may be caused by the shape of the potential energy surface in the vicinity of the saddle point. This may cause some of the energy to pass into the 'stable' vibrational modes of the complex which otherwise would appear in the reactive mode and so lead to reaction. The result is that the reaction path, instead of being directly over the 'pass' as shown in Fig. 1–3, may 'skate-around' the saddle point and return in a zigzag fashion back to the initial state. The returning zigzag motion represents vibrational energy generated in the molecular reactant. In these circumstances, the system, though possessing the critical energy ε_c, does not react. Allowance for such an effect, which constitutes a retraction of the assumption that $(q_v^* \upsilon^*)$ is 'separable' from the other degrees of freedom of the complex, is difficult to make; and for lack of knowledge on this account, κ is usually set equal to 1. Some further points relating to the influence of the shape of the potential energy surface will be found in Section 5–1–3.

The value of κ may undoubtedly be quite small, however, when the reaction involves a change in electronic state. This is particularly the case when the same total electronic spin quantum number S cannot be obtained by vector addition of the spins of the reactants on the one hand and of the products on the other; that is, when it is not possible to construct a potential energy surface which 'correlates' with both reactions and products. Such a reaction is said to be *spin forbidden* or, less aptly, *non-adiabatic*. It involves a violation of the spectroscopic correlation rule[31] which regulates the spin multiplicity $(2S+1)$ of the states of molecules formed from fragments of known multiplicity and vice versa. For such reactions, values of κ as low as 10^{-6} can be expected on theoretical grounds.[32] An example is the reaction

$$O(^3P)+CO(^1\Sigma)+M \rightarrow CO_2(^1\Sigma)+M \qquad \text{1–XXIII}$$

which requires a change from $S=1$ to $S=0$. This reaction is several thousand times slower than the comparable reaction

$$O(^3P)+SO(^3\Sigma)+M \rightarrow SO_2(^1A)+M \qquad \text{1–XXIV}$$

for which a potential energy surface is available which correlates with both reactants and products ($S_O - S_{SO} = S_{SO_2} = 0$).

When there is no reason to believe on the grounds just discussed that the transmission coefficient is significant, it is assumed to be unity. With this reservation, the general transition-state expression for the rate constant in molar units for the reaction

$$\alpha+\beta+\cdots \rightarrow \text{Products} \qquad \text{1–XXV}$$

is

$$k = l^{\ddagger}\frac{\bar{k}T}{h}\frac{N^{x-1}Q^{\ddagger}_{\alpha\beta\ldots}}{Q_{\alpha}Q_{\beta}\cdots}\exp[-E_0/RT] \qquad \text{1–61}$$

where N is Avogadro's number, x is the molecularity of the reaction, and the Q's are calculated in molecular units.

1–4–2 Comparison of theoretical and experimental rate expressions

Equation 1–61 is of the form

$$k = C(T)\exp(-E_0/RT) \qquad \text{1–62}$$

and agrees with experiment in so far as $C(T)$ is a shallow function of T. The following considerations show that generally this is so and, further, that the full expression (1–61) provides a theoretical basis for the different absolute values of the pre-exponential factor A found experimentally for different types of reaction. First we note that the experimental Arrhenius expression (equation 1–9) and equation 1–62 are mathematically incompatible if it is insisted that A and E_a are both independent of temperature; but the difficulty vanishes if E_a is simply defined as the slope of the Arrhenius plot at the temperature T:

$$E_a = -R\,d\ln k/d(1/T) \qquad 1\text{–}63$$

Secondly, of its nature, E_0 is independent of T but E_a is not (though it appears to be). In fact, E_a represents the temperature-dependent difference between the average energy of the activated complex and that of the reactants. This can be seen intuitively from Fig. 1–5, but it can also be demonstrated rigorously.[33] From equations 1–62 and 1–63 it follows that

$$E_a = E_0 + RT\,d\ln C(T)/d\ln T \qquad 1\text{–}64a$$

and

$$A = C(T)\exp\left[d\ln C(T)/d\ln T\right] \qquad 1\text{–}64b$$

The first term on the right of equation 1–64a is equal to $(E_c + \Delta E_z)$ and is generally the larger.

For *bimolecular reactions*, it is convenient to factorize $C(T)$ into the separate partition functions q_t, q_r, etc., for each degree of freedom of complex and reactants. To within an order of magnitude, values of q_t and q_r do not vary from one molecule to another and, with a proviso to be mentioned later, the same is true of q_v. Substituting $Q_t = q_t^3$, etc., in equation 1–61 and cancelling where appropriate enables $C(T)$ to be written in shorthand form:

$$C(T) \approx Nl^{\ddagger}\left[\frac{kT}{h}\frac{q_r^2}{q_t^3}\right]\left(\frac{g_{\ddagger}}{g_\alpha g_\beta}\right)\left(\frac{q_v}{q_r}\right)^{\Delta} \qquad 1\text{–}65$$

The exponent Δ is equal to the number of new internal degrees of freedom brought into being with the activated complex. It depends on the geometries of the reacting species and the complex and varies from 0 for a reaction between free atoms to 5 for a reaction between non-linear molecules. The factors q_t and q_r both depend on $T^{1/2}$ and the temperature dependence of q_v varies from T^1 for a very low frequency or high temperature to T^0 for the contrary conditions.§

For an internal free rotation, which may occur instead of a vibration, $q_{ir} \propto T^{1/2}$. Hence, at an unrealistic worst, $C(T) \propto T^3$ and therefore, from equation 1–64, $E_a = E_0 + 3RT$. For a reaction with $E_a = 10\,\text{kcal mole}^{-1}$, which typically might be studied between 450 and 550 K, this would produce about 6 per cent change in the slope of the Arrhenius plot over the temperature range. The curvature would almost certainly not be detected by current techniques. Bearing in mind that low activation energies are usually the most difficult to measure accurately, the absence of detectable curvature in the generality of Arrhenius plots is not surprising. This is not to say, however, that the absolute mean value of E_a

§ At 600 K the q_v's for frequencies of 300 and 800 cm^{-1} (typical values for bending modes in an activated complex) are effectively proportional to $T^{0.65}$ and $T^{0.3}$ respectively.

does not contain a significant contribution from $C(T)$. In the example just given, the contribution would amount to 30% of the measured value of E_a.

The term in square brackets in equation 1–65 is the transition-state expression for a hypothetical reaction between two free atoms and, apart from the approximation regarding the equivalence of q_t's, is *exactly* equivalent to the collision frequency. This can be verified by substituting the appropriate expressions for the partition functions in equation 1–61. Hence the remaining terms in equation 1–65 are equivalent to the steric factor P in a collision-theory expression for $C(T)$ (or, somewhat less exactly, for A). If the relatively small effect of the g factors is ignored, the term $(q_v/q_r)^{\Delta}$ becomes the measure of P, which therefore is related to the type of reaction through Δ. Since values of q_v and q_r are usually in the range 1–5 and 10–100 respectively, the fact that P is usually less than unity is explained. Further, since Δ increases with an increase in the number of rotational degrees of freedom possessed by the reactants, the corresponding values of P for otherwise analogous reactions should show a large decrease. An example of this can be obtained by comparing reactions of the types

$$A+B-R \rightarrow [A \cdots B \cdots R]^{\ddagger} \rightarrow A-B+R \qquad \text{1–XXVI}$$

and

$$R'+B-R \rightarrow [R' \cdots B \cdots R]^{\ddagger} \rightarrow R'-B+R \qquad \text{1–XXVII}$$

where A and B are atoms and R and R′ are non-linear radicals. The experimental values of P for the generality of reactions of type 1–XXVI, for which $\Delta = 2$, are about 10^{-1}, whereas for type 1–XXVII, for which $\Delta = 5$, they are about 10^{-3} (see Table 1–1). The difference, if not the absolute values, is in conformity with equation 1–65. Equation 1–65 cannot, of course, indicate more than trends to be expected in the values of P because q_v and q_r are not actually constant and, in particular, because vibrations in the reactants may be converted to more or less free internal rotations in the activated complex or vice versa. Values of q_{ir} for internal free rotations are of similar magnitude to those of q_r. It is probable, for example, that the difference between the experimental values of P for reactions 1–XXVI and 1–XXVII would be larger than it is if it were not for the occurrence of an extra quasi-free rotation in the transition state of reactions of type 1–XXVII. An example of the complete failure of equation 1–65, however, occurs with the bimolecular combination of polyatomic free radicals, for example CH_3+CH_3, for which P is close to 1, not about 10^{-5} or less as would be expected. To explain this it is necessary to suppose, plausibly but quite *ad hoc*, that the combining forces in the transition state are sufficiently weak to give rise to abnormally low deformation frequencies and to allow free rotation about the incipient bond.

Turning to *unimolecular reactions*, since translational and, to a first approximation, rotational degrees of freedom are cancelled in the rate expression for reactions of this type, $C(T)$ reduces to

$$C(T) \approx l^{\ddagger} \frac{\bar{k}T}{h} \cdot \frac{g^{\ddagger}}{g_{\alpha}} \frac{\prod^{3N-7} (q_v)^{\ddagger}}{\prod^{3N-6} (q_v)_{\alpha}} \qquad 1\text{–}66$$

(for a non-linear reactant and complex). Both the temperature dependence of $C(T)$ and its absolute value depend on whether the internal motions of the reactant molecule change to lower or higher frequencies in the activated complex and whether free rotations are created or destroyed; that is, whether the activated complex is 'loose' or 'tight' compared with the reactant molecule. For similar reasons to those given for bimolecular reactions, such changes cannot make $C(T)$ highly temperature dependent, but they must be invoked to explain the large range of absolute values of the pre-exponential factor which does in fact exist. When the changes are small and the reactant and complex have the same statistical weight, $C(T)$ becomes equal to $\bar{k}T/h$, which, at say 600 K, has the value $1.25 \times 10^{13}\,s^{-1}$. Values of A of about $10^{13}\,s^{-1}$ are commonly found and formerly were regarded as 'normal'; but this term is now better applied, if at all, to the range 10^{10} to $10^{17.5}\,s^{-1}$ which covers most of the values known at present (and of which 10^{13} is about the 'central' value). Arrhenius parameters for some unimolecular reactions are given in Table 1–3.

Table 1–3 Kinetic parameters for some unimolecular reactions
($p_{0.5}$ is the pressure at which the rate constant k_U has 'fallen off' to half its value at infinite pressure $(k_U)_{\infty}$ at T K)

Reaction	$\log A$ (s^{-1})	E_a (kcal mole^{-1})	$p_{0.5}$ (torr)	T K
Isomerizations				
$CH_3NC \rightarrow CH_3CN$	13.6	38.4	48	474
$CH_3NC \rightarrow CH_3CN$	—	—	63	504
$C_2H_5NC \rightarrow C_2H_5CN$	13.8	38.2	0.6	504
cyclo-$C_3H_6 \rightarrow C_3H_6$	15.4	65.6	5	765
cyclo-$CH_3C_3H_5 \rightarrow C_4H_8$	15.5	65.0	0.2	764
cyclo-$C_4H_6 \rightarrow C_4H_6$	13.4	32.7	0.6	423
$C_4H_7OC_2H_3 \rightarrow C_5H_9CHO$	11.2	29.1	—	—
Bond fissions				
$O_3 \rightarrow O_2 + O$	13.4	25	$10^{4.5}$	300
$N_2O_5 \rightarrow NO_2 + NO_3$	14.8	21	$10^{2.5}$	300
$C_2H_5 \rightarrow C_2H_4 + H$	14.4	40.9	$10^{2.7}$	959
$C_2H_6 \rightarrow 2CH_3$	17.1	90.4	20	959
$(CH_3CO)_2 \rightarrow 2CH_3CO$	16.0	67.2	<1	780
$(CH_3)_3CO \rightarrow (CH_3)_2CO + CH_3$	13.4	16.8	80	482
Molecular eliminations				
$(COOH)_2 \rightarrow HCOOH + CO_2$	11.9	30.0	—	—
$C_2H_5Cl \rightarrow C_2H_4 + HCl$	14.0	58.4	1.5	675
n-$C_3H_7Cl \rightarrow C_3H_6 + HCl$	13.5	55.0	0.05	720
cyclo-$C_4H_8 \rightarrow 2C_2H_4$	15.6	62.5	0.2	722

It is possible, in some cases, immediately to correlate the order of magnitude of A for a unimolecular reaction with expected properties of the activated complex. For reactions that open a ring structure, for example, the complex can reasonably be expected to be looser than the original molecule and the value of A should therefore be greater than $10^{13}\ s^{-1}$. This is evidently the case with the isomerizations of *cyclo*propane and its homologues to olefines,[34] for which A is about $10^{15.5}\ s^{-1}$. On the other hand, for the cyclization of linear hexatrienes to their cyclic isomers A is about $10^{11}\ s^{-1}$.[34]

Quantitative evaluation of A for any type of reaction via equation 1–64b depends on acquiring sufficient knowledge of the activated complex to evaluate $Q^{\ddagger}$. Much effort, therefore, has been put into determining the properties of activated complexes, particularly the vibration frequencies, which are the most difficult (and vital) to evaluate. It is here that transition-state theory makes its own particular use of the potential energy surface. The position of the saddle point automatically fixes the interatomic distances in the complex and hence its moments of inertia. Similarly the vibration frequencies can be evaluated from force constants which are calculated from the variation in potential energy with small displacements from the saddle point in appropriate directions. Details of this procedure are given in references 35 and 36.§ Inevitably, a quantitative test of the theory is bound up with the reliability or otherwise of the method used to calculate the potential energy surface. An outline of current methods for performing such calculations is given in Chapter 5, and we shall therefore postpone an account of tests of the theory based on potential energy surfaces until that chapter.

Another, more empirical, approach to calculating pre-exponential factors dispenses with the potential energy surface (which in any case can only be constructed for simple systems) and assigns dimensions and vibration frequencies to the activated complex by analogy with known bond lengths and angles and internal motions of similar stable molecules. For example the activated complex for the following reaction is assumed to have the a planar structure as shown:

$$O_3 + NO \rightarrow \left[O\text{—}O\cdots O\cdots N\text{—}O \right]^{\ddagger} \rightarrow O_2 + NO_2 \qquad \text{1–XXVIII}$$

and its frequencies are derived directly from those of similar vibrational modes in O_3, NO_2, and N_2O_4. The reaction coordinate is identified (as

§ It may be helpful to mention that for triatomic ABC, four force constants F_{11}, F_{22}, F_{12}, and F_{θ} are usually derived: they refer respectively to the separations of A from BC and AB from C, to the interaction these have upon each other, and to a change in the angle ABC. In Fig. 1–4 the curvatures ($\partial^2 V/\partial r^2$) of the surface along the directions A′B′, B′C′, and PQ close to the saddle point relate to F_{11}, F_{22}, and $\{\frac{1}{2}(F_{11}+F_{22})+F_{12}\}$ respectively. F_{θ} can be obtained by calculating the potential energy as a function of the angle ABC or, for a linear complex, more simply from the relation $F_{\theta} = -(r_{AB}r_{BC}/r_{AC})(\partial V/\partial r_{AC})$. There are two stretching vibrations and one bending; one of the former, identified with the antisymmetric stretch, is imaginary; and the latter is doubly degenerate in a linear complex.

usual) with the mode associated with antisymmetric stretching of the two central bonds of the complex. A certain amount of informed guesswork is required and, in expert hands, the method can yield values of A within an order of magnitude of the experimental. The paper by Herschbach *et al.*,[37] from which the above example is taken, gives a clear account of the quantitative effects of bond lengths, vibration frequencies, internal rotations, and so forth on the value of $C(T)$.

The main problem is still to determine the vibration frequencies. The frequencies associated with parts of the complex remote from the centre of reaction are unlikely to be much different from those of the same structures in the original reactants; their effects on $C(T)$ therefore cancel out. But the frequencies associated with the partial bonds intimately involved in the reaction and other bonds close to the site of reaction are less easily assessed. This applies particularly to the frequencies of bending modes. These are low and consequently introduce large errors in $Q_v^\ddagger$ if they are imprecisely known. Furthermore the bond lengths required to calculate the moments of inertia of $Q_r^\ddagger$ are uncertain. In order to make full use of the means transition-state theory offers for systematic discussion of the mass of kinetics data, the viewpoint widely adopted in recent times is to accept these difficulties and use the theory in reverse to explore the properties of the transition state. This is facilitated by the fact, which will emerge more clearly in subsequent chapters, that there is a limited number of types of elementary reaction and therefore a limited number of basic structures for the transition state. Since this type of exercise is usually conducted in terms of the thermodynamic formulation of transition-state theory, we shall postpone further discussion until this is set forth.

1–5 Thermodynamic formulation of activated-complex theory

The general rate expression (1–61) contains the pseudo-equilibrium constant $K^\ddagger$ of equation 1–58 expressed in statistical form. However $K^\ddagger$ can equally well be expressed in terms of standard thermodynamic functions:

$$-RT \ln K^\ddagger = \Delta G^\ddagger = \Delta H^\ddagger - T\Delta S^\ddagger \qquad 1\text{–}67$$

Here $\Delta G^\ddagger$, $\Delta H^\ddagger$, and $\Delta S^\ddagger$ are the *free energy*, *enthalpy* (*'heat'*), and *entropy of activation* respectively, referred to standard states of the reactants and activated complex. (For convenience, the conventional superscript $^\circ$ is omitted from the symbols.) Hence, from equation 1–58,

$$k = \frac{\bar{k}T}{h}\exp(-\Delta G^\ddagger/RT) = \frac{\bar{k}T}{h}\exp(\Delta S^\ddagger/R)\exp(-\Delta H^\ddagger/RT) \qquad 1\text{–}68$$

When this expression is compared with the Arrhenius equation, bearing in mind that $K^\ddagger$ is actually expressed in terms of concentration, it will be found that

$$E_a = \Delta H^\ddagger + xRT_m \qquad 1\text{–}69$$

and that

$$A = e^x (RT_m)^{x-1} \frac{kT_m}{h} \exp(\Delta S^\ddagger/R) \qquad 1\text{–}70$$

where T_m refers conveniently to the mean temperature of the measurements and x is the molecularity of the reaction. The term $(RT_m)^{x-1}$ with R in units of cm^3 atm $mole^{-1}$ deg^{-1} occurs in equation 1–70 because $\Delta S^\ddagger$ conventionally refers to ideal gases at 1 atm pressure as standard states.

For a unimolecular reaction, equation 1–70 becomes

$$A = 5.66 \times 10^{10}\, T_m \exp(\Delta S^\ddagger/R)\, s^{-1} \qquad 1\text{–}71$$

and for bimolecular reaction,

$$A = 1.26 \times 10^{13}\, T_m^2 \exp(\Delta S^\ddagger/R)\, cm^3\, mole^{-1}\, s^{-1} \qquad 1\text{–}72$$

Since $\Delta S^\ddagger$ conventionally is expressed in cal $mole^{-1}$ deg^{-1}, R here has the same units. Values of $\Delta S^\ddagger$ derived from the experimental values of A via these equations naturally are subject to the approximations on which equation 1–58 is based, including omission of the quantum (tunnelling) correction Γ^* on the reaction coordinate. Since

$$S_i = R\{\ln Q_i + d \ln Q_i / d \ln T\} \qquad 1\text{–}73$$

$\Delta S^\ddagger$ can be separated into components for the various degrees of freedom:

$$\Delta S^\ddagger = \Delta S_e^\ddagger + \Delta S_t^\ddagger + \Delta S_r^\ddagger + \Delta S_v^\ddagger \qquad 1\text{–}74$$

Of these in general only $\Delta S_t^\ddagger$ can be calculated *ab initio* without detailed knowledge of the transition state; in doing so, since S_t refers to unit pressure, Q_t in equation 1–46 must be multiplied by kT_m/P (in absolute units) before being substituted in equation 1–73.

1–5–1 Applications of activated-complex theory in thermodynamic form

Estimation of Arrhenius parameters from those of the reverse reaction
Microscopic reversibility requires that when two opposing reactions are at equilibrium they are in equilibrium with the same activated complex. Since at equilibrium $k/k_{rev} = K$, the following relations readily follow from equations 1–68 and 1–69:

$$\Delta S = \Delta S^\ddagger - \Delta S_{rev}^\ddagger \qquad 1\text{–}75$$

$$\Delta H = \Delta H^\ddagger - \Delta H_{rev}^\ddagger = E_a - E_{a,rev} - (x - x_{rev})RT \qquad 1\text{–}76$$

$$\Delta E = E_a - E_{a,rev} \qquad 1\text{–}77$$

where ΔS, ΔH, and ΔE are the standard entropy, enthalpy, and internal energy changes for the reaction. When the system is not at equilibrium, it is consistent with the equilibrium approximation on which transition-state theory is based to apply these relations provided ΔS, $\Delta S^\ddagger$, etc., refer to the same temperature. This is the justification for the relation between E_a, $E_{a,rev}$, and ΔE at absolute zero indicated in Fig. 1–5. Since values of

ΔS and ΔH are frequently available from spectroscopic and thermochemical data or can be estimated with sufficient accuracy by group additivity methods,[38] equations 1–75 and 1–76 enable the Arrhenius parameters of a reaction to be obtained from those of its reverse. When, as not infrequently happens, the reaction is much more accessible experimentally in one direction than in the other, the relations are of much practical value, though, as indicated in Section 1–3–4, the precision with which the equilibrium relation can be validly applied and in what circumstances is still debatable.

Exploration of the transition state

The thermodynamic version of activated-complex theory is much used to investigate the properties of activated complexes, that is their geometry, dimensions, and vibration frequencies. Applied to any particular reaction, the usual procedure is first to postulate the basic geometry and dimensions of the activated complex on more or less intuitive grounds based on the nature of the products and then to compare $(\Delta S_t^{\ddagger} + \Delta S_r^{\ddagger} + \Delta S_e^{\ddagger})$ calculated by equation 1–73 with $\Delta S_{exp}^{\ddagger}$ derived from the experimental value of A via equation 1–70. The difference is $\Delta S_v^{\ddagger}$, from which vibration frequencies can be assigned to the 'new' structures in the complex. Sometimes alternative geometries are able to be ascribed to the complex, for example linear or cyclic, and in such cases the reasonableness or otherwise of the corresponding values of $\Delta S_v^{\ddagger}$ may help to decide between them. Such investigations, of course, do not constitute tests of transition-state theory; their proper aim is to use it to systematize the experimental facts by building up a consistent body of empirical knowledge of the properties of activated complexes. In this way exceptional—and therefore interesting—or erroneous experimental data can be recognized.

Two elementary examples of this type of calculation will be considered. The first concerns the activated complex in the reaction[38]

$$H + C_2H_6 \rightarrow [H \cdots H \cdots CH_2CH_3]^{\ddagger} \rightarrow H_2 + C_2H_5 \qquad \text{1–XXIX}$$

The experimental value of A is $10^{14.1}$ and the mean temperature of the experiments 700 K. Thus, from equation 1–71, $\Delta S_{exp}^{\ddagger} = -21.6$ cal mole^{-1} deg^{-1}. Addition of a hydrogen atom makes little difference to the mass or moment of inertia of the ethane molecule but destroys its sixfold symmetry, that is, σ in equation 1–47 is reduced from 6 to 1. Therefore, apart from $\Delta S_v^{\ddagger}$, the net gain of entropy in the transition state is given by $R \ln 6 = +3.6$ cal mole^{-1} deg^{-1} (from equations 1–47 and 1–73) less the loss of the translational entropy of the hydrogen atom. For the latter, equation 1–73 becomes

$$S_t(\mathrm{H}) = (1.5R \ln 1 + 2.5R \ln 700 - 2.31) = 30.2 \text{ cal mole}^{-1} \text{ deg}^{-1} \qquad \text{1–78}^{(39)}$$

The electronic degeneracy g of the transition state is assumed to the same as that of H (namely 2). Hence,

$$\Delta S_{calc}^{\ddagger} = +3.6 - 30.2 + \Delta S_v^{\ddagger} = \Delta S_{exp}^{\ddagger} = -21.6$$

and therefore $\Delta S_v^{\ddagger} = +5.0$ cal mole^{-1} deg^{-1}. Apart from the reaction coordinate, the complex has two extra vibrational modes over and above those of ethane. Of the latter only the modes associated with the C—H bond attacked are likely to be affected by the presence of the extra H atom. Since stretching frequencies involving H atoms are generally too high to contribute significantly to the entropy, $\Delta S_v^{\ddagger}$ is reasonably ascribed to the doubly degenerate bending mode of the H ··· H ··· R group. Recourse to equations 1–48 and 1–73, or more conveniently to the table relating S_v to frequency and temperature in Benson's *Thermochemical Kinetics*[38] will show the corresponding frequency to be about 400 cm^{-1}. This straightforward result can be used to predict the values of A for other H+HR reactions.

Our second example is the much-discussed reaction:

$$C_2H_6 \rightarrow [H_3C \cdots CH_3]^{\ddagger} \rightarrow 2CH_3 \qquad \text{1–XXX}$$

for which $A = 10^{17.0 \pm 0.4}$ at 1000 K.[34] Application of equation 1–71 shows that $\Delta S_{exp}^{\ddagger}$ is strongly positive, namely 14.9 cal mole^{-1} deg^{-1}. With a unimolecular reaction $\Delta S_t^{\ddagger} = 0$, and assuming that reaction proceeds by elongation of the C—C bond, there is no change in σ or the centre of mass of the molecule or the moment of inertia about the C—C bond. In these circumstances, substitution of equation 1–47 in 1–73 yields $\Delta S_r^{\ddagger} = 2R \ln (r_{C-C}^{\ddagger}/r_{C-C})$, where r_{C-C} and $r_{C-C}^{\ddagger}$ are the C—C bond lengths in ethane and the complex respectively. The latter is unknown. In view of the qualitative fact noted previously that the complex must be very 'loose', a plausible value is the mean of the bond length in ethane (r_{C-C}) and the van der Waals diameter of the methyl radical, that is, about 2.7 Å. Hence $\Delta S_r^{\ddagger} = 2.1$ and therefore $\Delta S_v^{\ddagger} = 14.9 - 2.1 = 12.8$ cal mole^{-1} deg^{-1}. Conversion of the torsional vibration (350 cm^{-1}) about the C—C bond in ethane to free rotation in the complex would contribute about 4.1 − 3.4[38] = 0.7 cal mole^{-1} deg^{-1}; and the remaining 12.1 cal mole^{-1} deg^{-1} is most reasonably ascribed to a 'softening' of the four vibrational modes which correspond to rocking of the methyl groups against each other. In ethane, the frequencies of these are roughly 1000 cm^{-1} and contribute a total of about 5.3 cal mole^{-1} deg^{-1} to its entropy at 1000 K. Therefore the same modes in the complex must contribute 12.1 + 5.3 = 17.4 cal mole^{-1} deg^{-1} to $S_v^{\ddagger}$. This corresponds to a reduction of the mean frequency to about 200 cm^{-1}. Although the choice of $r_{C-C}^{\ddagger}$ was arbitrary, any reasonable value would lead to much the same result. The 'looseness' of the transition state in this and presumably in similar bond dissociations therefore consists largely in a relaxation of bending vibrations associated with the breaking bond.

The above examples were chosen for their simplicity. More subtle, though basically similar, considerations are needed to explore more complicated activated complexes. In such cases the assignment of frequencies to fit the experimental data becomes more arbitrary though naturally it must be consistent with experience gained from simpler reactions. It is fair to add, however, that the uninitiated reader of the literature may sometimes find it difficult to discover whether such an exploration is drawing upon accumulated experience, or adding to it. This perhaps can be accepted as an indication of the 'present state of the art' (of which more comprehensive accounts will be found in references 34 and 38). The main function of the discipline at present—and in this it is unique—is to provide the concepts necessary for meaningful discussion of the mass of kinetic data, particularly those relating to unimolecular reactions of complex molecules.

Bibliography

(A) M. H. Back and K. J. Laidler (eds.), *Selected Readings in Chemical Kinetics*, Pergamon Press, 1967.
(B) E. A. Guggenheim, *Elements of the Kinetic Theory of Gases*, Pergamon Press, 1960.
(C) R. Fowler and E. A. Guggenheim, *Statistical Thermodynamics*, 3rd impression, C.U.P., 1952.
(D) H. Eyring, The activated complex in chemical reactions, *J. Chem. Phys.*, **3,** 107, 1935.
(E) S. Glasstone, K. J. Laidler, and H. Eyring, *The Theory of Rate Processes*, McGraw-Hill, 1941.
(F) V. N. Kondrat'ev, *Chemical Kinetics of Gas Reactions*, trans. J. M. Crabtree, S. N. Carruthers, and N. B. Slater, Pergamon Press, 1964.
(G) H. S. Johnston, *Gas Phase Reaction Rate Theory*, Roland Press Co., New York, 1966.
(H) S. W. Benson, *Thermochemical Kinetics*, Wiley, 1968.

References

1 W. R. Schulz and D. J. LeRoy, *Can. J. Chem.*, **42,** 2480, 1964; *J. Chem. Phys.*, **42,** 3869, 1965.
2 J. O. Hirschfelder, C. F. Curtiss, and R. B. Bird, *Molecular Theory of Gases and Liquids*, Wiley, 1954.
3 R. N. Porter and M. Karplus, *J. Chem. Phys.*, **40,** 1105, 1964.
4 H. S. Johnston, *Gas Phase Reaction Rate Theory*, p. 142, Roland Press Co., New York, 1966.
5 F. W. Schneider and B. S. Rabinovitch, *J. Am. Chem. Soc.*, **84,** 4215, 1962.
6 E. Meyer, H. A. Olschewski, J. Troe, and H. Gg. Wagner, 12*th Symposium* (*Internat.*) *on Combustion*, p. 346, The Combustion Institute, 1969.
7 K. J. Hole and M. F. R. Mulcahy (unpublished).
8 M. C. Lin and K. J. Laidler, *Trans. Faraday Soc.*, **64,** 79, 1968.
9 K. M. Maloney and B. S. Rabinovitch, *J. Phys. Chem.*, **73,** 1652, 1969.
10 O. K. Rice, *Statistical Mechanics, Thermodynamics and Kinetics*, Chap. 17, Freeman & Co., 1967.
11 M. Volpe and H. S. Johnston, *J. Am. Chem. Soc.*, **78,** 3903, 1956.
12 A. B. Trenwith, *Trans. Faraday Soc.*, **63,** 2452, 1967.
13 F. J. Fletcher, B. S. Rabinovitch, K. W. Watkine, and D. J. Locker, *J. Phys. Chem.*, **70,** 2823, 1966; Y. N. Lin and B. S. Rabinovitch, *ibid.*, **72,** 1726, 1968.
14 H. O. Pritchard, R. G. Sowden, and A. F. Trotman-Dickenson, *Proc. Roy. Soc.*, **A217,** 563, 1953.
15 Y. N. Lin, S. C. Chan, and B. S. Rabinovitch, *J. Phys. Chem.*, **72,** 1932, 1968.

16 R. Fowler and E. A. Guggenheim, *Statistical Thermodynamics*, Chap. 12, 3rd impression, C.U.P., 1952.
17 J. Troe, *Ber. Bunsenges. physik Chem.*, **72,** 908, 1968.
18 S. W. Benson and W. B. De More, *Ann. Rev. Phys. Chem.*, **16,** 397, 1965.
19 K. J. Hole and M. F. R. Mulcahy, *J. Phys. Chem.*, **73,** 177, 1969.
20 F. Kaufman, *Ann. Rev. Phys. Chem.*, **20,** 45, 1969.
21 N. B. Slater, *Theory of Unimolecular Reactions*, p. 18, Cornell Univ. Press, 1959.
22 S. H. Bauer, *Ann. Rev. Phys. Chem.*, **16,** 245, 1965.
23 H. S. Johnston, *Gas Phase Reaction Kinetics of Neutral Oxygen Species*, NSRDS–NBS–20, U.S. Govt. Printing Office, 1968.
24 T. C. Clark, M. A. A. Clyne, and D. H. Stedman, *Trans. Faraday Soc.*, **62,** 3354, 1966.
25 E. W. Montroll and K. E. Shuler, *Adv. Chem. Phys.*, **1,** 361, 1958; R. D. Present, *J. Chem. Phys.*, **48,** 4875, 1968.
26 D. J. Le Roy, B. A. Ridley, and K. A. Quickert, *Disc. Faraday Soc.*, **44,** 92, 173, 1967.
27 R. Fowler and E. A. Guggenheim, *Statistical Thermodynamics*, Chaps V and III, 3rd impression, C.U.P., 1952.
28 D. M. Bishop and K. J. Laidler, *J. Chem. Phys.*, **42,** 1688, 1965; J. N. Murrell and K. J. Laidler, *Trans. Faraday Soc.*, **64,** 371, 1968; or see K. J. Laidler, *Theories of Chemical Reaction Rates*, McGraw-Hill, 1969.
29 H. S. Johnston, *Gas Phase Reaction Rate Theory*, p. 24, Roland Press Co., New York, 1966.
30 I. Shavitt, *J. Chem. Phys.*, **49,** 4048, 1968.
31 G. Herzberg, *Molecular Spectra and Molecular Structure*, Vol. I, *Spectra of Diatomic Molecules*, 2nd edn, p. 318, Van Nostrand, 1950.
32 K. J. Laidler, *The Chemical Kinetics of Excited States*, Clarendon Press, 1955.
33 R. Fowler and E. A. Guggenheim, *Statistical Thermodynamics*, pp. 501, 521, 3rd impression, C.U.P., 1952.
34 H. M. Frey and R. Walsh, *Chem. Rev.*, **69,** 103, 1969.
35 H. S. Johnston, *Gas Phase Reaction Rate Theory*, pp. 333, 339, Roland Press Co., New York, 1966.
36 G. Herzberg, *Infrared and Raman Spectra*, p. 186, Van Nostrand, 1945.
37 D. R. Herschbach, H. S. Johnston, K. S. Pitzer, and R. E. Powell, *J. Chem. Phys.*, **25,** 736, 1956. See also D. J. Wilson and H. S. Johnston, *J. Am. Chem. Soc.*, **79,** 29, 1957.
38 S. W. Benson, *Thermochemical Kinetics*, Wiley, 1968.
39 E. A. Moelwyn-Hughes, *Physical Chemistry*, p. 363, Pergamon Press, 1957.

Problems

1–1 The activation energy for a bimolecular reaction $A+B_2 \rightarrow AB+B$ is 10 kcal (42 kJ) mole^{-1}. What are the temperatures at which 0.01, 1, 10, and 90 per cent of 'hard-sphere' collisions between the reactant molecules occur with sufficient relative kinetic energy along their line of centres for reaction? If the reaction is exothermic to the extent of 5 kcal (21 kJ) mole^{-1} and the molecular weight of B_2 is 10 times that of A, use plausible assumptions about the collision diameters and steric factors to estimate the temperature at which the reaction of an equimolecular mixture of $A+B_2$ is 90 per cent complete at equilibrium.

1–2 The following rate constants for the bimolecular reaction $NO_2+CO \rightarrow NO+CO_2$ were obtained by Crist and Brown (*J. Chem. Phys.*, **9,** 840, 1941):

k cm^3 mole^{-1} s^{-1}	0.312	0.675	1.20	3.08	13.8
Temperature K	498.4	510.3	522.7	536.2	563.3

Determine graphically the Arrhenius activation energy (E_a). What fraction of the collisions between NO_2 and CO molecules are potentially reactive at 520 K? Calculate the steric factor P at 520 K (a) assuming 'hard-sphere' molecules and (b) allowing for van der Waals attraction by means of the formula on p. 12. What would be the error in P if E_a had been estimated too low by 10 per cent? ($10^8\sigma_{NO_2} = 4.0$ cm; $10^8\sigma_{CO} = 3.6$ cm; $\varepsilon/k = (\varepsilon_{NO_2}\varepsilon_{CO})^{1/2}/k = 160$).

1–3 (a) Show that, on the basis of the Lindemann–Hinshelwood mechanism, the pressure at which the rate constant of a unimolecular reaction has 'fallen-off' to one half its high pressure value (k_∞) is given by $p_{0.5} = (RTk_\infty/k_A)$ where k_A is the bimolecular rate constant for activation. (b) k_∞ for the unimolecular decomposition of ethane at 900 K is 1.3×10^{-5} s^{-1} and E_a is 90 kcal (380 kJ) mole^{-1}. Calculate $p_{0.5}$ on the assumption that the critical energy for reaction must be supplied by collisions with relative kinetic energy along the line of centres greater than E_a. The observed value is about 10 torr. What is the origin of the discrepancy? ($Z = 10^{14}$ cm^3 mole^{-1} s^{-1}).

1–4 Reverse the procedure of Problem 1–3(b) to calculate k_A from $p_{0.5}$ and hence, by applying trial and error to equation 1–31, estimate the number of classical oscillators that must be excited to account for the rate of activation; (to nearest whole number). What fraction is this of the maximum possible? (Assume $E_c = E_a$; $\lambda = 1$).

1–5 An estimate is required of the half-life of H_2O_2 vapour in low concentration in pure air (an inert medium) at 1 atm pressure at 298 K, but no measurements of the rate of the homogeneous decomposition are available for this temperature. Assuming the following reaction steps with reaction 1 rate-determining,

$$H_2O_2 + M \rightarrow OH + OH + M \quad (1)$$

$$OH + H_2O_2 \rightarrow H_2O + HO_2 \quad (2)$$

$$HO_2 + HO_2 \rightarrow H_2O_2 + O_2 \quad (3)$$

estimate $t_{0.5}$ from the following data:

$\Delta G_f^\circ(H_2O_2) = -25.21$ kcal (-105.5 kJ) mole^{-1};

$\Delta G_f^\circ(OH) = 8.19$ (34.27);

$k_{-1} = 1.3 \times 10^{18}$ cm^6 mole^{-2} s^{-1} at 298 K

(Caldwell and Back, *Trans. Faraday Soc.*, **61**, 1939, 1965).
Direct measurements of the decomposition in the range 720–950 K gave

$$2k_1 = 1.7 \times 10^{17} \exp(-46.3\ \text{kcal}/RT)$$

(Baldwin and Brattan, *8th Combustion Symp.*, 1962, p. 110). How well does $t_{0.5}$ obtained *via* a 420° extrapolation of these measurements compare with your previous result?

1–6 (a) Write down the activated-complex theory expression for the rate constant of the reaction $Br + H_2 \rightarrow HBr + H$ using the same assumptions as are embodied in equation 1–61. (Assume a linear complex and $g^\ddagger_{HHBr} = 2$. Note that $g_{Br} = 4$). (b) If the pre-exponential factor is written as Constant $\times T^n$, what are the greatest and least possible values of n? (The fundamental vibration frequency of H_2 (ω_{H_2}) is 4395 cm^{-1}.)

1–7 In the expression for the partition function q_v for a harmonic oscillator, $(h\upsilon/\bar{k}) = 1.44\omega$ with the frequency ω expressed in cm^{-1}. Plot a graph of q_v (without zero-point energy) as a function of ω/T over the range 0–4 for ω/T. Hence evaluate q_v (approximately) for the following combinations of frequencies and temperatures. 200 cm^{-1}: 300, 1000 K; 500 cm^{-1}: 300, 1000 K; 2000 cm^{-1}: 1000, 2000 K.

1–8 Show that each vibrational partition function q_v in the activated-complex theory rate expression makes a contribution $\pm\delta E_a = 1.44R\omega(q_v - 1)$ to the Arrhenius activation energy (apart from zero-point energy). Use this result and the graph plotted for problem 1–7 to predict the change in E_a between 200–1000 K for the reaction $H + HCl \rightarrow H_2 + Cl$. [$\omega_{HCl} = 2990\ cm^{-1}$; use the sym. stretching ($\omega_s^\ddagger = 1360$) and bending ($\omega_b^\ddagger = 720\ cm^{-1}$) frequencies calculated from the potential energy surface for the linear complex by Westenberg (*J. Chem. Phys.*, **48,** 4405, 1968)]. What is the contribution to E_a from zero-point energies?

1–9 Calculate the entropies of activation $\Delta S^\ddagger$ of the following unimolecular reactions from the experimental Arrhenius pre-exponential factors at the reaction temperatures T_r.

Reaction	$\log A$ (s^{-1})	T_r (K)
$(CH_3CO)_2 \rightarrow CH_3CO + CH_3CO$	16.0	600
$(CH_3CO)_2O \rightarrow CH_2CO + CH_3COOH$	12.0	600
$C_6H_5CH_2CH_3 \rightarrow C_6H_5CH_2 + CH_3$	14.6	950
$C_6H_5CH_2Br \rightarrow C_6H_5CH_2 + Br$	13.0	950
$\underline{CH_2CH_2OCH_2} \rightarrow HCHO + C_2H_4$	14.8	750
$CH_2{:}CHOC_2H_5 \rightarrow CH_3CHO + C_2H_4$	11.4	750

Discuss qualitatively the differences between the values for the bracketed reactions in terms of likely structures and properties of the activated complexes. What value of $\log A$ corresponds to $\Delta S^\ddagger = 0$ for a unimolecular reaction at 750 K? at 1750 K?

1–10 By means of equation 1–73 derive the following expressions for the translational, rotational, and vibrational entropies of activation for a bimolecular reaction between linear species X and Y involving a linear activated complex.

$$\Delta S_t^\ddagger = R\{1.5 \ln[(M_x + M_y)/M_x M_y] - 2.5 \ln T + 1.16\} \qquad (M_O = 16)$$

$$\Delta S_r^\ddagger = R\{\ln(I_\ddagger / I_x I_y) + \ln(\sigma_x \sigma_y / \sigma_\ddagger) - \ln T - 89.39\} \qquad (I \text{ in g cm}^2)$$

$$\Delta S_v^\ddagger = \left\{\sum_\ddagger S_v - \left(\sum_x S_v + \sum_y S_v\right)\right\}$$

where

$$S_{v,i} = R\{u_i[\exp(u_i) - 1]^{-1} - \ln(1 - \exp(-u_i))\} \qquad (u_i = 1.44\omega_i/T)$$

1–11 The Arrhenius parameters for the reaction $HO + CO \rightarrow CO_2 + H$ are uncertain at present (1972). Predict the value of $\Delta S^\ddagger$ at 300 K, and hence the pre-exponential factor, on the basis of a linear activated complex assumed to have $I^\ddagger = 85 \times 10^{-40}\ g\ cm^2$ and the following frequencies:

	2x stretching	2x bending	2x bending
ω (cm^{-1})	1600	1000	500

($I_{OH} = 1.5 \times 10^{-40}$; $I_{CO} = 15 \times 10^{-40}\ g\ cm^2$. $\omega_{OH} = 3735$; $\omega_{CO} = 2170\ cm^{-1}$).

2 Experimental methods I: Investigation of overall kinetics

The basic facts required from an experimental investigation of the kinetics of a reaction are the rates at which the reactants are consumed and the products formed and how these rates depend on the concentrations of the reactants and the temperature. It is also necessary to determine to what degree the rate of reaction is sensitive to the presence of inert gases and how it depends on the nature and extent of the surface of the containing vessel. Another point to be ascertained is whether the reaction is subject to autocatalysis or self-inhibition, that is, whether it is accelerated or retarded by its products.

It was noted in Chapter 1 that 'rate of reaction' is an ambiguous term: it may refer variously to the rate of disappearance of a particular reactant or to the appearance of one of perhaps several products. The relations between these rates are expressed by the reaction stoichiometry and it is important to note that this may vary with the reaction conditions. To quote an example: the main products of the thermal decomposition of *n*-butane at 790 K are methane, ethane, ethylene, and propylene.[1] When the initial pressure of *n*-butane is 150 torr, the ethylene and ethane are formed in the proportion 1.5:1 whereas at 15 torr the proportion is 2.3:1. The ratio of methane to propylene, on the other hand, is independent of the pressure. Similarly, lowering the reaction temperature to 739 K reduces the ethylene: ethane ratio at 150 torr to 1:1 but does not affect the ratio of methane to propylene. Thus rather extensive measurements of the rates of change of several species are often required before it can be stated confidently that the basic experimental facts relating to a particular reaction system are known, even for a fairly restricted range of conditions. This is by no means to say that stoichiometrically simple reactions do not exist, but they are less common than might be supposed. An alteration in stoichiometry observed when a reactant concentration or some other factor is changed may be caused by the presence of two or more independent reactions which respond differently to the change. But more frequently it reflects an alteration within the mechanism of a single complex reaction. Systematic investigation of the rates of change of the concentrations of individual reactants and products (and intermediate species) is therefore not only necessary to define what in fact is being investigated but also constitutes a principal means of elucidating the mechanism of the reaction. Nor should the kinetic behaviour of 'minor' or 'trace' constituents of the products be neglected, since this also not infrequently provides important insight into the reaction mechanism.

Gas kinetics as a discipline, therefore, relies heavily on chemical analysis. The range of analysis required naturally includes the analysis of substances that are not gaseous at ordinary temperatures and is almost as wide as analytical chemistry itself. Nevertheless, the specificity, sensitivity, and convenience of gas chromatography make it the most generally useful for the kinetic methods described in this chapter. Indeed, in retrospect, it seems clear that before the introduction of gas chromatography in the 1950s, serious study of complex reaction mechanisms involving any but the simplest molecular species was scarcely practicable. That this is far from being the case at present it is hoped to show in Chapter 3.

There are four general methods by which measurements of reaction rate are usually performed. These are the *static*, *linear flow*, *stirred flow*, and *shock-tube* methods. What follows is an outline of these methods as applied to reactants that are stable at room temperature. Applications to labile species are considered in Chapter 4. The treatment is selective and for the most part only the basic features of the experimental techniques are presented. The reader should consult the standard works listed in the Bibliography at the end of this chapter for further details of construction of apparatus, instrumentation, and experimental procedures.

2–1 The static method

This is the simplest and most widely used method. In its usual form it is suitable for reaction times greater than about one minute. The reactants are admitted in determined proportions to an evacuated vessel in a thermostat, the initial total pressure is measured, and the composition of the mixture is determined as a function of time. A typical apparatus is shown diagrammatically in Fig. 2–1.

The volume chosen for the reaction vessel is often the result of a compromise. A large vessel is desirable to minimize wall effects and dead space (see below) and to provide a sufficient quantity of products for accurate analysis; but the trouble and expense involved in constructing the thermostat usually increases with the size required. In addition, errors due to the time taken to admit the reactants and to lack of temperature equilibration during the reaction increase with increasing volume of the vessel. (Further reference to these errors will be made later.) A few hundred cubic centimetres is a common size though vessels larger than 20 litres have been used for work at low pressures.[(3)] Borosilicate glass and fused quartz are the usual materials for use up to about 550 and 1100 °C respectively. To test the homogeneity of the reaction the surface-to-volume ratio can be increased by filling the vessel with glass wool or pieces of tubing; when, however, more definite geometry is required, vessels can be constructed from sections of tubing of any desired diameter joined together in parallel.

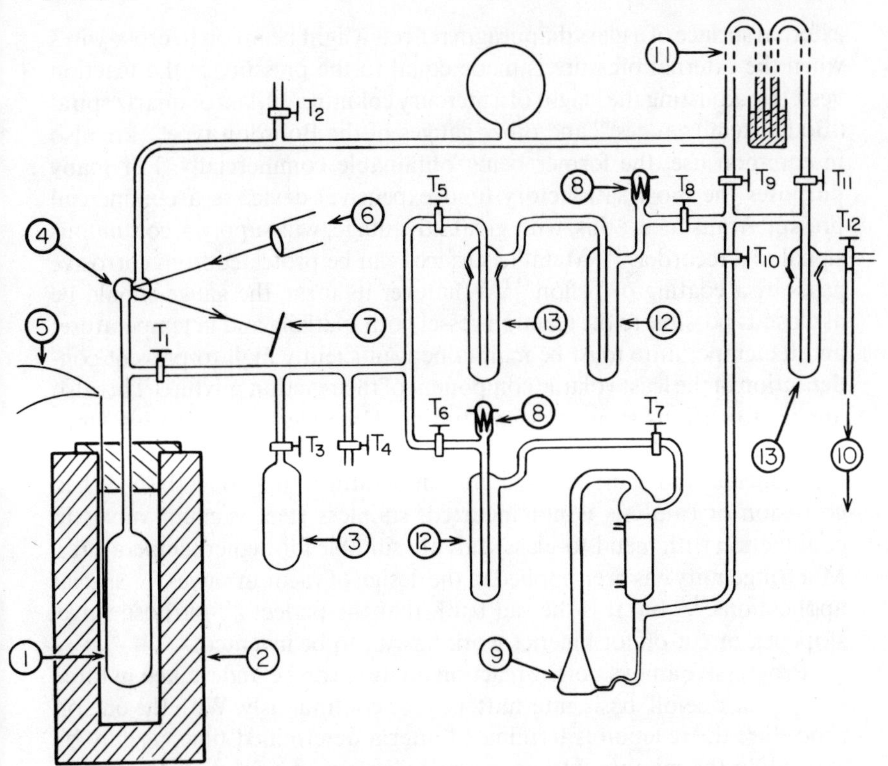

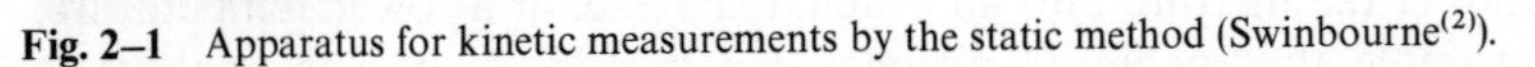

Fig. 2–1 Apparatus for kinetic measurements by the static method (Swinbourne[2]). 1, Reaction vessel; 2, constant temperature furnace; 3, reactant storage; 4, glass diaphragm gauge; 5, thermocouple; 6, lamp; 7, screen; 8, Pirani gauge heads; 9, mercury diffusion pump; 10, to rotary oil pump; 11, mercury manometers; 12, 13 liquid nitrogen traps; T, stopcocks.

The common form of thermostat for work at high temperatures is an electrically heated hollow copper or aluminium block cast in two or more pieces so as to admit the reaction vessel. It is well insulated thermally and provided with a temperature controller. Alternatively, precise temperature control (± 0.1 °C) can be obtained with an air thermostat of suitable design.[4] For lower temperatures a liquid thermostat has several advantages. With efficient stirring, very uniform temperature can be obtained and dead space can be minimized; in addition, several reaction vessels—for example with different volumes or surface treatments—can be introduced together. A eutectic mixture of potassium nitrate and sodium nitrite (55:45 by weight), which melts at 140 °C, is a suitable medium for temperatures up to 450 °C.[5] Low-melting alloys such as bismuth solder[6] (m.p. 111 °C) have also been used.

In measuring the pressure in the reaction vessel it is generally desirable to avoid contact between the reaction mixture and mercury or other manometric fluid. In the apparatus shown in Fig. 2–1, this is accomplished by one of the several available forms of diaphragm gauge: the silvered external surface of a glass diaphragm reflects a light beam on to cross-wires when the external pressure is made equal to the pressure in the reaction vessel by adjusting the height of a mercury column.[7] Glass or quartz spiral ('Bodenstein') gauges[8] and other gauges of the Bourdon type[9] are also in common use, the former being obtainable commercially. For many purposes the most satisfactory (and expensive) device is a commercial pressure transducer; this, with great advantage, will supply a continuous signal to a recorder.[10] Metallic surfaces can be protected from corrosive gases by a coating of Teflon.[11] Whatever its form, the gauge should be as close as possible to the reaction vessel both spatially and in temperature; and its temperature must be maintained sufficiently high to prevent condensation of the least volatile component of the reaction mixture. This also applies to various stopcocks and tubing. Greaseless high-vacuum stopcocks are available commercially which incorporate Viton or Teflon components and can be used at temperatures up to 200 °C. When corrosion or catalysis is not incurred, stainless steel vacuum valves in conjunction with metal-to-glass seals are suitable for higher temperatures. Much ingenuity has been applied to the design of vacuum valves for special applications,[12] but it is the sad truth that the perfect all-purpose valve, stopcock or cut-off for kinetics work has yet to be invented.

Progressive analysis of the reaction mixture can be undertaken in three ways: on a 'one-off' basis, intermittently, or continuously. With the one-off procedure the reaction is terminated after a determined time, usually by expanding the mixture into an evacuated vessel at a lower temperature, for example that of liquid nitrogen (as in Fig. 2–1). The analysis is performed at leisure and the experiment is repeated for various reaction times. With the intermittent procedure the reaction mixture is sampled at successive intervals, the amounts withdrawn being sufficiently small to leave the concentrations in the reaction vessel effectively unchanged. The high sensitivity of gas-chromatographic analysis can render this method feasible with a reaction vessel of volume as low as 200–300 ml. Special quick-acting valves have been designed for transferring such samples directly to the chromatographic column.[13] By the use of such devices, the highly desirable characteristics of frequent, accurate, and substantially complete analysis can often be achieved.

In principle, continuous monitoring of the reaction mixture can be obtained by measuring any property which changes with the progress of the reaction. When a change in the number of molecules occurs, the simplest measurement is that of pressure. Such measurements are widely used in studies of pyrolytic reactions. In principle the concentration of the reactant (A) at any time can be calculated from the total pressure P by one or other form of the theoretically equivalent relations

$$[\mathrm{A}] = \frac{(qP_0 - P)}{RT(q-1)} = \frac{(P_\infty - P)}{RT(q-1)} \qquad \text{2–1(a), (b)}$$

where P_0 and P_∞ are the initial and final pressures and q is the number of molecules into which the reactant molecule decomposes. It is obvious that this procedure presupposes knowledge of the reaction stoichiometry. When this has been determined for a range of conditions the method can be used very conveniently to increase the number of kinetic measurements within that range.

If the absorption spectra of components of the reaction mixture are sufficiently separate, the reaction can be followed photometrically *in situ*. Typical apparatus using absorption in the visual–ultraviolet region is shown in Fig. 2–2. Infrared spectrophotometry can also be used in

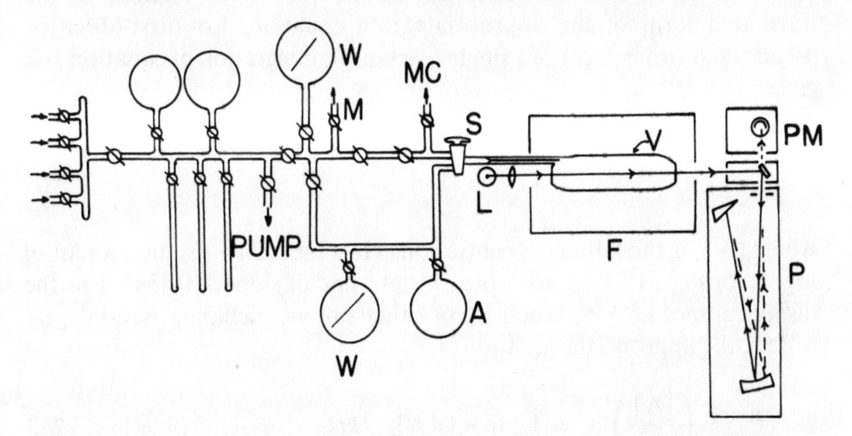

Fig. 2–2 Apparatus for kinetic measurements by the static method using optical absorption (Kaufman and Decker[14]). (Copyright 1962 The Williams and Wilkins Co., Baltimore; reproduced by permission.)

V, Quartz reaction vessel; F, furnace; PM, photomultiplier tube; P, quartz monochromator; L, light source; S, three-way stopcock; MC, to McLeod gauge; M, to monometer; W, dial pressure gauges; A, gas storage.

appropriate circumstances.[15] When there is strong absorption, the exponential nature of Beer's law makes it desirable to use a detector–display circuit that relates the read-out signal logarithmically to the light absorbed.[16] By means of an appropriate optical system[17] reactions at high temperature can be followed with a standard spectrophotometer and quasi-simultaneous determination of several species may be obtained by using a fast scanning instrument.

Continuous analysis can also be carried out by mass spectrometry.[18,19] A pin-hole probe communicating with the ionization chamber of the mass spectrometer is incorporated in the reaction vessel and, with a suitable design, a leak rate of 3% per hour can be sufficient for frequent and complete analysis of 300 torr of reaction mixture.[19] Unfortunately, the method is likely to suffer from the lack of discrimination which is endemic to mass spectrometric analysis of mixtures of related compounds. An advantage of any continuous method is that the signals can be displayed by a pen-recorder, stored in a data logger or transmitted directly to a computer for processing.

2–1–1 Derivation of kinetic parameters

Methods of deriving rate parameters from experimental data obtained by the static and other methods are described in detail in a recent book by Swinbourne (see Bibliography ref. (D)). A few elementary aspects only are noted here, mainly for the purpose of recording the integrated forms of standard rate equations for future reference.

The static method is much used for determining reaction order (n). This is often carried out by fitting the concentrations of the reactants, measured directly or indirectly during the course of reaction, to the integrated form of the appropriate rate equation. For first-order or pseudo-first-order decay of a single reactant A, integration of equation 1–2 gives

$$f = \frac{[A]}{[A]_0} = \exp[-kt] \qquad 2\text{–}2$$

where $[A]_0$ is the initial concentration. Thus for such a reaction a plot of $\log_{10} f$ or $\log_{10}[A]$ against t is a straight line of slope $-0.434k$. For the disappearance of A by reactions of other orders, including non-integral orders, the appropriate equation is

$$f^{1-n} = \left(\frac{[A]}{[A]_0}\right)^{1-n} = 1+(n-1)[A]_0^{n-1}kt \qquad (n \neq 1) \qquad 2\text{–}3$$

For such a reaction a plot of f^{1-n} against t is a straight line of slope $(n-1)[A]_0^{n-1}k$. It will be seen that, with a first-order reaction, only relative concentrations of A are required to determine the rate constant; but with reactions of other orders, at least one determination of the absolute value of $[A]$ must be made (for example, $[A]_0$.) When two or more reactants are involved the equations become more complicated; thus, for a reaction $A+B \rightarrow$ Products, that is first order in each reactant, the expression is

$$\ln([A]_0/[A]) - \ln([B]_0/[B]) = ([B]_0 - [A]_0)kt \qquad 2\text{–}4$$

When $[B]_0$ is in large excess over $[A]_0$ or vice versa, equation 2–4 becomes equivalent to equation 2–2. This fact can be useful when the absolute concentration of one reactant, say B, is more easily determined than that of the other; with B in excess, the pseudo first-order rate constant $[B]_0k$, and hence k, can be found without the need to measure $[A]$ absolutely.

When the experimenter is unaware that he is dealing with a reaction of complex mechanism, fitting concentrations to the above equations can lead to deceptively simple or otherwise erroneous results, particularly if the concentrations are measured indirectly. For this reason, the 'initial rate' procedure is commonly adopted: a plot is made of the relevant concentration against time and the tangent is taken at the origin. In principle, this measures the rate when the concentrations of the products are too low for the effect of secondary reactions to be appreciable. Thus for the reaction $A + B \rightarrow yC$,

$$\frac{1}{y}\frac{d[C]}{dt} = \left(\frac{-d[A]}{dt}\right) \approx k[A]_0^{n_A}[B]_0^{n_B} \qquad 2\text{–}5$$

The reaction orders for A and B are determined by measuring the initial rate as a function of the initial partial pressure of A with a constant pressure of B and vice versa. Concentrations of products corresponding to a few per cent reaction and frequently considerably less can usually be measured by gas chromatography.

2–1–2 Potential errors

Like other methods the static method has potential errors peculiar to itself. One such arises from the time required for the pressure and temperature to reach their required values after the reactants are admitted to the reaction vessel. In practice this usually is less an error than a limitation on the magnitude of the rate which can be measured. The time involved is normally a few tenths of a second, depending upon the diameter of the admitting stopcock.[20] During this period the gas heats up adiabatically as it rushes into the evacuated vessel and its temperature can rise transiently to more than 200 °C above its final equilibrium value.[21] In most cases the effect is too short-lived to influence the kinetics, but this may not be the case with certain autocatalytic reactions that are initiated by the production of a trace of an active intermediate.

A more serious error is likely to be caused by *dead space*, that is, by any volume containing reactants which is not heated to the reaction temperature. This often comprises the pressure gauge and the tubing between the reaction vessel and the stopcock by which the reactants are admitted. It may easily amount to several per cent of the total volume, though frequently it can be reduced substantially and sometimes eliminated altogether by careful design. The effect depends upon the circumstances of measurements and only a few illustrative examples can be given here. It is convenient to define a dead space factor δ which allows for the difference in temperature between the dead volume and the reaction vessel. Thus

$$\delta = V_d/[V_d + V_r(T_d/T_r)] \qquad 2\text{–}6$$

where V_d, T_d, and V_r, T_r are the respective volumes and temperatures of the dead space and the reaction vessel. Typically, for a reaction vessel at 500 °C plus 2 per cent dead volume at room temperature, δ is 0.05.

For a reaction occurring without pressure change it is easily shown that the fraction of the original number of reactant molecules which would have remained at any time in a vessel without dead space (f_{true}) is related to that actually found by total extraction of the reaction mixture from the vessel and dead space (f_{obs}) by the following expression:

$$f_{true} = (f_{obs} - \delta)/(1 - \delta) \tag{2–7}$$

Clearly, the error increases with increasing progress of the reaction. More subtle effects occur when the reaction produces a pressure change.[22,23] When there is an increase in pressure, some of the reactant molecules are compressed into the dead space as the reaction proceeds; hence more reactant disappears from the reaction vessel in a given time than would be the case if there were no dead space. The contrary effect occurs when there is a pressure decrease. Thus if the reaction rate is measured by determining the concentration of a reactant *in situ*, for example by photometry, the rate will appear to be faster than the true rate in the first case and slower in the second. For a first-order reaction giving an increase in pressure, it can be shown that

$$\frac{[A]}{[A]_0} = f_{obs} = \frac{1}{(q-1)}\{qz^{-\delta} - z^{1-\delta}\} \tag{2–8}$$

where

$$z = q - (q-1)\exp[-kt] \tag{2–9}$$

This may be compared with equation 2–2 which applies to an ideal system without dead space. Solution of equation 2–8 will show, for example, that, for a first-order pyrolysis of a compound yielding three product molecules studied in an apparatus with $\delta = 0.05$, a value of $f_{obs} = 0.5$ recorded in the reaction vessel would be 3.5% too low. Neglect of the effect of dead space would cause an error of +5% in the value of the rate constant deduced from this measurement.

As might be expected, kinetic measurements based on measurements of pressure increase are particularly susceptible to dead-space errors. For a first-order pyrolysis the ratio of the total pressure at any time to its initial value is given by[22]

$$(P/P_0) = z^{1-\delta} \tag{2–10}$$

and hence

$$(P_\infty/P_0) = q^{1-\delta} \tag{2–11}$$

(This equation is valid for any reaction order.) If the reaction mentioned above were to be studied in the same apparatus by pressure change instead of by photometry of the reactant concentration, the total pressure which appeared to indicate 50% reaction, namely $2P_0$, would be 3.5% too low, and the rate constant deduced from this measurement (alone) neglecting the dead space would be too low by 10%. Hence the dead-space error in this experiment would be twice as large and of the opposite sign from that attaching to the photometric experiment. The most accurate procedure for calculating the instantaneous values of [A] from total pressure measurements for substitution in equations 2–2 or 2–3 is by means of equation 2–1(b), using the value of P_∞ actually observed or calculated by equation 2–11.[23] §

A further possibility of error arises from the fact that the *evolution or absorption of heat* by the reaction causes the temperature of the gas in the centre of the reaction vessel to be different from that at the walls, that is, from the temperature of the thermostat.[24] Normally the effect is quite small, but it may become appreciable with strongly exothermic reactions or large vessels. Its magnitude can be checked by measuring the temperature difference (ΔT) between the centre and the wall of the reaction vessel during reaction. It has been shown by Boddington and Gray[24] that the errors in the Arrhenius parameters arising from the effect can virtually be eliminated if the observed rate is related not to the temperature of the thermostat T_s but to the temperature $(T_s + 2\Delta T/5)$ for a spherical vessel or to $(T_s + \Delta T/2)$ for an infinite cylinder.

2–1–3 Two investigations by the static method

A more realistic impression of the application of the static technique may be gained from a short account of the experimental side of investigations in which it has been used. From the vast number of such studies we have space only to consider two very briefly. Though relatively uncomplicated, they are reasonably representative of contemporary technique.

Frey[25] studied the isomerization of 3-methylcyclobutene to trans-penta-1,3 diene in a 150 ml vessel at 386–429 K and pressures from 0.4 to 23 torr.

$$C_4H_5CH_3 \rightarrow CH_3CH{:}CHCH{:}CH_2 \qquad \text{2–I}$$

The temperature during an experiment was held constant to $\pm 0.02°$ in an oil thermostat and was accurate to $\pm 0.1°$; greaseless valves were used and the dead volume was 0.2‰, that is, negligible. The reactant was admitted to the reaction vessel from a calibrated gas pipette, thus fixing the initial pressure; the partly reacted gas, after total withdrawal into a liquid nitrogen trap, was analysed by gas chromatography. No product other than the main product was detected even at 90% conversion. The reaction was found to be accurately first order at each of nine temperatures, as exemplified by the values of the rate constant calculated from equation 2–2 given in Table 2–1. Packing the reaction vessel with glass tubing to multiply the surface-to-volume ratio by 10 had no effect on the rate constant. This, however, decreased uniformly with decreasing pressure over the pressure range mentioned above. The value at infinite pressure k_∞ was obtained by extrapolating a plot of $1/k$ against $1/P$ (see Chapter 5). From these and other results, together with reference to the known behaviour of analogous compounds, it was concluded that the reaction is

§ In this case, the value of the rate constant is given by $k = \{[(P_\infty/P_0) - 1]/(q-1)\}^n S_0/(1-\delta)$, where S_0 is the initial slope of the plot of the appropriate integrated concentration function against time for $n = 1$ or $n \neq 1$ as described in Section 2–1–1.

Table 2–1 Experimental data for the isomerization of 3-methylcyclobutene at 409.6 K (Frey[25])

(initial pressure of 3-methylcyclobutene (MCB) 7 torr)

Time (s)	360	540	780	960	1140
% MCB remaining	83.8	77.0	68.6	62.5	56.9
$10^4\,k\,s^{-1}$	4.90	4.85	4.84	4.90	4.94
Time (s)	1320	1500	1680	1920	2040
% MCB remaining	52.5	48.1	43.85	39.1	36.8
$10^4\,k\,s^{-1}$	4.88	4.88	4.91	4.93	4.90

an elementary unimolecular transformation. An Arrhenius plot of the rate constants at 7 torr pressure gave a good straight line which, when corrected for the pressure dependence, yielded the relation

$$k_\infty = 10^{13.53} \exp[-31.55/RT]\,s^{-1} \qquad 2\text{–}12$$

The Arrhenius parameters are subject to negligible statistical error and their absolute accuracy was estimated to be about $\pm 1\%$.

In subsequent work,[25b] a vessel of 1 litre capacity was used to study the reaction at pressures down to 0.01 torr. At the lowest pressure the rate constant was found to be less than a third the value at infinite pressure (at 422 K) and, in accordance with the theory of unimolecular reactions, was partly restored to this value when the reaction was conducted in the presence of carbon dioxide or another inert gas.

Our second example comes from the work of Whittle and collaborators[26] on the reaction between bromine and fluoroform at 634–738 K. The stoichiometry:

$$Br_2 + CF_3H \rightarrow HBr + CF_3Br \qquad 2\text{–II}$$

was established by analyses of known volumes of the partly reacted gas withdrawn from the reaction vessel: bromine was removed by freezing out at $-80\,°C$ and hydrogen bromide by dissolving in water; and the residual $CF_3H + CF_3Br$ mixture was analysed by gas chromatography. The kinetics were determined photometrically by measuring the rate of disappearance of bromine from a quartz reaction vessel in an oven thermostated to $\pm 0.2\,°C$, total pressures being in the range 50–450 torr. A logarithmic photometer was used to give an output to a pen-recorder and thus effectively to produce a plot of bromine concentration against time. The photometer was calibrated with known pressures of bromine measured by a spiral gauge and the light beam was shown to be too weak to initiate photochemical reaction. Reaction orders in bromine and fluoroform were obtained at various temperatures by the initial rate method. Plots of log (initial rate) against log (concentration) were straight lines with (typically) slopes 0.51 ± 0.02 for bromine and 0.99 ± 0.01 for fluoroform. Hence, the rate equation is

$$-d[Br_2]/dt = k[Br_2]^{1/2}[CF_3H] \qquad 2\text{–}13$$

At longer reaction times the reaction was found to be retarded by the hydrogen bromide produced and this was verified by initial addition of this gas. However, the formation of hydrogen bromide did not affect the initial rate measurements. Equation 2–13 is characteristic of thermal bromination reactions (see Section 3–5–5) and indicates that the rate is determined by a reaction between fluoroform and bromine atoms. The latter are in equilibrium with molecular bromine at the reaction temperature:

$$Br_2 \rightleftharpoons 2Br \qquad \text{2–III}$$

$$Br + CF_3H \rightarrow HBr + CF_3 \qquad \text{2–IV}$$

In this case $k_4 = k/K_3^{1/2}$ and the authors used known values of K_3 to obtain k_4 at several temperatures from their results. An Arrhenius plot yielded the relation

$$k_4 = (13.107 \pm 0.035) \exp\left[-(22.32 \pm 0.11)/RT\right] \text{ cm}^3 \text{ mole}^{-1} \text{ s}^{-1} \qquad \text{2–14}$$

The inhibition by hydrogen bromide, which is due to the reverse of reaction 2–IV, was also studied but will not be discussed here.

2–2 The linear flow method

In the usual form of this technique the reactants are passed at a constant measured rate of flow through a tube at constant temperature and the amount of reaction that occurs in the time elapsing between the entrance and the exit is determined. The reaction time is varied by changing the rate of flow. The reaction tube is usually 2–3 cm in diameter and 20–30 cm long, though longer (coiled) tubes have been used for slower reactions. The usual practice is to collect the products and unreacted reactants for a measured time after steady conditions have been established and to analyse them subsequently, although, when relatively involatile reactants produce a permanent gas, the accumulation of the latter can be monitored continuously. For work at atmospheric pressure, the reactants are usually diluted with an excess of inert carrier gas before entering the reaction tube.[27] It is frequently more convenient, however, to work at pressures of about 5–50 torr; the flow can then be maintained by a mercury ejector pump or simply by cold traps or a large evacuated flask downstream. Figure 2–3 is a diagram of an apparatus suitable for such pressures. The pressure in the reaction tube is controlled by a thermostated capillary. The apparatus can also be constructed in circulatory form: a carrier gas is used which, after the products and unreacted reactants have been trapped out, is returned with fresh reactant to the reaction vessel.[28]

The flow method is less flexible than the static method, but it can be used at reaction times down to about 1 second or somewhat less. It is therefore suitable for faster reactions. Since an indefinitely large amount of the products can be collected for analysis, it is also suitable when the reaction is to be studied at a small degree of reaction.

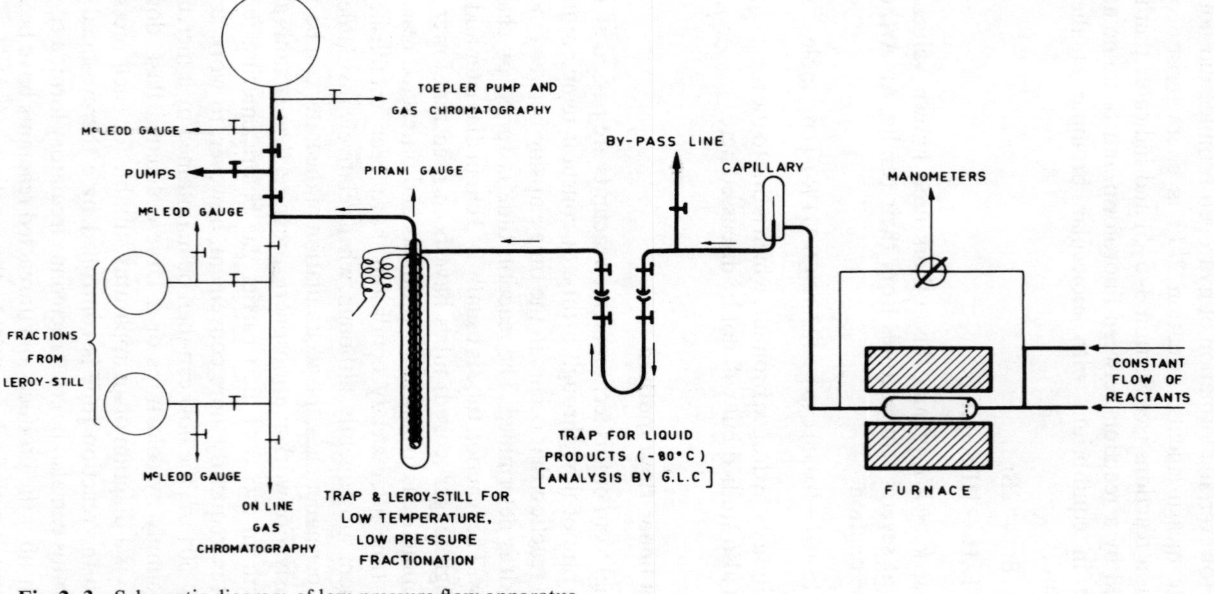

Fig. 2–3 Schematic diagram of low-pressure flow apparatus.

2–2–1 Derivation of kinetic parameters

The linear flow method is not very suitable for the study of reactions with complex kinetics, and indeed in recent times has been mostly applied to first-order or pseudo-first-order reactions. It is nearly always assumed that all the reactant molecules spend the same time in the reaction tube, apart, of course, from those disappearing by reaction. The residence or 'contact' time is related to various quantities by the expressions:

$$t_c = \frac{L}{\bar{u}} = \frac{\pi \mathbf{R}^2 L}{v} = \frac{V}{v} = \frac{PV}{RT\sum N_{i,0}} \qquad \text{2–15(a),(b),(c),(d)}$$

where L, $\mathbf{R}$, and V are the length, radius, and volume of the reaction tube; v and $\bar{u}$ are respectively the volumetric and average linear flow velocities measured at the reaction temperature T and pressure P; and $\sum N_{i,0}$ is the total number of moles entering the reaction tube per second. Rate constants are obtained by applying the appropriate integrated rate expression. If the reaction produces no change in volume these are directly analogous to equations 2–2 and 2–3. Thus, for a first-order reaction,

$$f = \frac{N_A}{N_{A,0}} = \frac{[A]}{[A]_0} = \exp(-kt_c) \qquad \text{2–16}$$

or

$$kt_c = -\ln f \qquad \text{2–17}$$

When the change in volume brought about by the reaction is appreciable relative to the total volume of reactants and carrier gas (if any), the actual residence time is shorter or longer because of the expansion or contraction. Equation 2–17 then becomes

$$kt_c = -x_A(q-1)(1-f)-[1+(q-1)x_A]\ln f \qquad \text{2–18}$$

where q is as defined in equation 2–1, x_A is the mole fraction of the reactant in the original gas, and t_c is calculated without reference to the volume change, that is by equation 2–15. The derivations of this expression and the corresponding expressions for second-order reactions are given by Harris.[29] As the degree of reaction approaches zero, these expressions all approximate to differential rate expressions analogous to equation 2–5:

$$kt_c = \frac{-\Delta[A]}{[A]_0^{n_A}[B]_0^{n_B}} \qquad \text{2–19}$$

2–2–2 Heat and mass transfer in the reacting gas (potential errors in the linear flow method)

The above equations are based on assumptions regarding heat and mass transfer in the reaction tube. In addition to the assumption that all the molecules spend the same time in the tube, it is assumed that they attain the reaction temperature immediately on entering it and that, at the exit, they are immediately cooled sufficiently to bring the reaction to a halt. The following discussion is presented partly to indicate the errors these assumptions incur in rate measurements and partly as a brief introduction to the interrelation between the rate of chemical reaction and heat and mass transfer (particularly the latter) in flowing gas. This general topic is becoming increasingly important as studies in gas kinetics are extended to faster reactions and higher temperatures. Some further aspects, related to the kinetics of combustion, will be found in Chapter 7.

The assumption of *thermal equilibration* in the reaction tube means in practice that the axial temperature profile of the reacting gas is taken to be identical to that of the tube. The latter is usually determined by passing a thermocouple through the tube, and adjustments are made to the furnace or other thermostat until the temperature profile is 'flat' with abrupt falls at the entrance and exit of the tube. Thereafter the reaction temperature is taken to be that recorded by a thermocouple situated either inside a central thermometer well or in contact with the outer wall of the tube. The significance of such measurements with regard to the gas temperature, however, is doubtful. It is more reasonable to assume that the measured profile relates to the temperature of the wall of the tube rather than to that of the gas. In general, the gas will pass some distance into the tube before reaching the wall temperature and again some distance out of it before the temperature falls sufficiently for the reaction to stop. For a first-order reaction, the apparent value of the rate constant k_{app}, obtained by substituting the initial and final concentrations in equation 2–17, is related to the true value by the following expression:[30]

$$\frac{k_{app}}{k} \approx 1 - \frac{\mathbf{R}^2}{3.66\kappa t_c}\left\{0.58 + \ln\left[\frac{0.82E(T_s - T_0)}{RT_s^2}\right]\right\} \qquad 2\text{–}20$$

where T_s is the temperature of the wall and T_0 that of the entering gas; E is the activation energy of the reaction (which need be known only approximately) and κ is the coefficient of thermal diffusivity of the gas. The value of κ can be taken as equal to the coefficient of molecular self-diffusion at the mean temperature of the gas and the wall. As expected, the expression approaches unity as t_c or κ becomes large or $\mathbf{R}$ becomes small.

Since κ is inversely proportional to the pressure, the error increases with increasing pressure (P). Some values of the error appropriate to the reaction of an organic vapour under various conditions are given in column 5 of Table 2–2. It can be shown from equation 2–20 that, as a rough general rule, if the error in k_{app} due to lack of thermal equilibrium is to be less than 10% the quantity $(t_c/\mathbf{R}^2P)$ should be greater than about 0.05 s cm^{-2} torr^{-1}. When a carrier gas is used, thermal equilibration can be greatly accelerated by heating the carrier gas to the reaction temperature separately before the reactants are injected into it at the entrance of the reaction tube.

Turning to *mass transfer*, the assumption that all the reactant molecules spend the same time in the reaction tube implies that they move through it by simple 'piston' or 'plug' flow. Although, as will be seen, this is often a valid approximation, the actual situation is a good deal more subtle than it may at first appear. In the first place, if the reactants enter the reaction tube from a narrower tube there is a tendency for them to form a jet and to leave an annular region of relatively stagnant gas near the wall.(31) This effect can be reduced by causing the reactants to enter at right angles to the reaction tube or through a number of small holes. Secondly, assuming they enter uniformly, the reactant molecules move forward not only by flow but also by diffusion. The latter is caused by the fact that, owing to the reaction, their concentration is less near the exit than near the entrance of the tube. It is necessary to consider this in a little more detail.

Table 2–2 Heat- and mass-transfer parameters and calculated errors in first-order rate constants for the reaction of an organic vapour in a tube 20 cm long and 2 cm in diameter ($E = 65$ kcal mole^{-1}).

ε_h and ε_m are the errors arising from the assumption of thermal equilibrium and piston flow respectively. Each is defined by the expression $k = (1 + 0.01\varepsilon)k_{app}$, k being the true and k_{app} the apparent rate constant (T_s = wall temperature).

T_s (K)	p (torr)	t_c (s)	κ (cm^2 s^{-1})	ε_h (%)	D (cm^2 s^{-1})	G (cm^2 s^{-1})	μ	ε_m (%)
1000	1.0	1.0	166	0.5	230	230	1.16	9
1000	1.0	0.1	166	6	230	235	0.12	1
1000	10	0.1	17	140	23	59	0.03	0
1000	10	2.0	17	3	23	23	0.23	3
890	760	20	0.18	40	0.24	0.33	0.03	0
750	760	20	0.13	60	0.18	0.30	0.03	0

Assuming constant temperature and neglecting any volume change caused by the reaction, the steady-state concentration of the reactant [A] at any point (z, r) in a cylindrical tube is given by the solution of the conservation equation,

$$-u(r)\frac{\partial[\mathrm{A}]}{\partial z} + D\left(\frac{\partial^2[\mathrm{A}]}{\partial r^2} + \frac{1}{r}\frac{\partial[\mathrm{A}]}{\partial r} + \frac{\partial^2[\mathrm{A}]}{\partial z^2}\right) = k[\mathrm{A}]^n \qquad 2\text{–}21$$

where $u(r)$ is the linear velocity of flow at a radial distance r from the axis of the tube, z is the axial distance from the entrance, and D is the coefficient of diffusion of the reactant molecules through the gaseous mixture. The terms on the left represent the difference between the number of molecules entering and leaving an element of volume per second by flow and diffusion; this must be equal to the number of molecules consumed by reaction. The concentration measured experimentally is normally the 'cup-mixing' effluent concentration $[\bar{\mathrm{A}}]$, that is, the concentration found when the total gas emerging at distance L in unit time is uniformly mixed in unit volume. This is given by

$$[\bar{A}] = \frac{2t_c}{\mathbf{R}^2 L}\int_0^{\mathbf{R}} [A]_{L,r} r u(r)\, dr \qquad 2\text{–}22$$

The piston flow assumption is equivalent to setting

$$D = 0 \qquad 2\text{–}23$$

and

$$u(r) = L/t_c = \text{constant} \qquad 2\text{–}24$$

Solution of equation 2–21 for the boundary conditions $[A] = [A]_0$ at $z = 0$ and $[A] = [\bar{A}]$ at $z = L$ then leads directly to the simple rate expressions such as equation 2–17 for a first-order reaction. The question is whether the assumptions 2–23 and 2–24 are sufficiently realistic for the given experimental conditions.

The mean distance diffused by a molecule in one residence time is about $(2Dt_c)^{1/2}$. Diffusion coefficients are inversely proportional to the pressure and at 10 torr are likely to be in the range 10–100 $cm^2\,s^{-1}$, depending upon the nature of the gas and the temperature. Hence if t_c is of the order of seconds or longer, diffusion distances will be commensurate with the length of the reaction tube. The effect of diffusion should therefore be considered.

It can be shown that, under any conditions likely to be realized, the gas as a whole will move by viscous flow. Thus $u(r)$ is given by the Poiseuille formula,

$$u(r) = 2\bar{u}(1 - r^2/\mathbf{R}^2) \qquad 2\text{–}25$$

and not by equation 2–24. This means that the gas near the axis of the tube flows faster than the gas near the wall; therefore along any radius at z the reaction has proceeded further at points near the wall than at the axis. Hence, there is a radial concentration gradient causing the reactant molecules to diffuse towards the wall. If diffusion were altogether absent, the reactant molecules would move with the bulk of the gas in purely viscous flow and it is possible to show that, for the same residence time t_c, this would result in a smaller degree of reaction than piston flow (see Fig. 2–4). The effect of radial diffusion is partly to offset the difference. The radial concentration gradient exists, of course, in addition to the axial concentration gradient which causes the reactant molecules to diffuse towards the exit (and consequently reduces their time for reaction in the tube).

In principle the value of $[\bar{A}]/[A]_0$ can be obtained by solving equations 2–21 and 2–22 after substituting for $u(r)$ from equation 2–25. However, the mathematical complexity can be reduced substantially by making use of a remarkable phenomenon discovered by Taylor.[32] When a solution is allowed to flow steadily into a tube after the pure solvent, under appropriate conditions the combined action of the parabolic flow profile

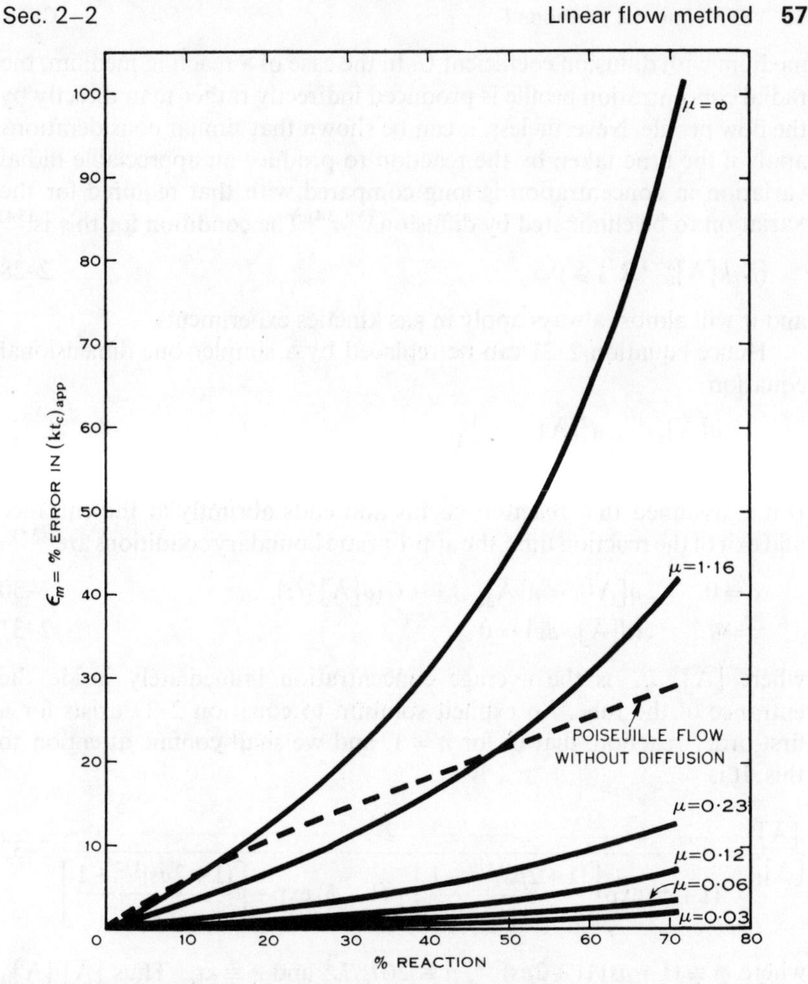

Fig. 2–4 The linear-flow method. Error in a first-order rate constant (k_{app}) calculated assuming piston flow without diffusion. The error ε_m is shown as a function of the degree of reaction for different values of μ ($= 2Gt_c/L^2$). The broken line shows the error incurred by the piston-flow assumption if (hypothetically) viscous flow were occurring without either radial or axial diffusion. ε_m is defined by the relation $(kt_c) = (1+\varepsilon_m/100)(kt_c)_{app}$ (Mulcahy and Pethard[30]).

and radial diffusion causes the solute to become axially dispersed about a radial plane moving with the mean speed of flow 'exactly as though it were being diffused by a process which obeys the same law as molecular diffusion'[32]—but with a diffusion coefficient G' given by

$$G' = \mathbf{R}^2\bar{u}^2/48D \qquad 2\text{–}26$$

When, in addition, axial diffusion is appreciable, the pseudo diffusion coefficient becomes[33]

$$G = (D+\mathbf{R}^2\bar{u}^2/48D) \qquad 2\text{–}27$$

The situation is equivalent to piston flow of the solute with velocity $\bar{u}$ in a medium with diffusion coefficient G. In the case of a reacting medium, the radial concentration profile is produced indirectly rather than directly by the flow profile. Nevertheless, it can be shown that similar considerations apply if the time taken by the reaction to produce an appreciable radial variation in concentration is long compared with that required for the variation to be eliminated by diffusion.[30,34] The condition for this is[34]

$$(D/k[\mathrm{A}]_0^{n-1}\mathbf{R}^2) \geqslant 0.5 \qquad 2\text{–}28$$

and it will almost always apply in gas kinetics experiments.

Hence equation 2–21 can be replaced by a simpler one-dimensional equation:

$$-\bar{u}\frac{d[\overline{\mathrm{A}}]_z}{dz} + G\frac{d^2[\overline{\mathrm{A}}]_z}{dz^2} = k[\overline{\mathrm{A}}]_z^n \qquad 2\text{–}29$$

If it is assumed that reaction begins and ends abruptly at the entrance and exit of the reaction tube, the appropriate boundary conditions are:[35]

$$x = 0, \qquad \bar{u}[\mathrm{A}]_0 = \bar{u}[\overline{\mathrm{A}}]_{z\to 0+} - G(d[\overline{\mathrm{A}}]/dz)_{z\to 0+} \qquad 2\text{–}30$$

$$x = L, \qquad (d[\overline{\mathrm{A}}]_z/dz) = 0 \qquad 2\text{–}31$$

where $[\overline{\mathrm{A}}]_{z\to 0+}$ is the average concentration immediately inside the entrance of the tube. An explicit solution to equation 2–29 exists for a first-order reaction, that is, for $n = 1$, and we shall confine attention to this. It is

$$\frac{[\overline{\mathrm{A}}]}{[\mathrm{A}]_0} = \frac{2}{(1+\phi)\exp\left[\dfrac{(1+2\mu s)^{1/2}-1}{\mu}\right] + (1-\phi)\exp-\left[\dfrac{(1+2\mu s)^{1/2}+1}{\mu}\right]} \qquad 2\text{–}32$$

where $\phi \equiv (1+\mu s)/(1+2\mu s)^{1/2}$, $\mu \equiv 2Gt_c/L^2$ and $s \equiv kt_c$. Thus $[\overline{\mathrm{A}}]/[\mathrm{A}]_0$ is a function of the dimensionless quantities μ and s; and if k is to be determined from the measured value of $[\overline{\mathrm{A}}]/[\mathrm{A}]_0$ it is equation 2–32 which, in principle, should be used rather than equation 2–17. Before comparing equations 2–32 and 2–17 generally, it will be useful to note two limiting cases. Equation 2–32 becomes identical with equation 2–17 when $\mu \to 0$ or $s \to 0$, that is, for example, when $G \to 0$ or $t_c \to 0$. Hence no error is incurred by using equation 2–17 under these conditions. When on the other hand $\mu \to \infty$, that is, principally when $D \to \infty$ equation 2–32 reduces to

$$\frac{[\overline{\mathrm{A}}]}{[\mathrm{A}]_0} = \frac{1}{kt_c + 1} \qquad 2\text{–}33$$

This is the case of perfect mixing in the reaction tube (brought about here by diffusion) to which further reference will be made in Section 2–3–1. It represents the maximum departure from the piston–flow situation *for the particular entrance and exit conditions chosen.*

The error in the observed rate constant k_{app} incurred by using equation 2–17 instead of equation 2–32 can be found by calculating k_{app} and k from the measured value of $[\text{A}]/[\text{A}]_0$ using the respective equations. This is shown in Fig. 2–4 as a function of the percentage reaction for values of μ corresponding to the conditions given in Table 2–2. The value of k_{app} is always too low; the error for a given residence time increases with increasing degree of reaction and, of course, increasing values of μ. Values at 25% reaction are given in column 9 of Table 2–2. Errors for other experimental conditions can be estimated from Fig. 2–4 by interpolating values of μ. For the range of conditions appropriate to the types of reaction considered in this chapter the Taylor dispersion term G' makes an appreciable contribution to the value of μ only when the absolute error is small. Hence in calculating μ it is usually sufficient to substitute D for G. As a rough rule, if the error in k_{app} is to be less than 10% in experiments involving up to 25% reaction, the experimental conditions should be such that t_c/L^2P is less than about 0.005 s cm^{-2} torr^{-1}. This may be compared with the statement on p. 54 regarding the error associated with lack of thermal equilibration.

Flow through the reaction tube implies a *pressure drop* along its length. When the flow rate is high, this can also affect the reaction rate. Normally the effect is negligible for the type of experiment considered in this chapter, but it may need to be considered in the fast-flow experiments described in Chapter 4. It is convenient to discuss it briefly here. In fast-flow experiments the extent of reaction is usually determined by measuring the concentration of a reactant *in situ* at two or more positions along the reaction tube, for example by optical absorption. For the present purposes it is sufficient to consider simply an upstream and a downstream position. With the *in situ* method of measuring the concentrations the effect of pressure drop is twofold. First, for a given upstream pressure, it would cause the concentration of reactant downstream to be smaller than the upstream value even if no reaction occurred; unless allowed for, this appears as a loss of reactant due to reaction. Second, the expansion due to the lower pressure causes the molecules to spend less time in the reaction tube, and therefore less reaction occurs.

Here again the relation between the observed degree of reaction and the true rate constant is obtained by solving the appropriate conservation equation. An important practical case is that of a reaction which effectively consumes only one reactant and is conducted in a large excess of carrier gas. The carrier gas may be inert or, alternatively, it may constitute a second reactant (B). Thus,

$$\text{A}(+\text{B}) \rightarrow \text{Products}$$

In these circumstances, consideration of the number of molecules of A entering and leaving an element of volume $\pi\mathbf{R}^2\,dz$ per second gives

$$-v\,d[\text{A}]-[\text{A}]\,dv = \pi\mathbf{R}^2 k[\text{A}]^{n_\text{A}}[\text{B}]^{n_\text{B}}\,dz \qquad 2\text{–}34$$

The volumetric flow velocity v is related to the pressure P at any position z by equation 2–15(c), (d) and P is related to z and the upstream pressure P_0 by the Poiseuille expression:

$$(P/P_0) = (1-bz)^{1/2} \qquad 2\text{–}35$$

where

$$b = \frac{16\eta RT}{\pi \mathbf{R}^2} \cdot \frac{\sum N_i}{P_0^2} \qquad 2\text{–}36$$

and η is the viscosity of the gas. Equations 2–35 and 2–36 enable equation 2–34 to be expressed exclusively in terms of [A] and P. Setting P/P_0 at the downstream position equal to y, integration gives

$$kt_c[\mathrm{B}]^{n_\mathrm{B}} = \frac{3+n_\mathrm{B}}{2}\left(\frac{1-y^2}{1-y^{3+n_\mathrm{B}}}\right)\ln\frac{y[\mathrm{A}]_0}{[\mathrm{A}]} \qquad 2\text{–}37$$

for a reaction first order in [A], that is for $n_\mathrm{A} = 1$, and

$$kt_c[\mathrm{A}]_0[\mathrm{B}]^{n_\mathrm{B}} = \frac{4+n_\mathrm{B}}{2}\left(\frac{1-y^2}{1-y^{4+n_\mathrm{B}}}\right)\left(\frac{y[\mathrm{A}]_0}{[\mathrm{A}]}-1\right) \qquad 2\text{–}38$$

for a reaction second order in [A]. Here t_c is defined, as usual, by equation 2–15. When B is inert or non-existent, $n_\mathrm{B} = 0$; and both equations reduce to the simple rate expressions, 2–2 and 2–3 as y tends to unity.

2–2–3 An investigation by the linear flow method: the pyrolysis of toluene

Among the great number of pyrolytic reactions studied by the linear flow method during the past two decades this reaction has been something of a *cause célèbre*. It has been investigated in at least four laboratories, the principal object being to discover how the observed rate of decomposition in the flow system is related to the rate of dissociation of the $C_6H_5CH_2$—H bond. The question is linked with the validity or otherwise of the *toluene carrier* technique which has been used extensively to study the unimolecular dissociations of numerous molecules to radicals. Further reference to this technique will be found in Chapter 3. In the meantime the work of Price[36] on the pyrolysis of toluene *per se* provides an interesting example of the linear flow method. Although the objective was to isolate the initial reaction step rather than to study the overall kinetics, the example is chosen as representative of the type of investigation to which the method has been most often applied.

The products of the pyrolysis are complex but, in the main, can be considered formally to be produced by the reactions:[37,38]

$$2C_6H_5CH_3 \rightarrow (C_6H_5CH_2)_2 + H_2 \qquad 2\text{–V}$$

$$3C_6H_5CH_3 \rightarrow (C_6H_5CH_2)_2 + C_6H_6 + CH_4 \qquad 2\text{–VI}$$

$$2C_6H_5CH_3 \rightarrow (C_6H_4CH_3)_2 + H_2 \qquad 2\text{–VII}$$

These equations are purely the stoichiometric expression of the final results of a complex free-radical mechanism. The flow system used by Price was basically similar to that shown in Fig. 2–3. Pressures were mainly between 10 and 20 torr and temperatures between 910 and 1140 K; a maximum of 5% of the toluene was decomposed. Rate measurements were based on the production of methane and hydrogen which were accumulated together for a given time. The combined volume was measured in a gas burette and analysed by gas chromatography. The fraction of methane proved to be constant to within ±6%. First-order rate constants were calculated on the assumptions that the initial reaction is

$$C_6H_5CH_3 \rightarrow C_6H_5CH_2 + H \qquad \text{2–VIII}$$

and that each hydrogen atom formed produces one molecule either of hydrogen or methane by reactions of the type:

$$H + C_6H_5CH_3 \rightarrow C_6H_5CH_2 + H_2 \qquad \text{2–IX}$$

$$H + C_6H_5CH_3 \rightarrow C_6H_6 + CH_3 \qquad \text{2–X}$$

$$CH_3 + C_6H_5CH_3 \rightarrow C_6H_5CH_2 + CH_4 \qquad \text{2–XI}$$

Values were reproducible to about ±5% and at 1053 K were found to be independent of the concentration of toluene. They decreased when the residence time was reduced to less than 1 second, probably because of lack of thermal equilibration. When a reaction vessel packed with quartz tubing was used, the rate increased in proportion to the increased surface and became much less dependent on temperature. These facts enabled the contribution from heterogeneous reaction in the empty vessel to be allowed for. Because of the large difference between the temperature coefficients of the homogeneous and heterogeneous reactions, the correction became negligible above 1050 K. Finally, an Arrhenius plot of the rate constants for the homogeneous reaction gave the expression

$$k = 10^{14.8} \exp(-85/RT)\,\mathrm{s}^{-1} \qquad \text{2–39}$$

for the temperature range mentioned above. Whether this relation can be fairly attributed to reaction 2–VIII depends on the validity of the assumed reaction mechanism. The observed stoichiometry provides reasonable though not conclusive evidence for this view. (Supporting evidence is provided by an independent determination of 84±3 kcal mole^{-1} for the enthalpy change of reaction 2–VIII;[39] since the activation energy of the reverse reaction is expected to be zero this value should be equal to the activation energy of reaction 2–VIII.)

2–3 The stirred-flow method

With this method, the reaction is conducted in a flow regime which may be considered the opposite of that usually assumed to occur in the linear-flow method. The reactants flow through a reaction vessel designed to mix reactants, products, and carrier gas (if any) completely and uniformly throughout its volume. In the language of chemical engineering, the vessel functions as a 'perfectly stirred reactor'. When a steady state is reached, the mixture that flows out of the vessel and is collected for analysis has exactly the same composition as the contents of the vessel. Until recently, the method has been applied rather rarely to gas-phase reactions; but it has several advantages over the more conventional linear-flow method. In the first place, in principle, mixing cannot be reduced below the level of diffusional interpenetration but it can be intensified at will by a suitable device; therefore the assumption of complete mixing is more capable of being substantiated than the assumption of no mixing which is usually made with the linear-flow method. Secondly, the agitation required to produce uniform composition also brings about rapid equilibration of the gas temperature. And, thirdly and most usefully, in the perfectly stirred condition, the concentrations of reactants, products, and any intermediate species in the reactor are all maintained constant. Therefore the rate of efflux of each product and intermediate species is precisely equal to the rate at which the species is produced by the reaction at stationary concentrations of all the other species, including, of course, the reactants. Since these concentrations can be determined in the effluent gas, this fact is of great value in studies of complex reaction mechanisms. On the other hand, like the linear flow technique the method lacks some of the control that is possible with the static method; it can be difficult to change one variable, such as pressure, without to some extent changing others, such as the degree of reaction. The relatively few investigations hitherto carried out by the method have been concerned with reactions with first-order kinetics, but there is no reason to prevent it from being applied to reactions of other orders provided effectively perfect mixing can be produced at the molecular level.[40]

A simple stirred-flow reactor suitable for operation at pressures of about 1 to 50 torr and residence times of about 0.1 to 100 s is shown in Fig. 2–5. The reactants with or without carrier gas enter through the central spherical perforated 'diffuser'; and rapid mixing occurs by a combination of vigorous convection and diffusion. The remainder of the apparatus is similar to that used for the linear flow method. If, however, the residence time is determined by measuring the total flow of gas, the flow meter should be included downstream of the reactor. The reason for this will appear in the following discussion.

2–3–1 Derivation of kinetic parameters

Rate expressions for stirred flow are very simply derived. In the stationary condition, the rate at which reactant molecules enter the reactor is exactly balanced by the rate at which they leave plus the rate of reaction. Thus, for a first-order reaction, $A \rightarrow qQ$:

$$v_0[A]_0 - v[A] - Vk[A] = 0 \qquad 2\text{–}40$$

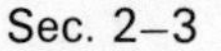

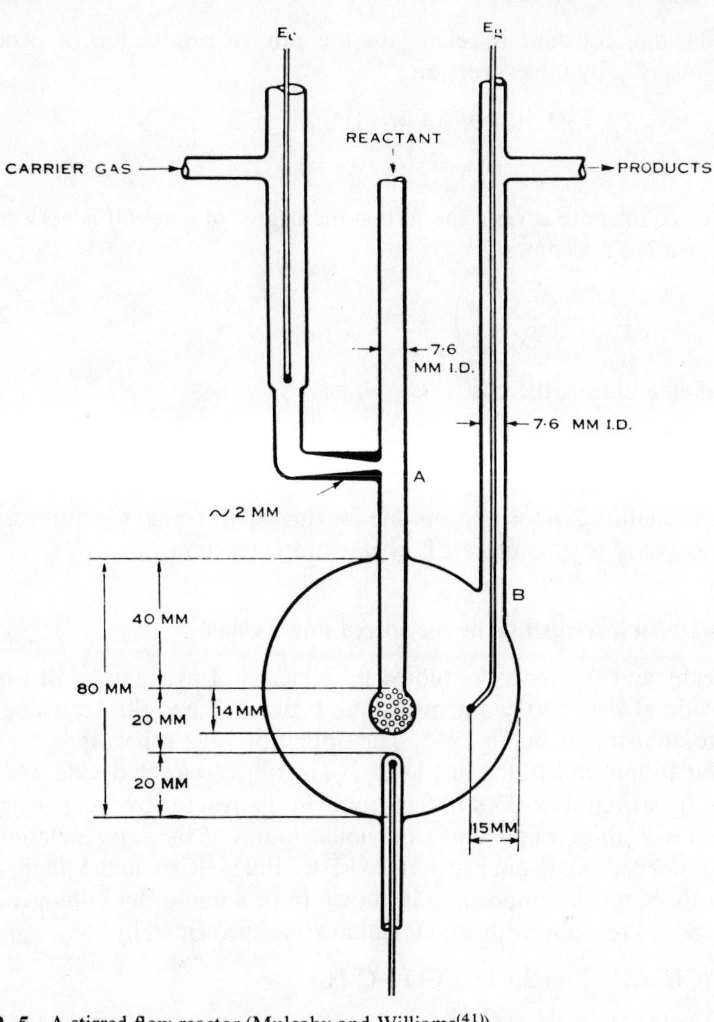

Fig. 2–5 A stirred-flow reactor (Mulcahy and Williams[41]).

where v_0, v are the volumetric flow rates in and out of the reactor (corrected to the reaction temperature) and [A] refers to the concentration in the reactor, which has the volume V. Hence

$$\frac{[\mathrm{A}]}{[\mathrm{A}]_0} = \frac{v_0}{Vk+v} \qquad 2\text{–}41$$

This expression is identical to the equation for linear flow with infinitely fast diffusion and, for the case of $q = 1$, is equivalent to equation 2–33. Equation 2–41, when expressed in terms of the molecular flow rates of the reactant ($N_\mathrm{A} = v[\mathrm{A}]$ mole s^{-1}), becomes:

$$\frac{N_\mathrm{A}}{N_{\mathrm{A},0}} = \frac{v}{Vk+v} = \frac{1}{kt'_\mathrm{c}+1} \qquad 2\text{–}42$$

where the residence time $t'_\mathrm{c} = V/v$ is defined by the rate of total efflux.

The rate constant is related to the rate of production of product molecules N_Q by the expression:[41]

$$k = \frac{N_Q RT}{qPV}\left\{\frac{q(N_{A,0}+N_M)+(q-1)N_Q}{qN_{A,0}-N_Q}\right\} \qquad 2\text{–}43$$

where N_M refers to carrier gas. When the degree of reaction is very small equation 2–43 becomes

$$k \approx \frac{N_Q RT}{qPV}\left(\frac{N_{A,0}+N_M}{N_{A,0}}\right) \qquad 2\text{–}44$$

and if, in addition, there is no carrier gas,

$$k \approx \frac{N_Q RT}{qPV} \qquad 2\text{–}45$$

When equation 2–45 is appropriate for the experimental conditions, it is not necessary to measure the flow rate of the reactant.

2–3–2 An investigation by the stirred-flow method

Mulcahy and Williams[41] studied the kinetics of pyrolysis of di-*t*-butyl peroxide at 430–550 K and pressures between 2 and 30 torr using the reactor illustrated in Fig. 2–5. The apparatus was otherwise generally similar to that illustrated in Fig. 2–3. The object was to decide whether perfectly stirred flow does in fact occur in the reactor by comparing the results with those obtained in previous studies of the same reaction by the static method. From previous work by Raley, Rust, and Vaughan[42] and others, the decomposition is known to be a first-order homogeneous free-radical reaction with the stoichiometry represented by

$$[(CH_3)_3CO]_2 \rightarrow 2(CH_3)_2CO + C_2H_6 \qquad 2\text{–XII}$$

The kinetic measurements were based on the rate of formation of ethane from the peroxide, either alone or in the presence of carbon dioxide used as a carrier gas. The flow rate of the peroxide was determined from the loss in weight of a thermostated reservoir of the liquid during the period of the experiment. After separation of the unreacted peroxide and acetone, the ethane and carbon dioxide (if present) were collected at 77 K and measured by fractionation into vessels of known volume provided with manometers. Values of the rate constant were calculated from equation 2–43 with $N_Q = 3N_{\text{ethane}}$ and $q = 3$. A few illustrative results are given in Table 2–3. When corrected to a common temperature, the standard deviation from the mean of these and other values not recorded in the table was ± 6 per cent. The Arrhenius relation found,

$$k = 10^{16.1} \exp(-38.3 \pm 0.4/RT)\ \text{s}^{-1} \qquad 2\text{–}46$$

agrees well with results obtained in several studies by the static method

Table 2–3 Kinetics of decomposition of di-*t*-butyl peroxide at 481 K

Some results obtained by the stirred-flow method. k is calculated from the data by equation 2–43; $k_{481} = k$ adjusted to 481 K using the measured activation energy (Mulcahy and Williams[41]).

Temp. (K)	*Con-version* (%)	*Total pressure* $\times 10^{-1}$ torr	$N_{A,0} \times 10^6$ (mole s^{-1})	$N_M/N_{A,0}$	t'_c (s)	$N_E \times 10^6$ (mole s^{-1})	$k \times 10^2$ (s^{-1})	$k_{481} \times 10^2$ (s^{-1})
481.5	0.9	0.82	8.82	48.5	0.17	0.0803	5.24	5.0
476.7	4.1	0.67	52.1	0	1.20	2.15	3.44	4.9
479.9	4.8	1.20	5.21	18.4	1.11	0.252	4.61	5.1
482.5	9.1	0.92	42.7	0	1.70	3.89	5.89	5.2
481.8	15.8	2.70	6.97	7.8	3.95	1.10	4.77	4.5
481.8	21.9	1.28	17.30	0	4.81	3.78	5.82	5.4
482.6	42.8	1.18	4.64	0	12.8	1.98	5.85	5.1
482.8	64.0	1.87	2.34_0	0	32.7	1.50_0	5.44	4.7

and thus indicates that the reactor functions genuinely as a perfectly stirred reactor under the specified conditions.

2–4 The shock-tube method

This technique uses a shock wave to heat the mixture of reactants to the reaction temperature. The gas is contained in a tubular vessel and the shock wave is generated by bursting a diaphragm as shown in Fig. 2–6. The temperature of the gas is thereby raised suddenly and uniformly to a value which depends on the intensity of the shock and the thermal properties of the gas. A rise of several thousand degrees is easily achieved. This takes place in the time corresponding to a few molecular collisions and, in the normal course of events, the temperature of the gas behind the advancing shock 'front' remains substantially constant at this level for a few hundred microseconds. With the development of instrumentation to study the progress of chemical change over intervals of this duration, the shock tube has become standard apparatus for the study of reactions at high temperatures. The advantage of almost instantaneous uniform heating throughout a finite volume is supplemented by the fact that the reaction is observed over a shorter time than is required for molecules to diffuse to or from the walls. Hence heterogeneous effects are eliminated. Very frequently, this combination of desirable conditions is difficult, if not impossible, to achieve by other means.

The theory and practice of the shock tube have been expounded in numerous reviews, two of which are cited in the Bibliography. The following account is purely qualitative. As shown in Fig. 2–6, a shock tube consists of a metal or heavy-walled glass pipe 2 to 10 cm in diameter and 2 to 10 m long, divided into two compartments by a metal or plastic diaphragm. (The large 'dump tank' shown in the figure is for a particular

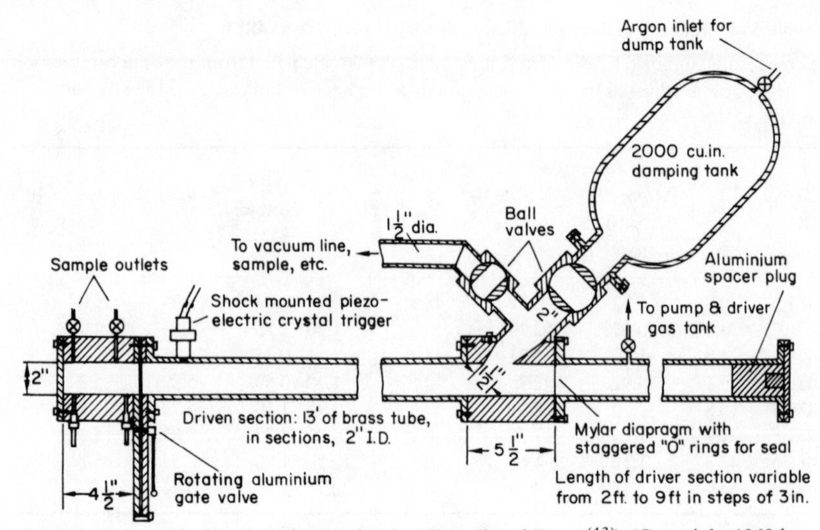

Fig. 2–6 Schematic drawing of a shock tube (Carroll and Bauer[43]). (Copyright 1969 by the American Chemical Society, reprinted by permission.)

application to be described later; it is not present in the basic form of the apparatus.) The left-hand compartment contains the reactants, which usually are heavily diluted by argon or other inert gas. The initial total pressure is typically 1 to 100 torr. The compartment on the right contains an inert 'driver' gas, normally hydrogen or helium, at a pressure of 1 to 10 atm; and the shock wave is produced by bursting the diaphragm either by puncturing it with an in-built stylus or simply by increasing the pressure in the driver compartment. A shock front travels at constant supersonic speed through the reactant gas and is followed at constant subsonic speed by the 'contact' surface of the expanding driver gas. This acts as a piston, pushing the shocked gas forward and compressing it. Between the shock front and the contact surface is the body of gas upon which the observations are made. When the shock front reaches the end of the tube, it is reflected back through the compressed gas, producing further heating and compression.

Measurements are carried out on the gas behind either the incident or the reflected shock, assuming, in the latter case, that it can be arranged for negligible reaction to occur in the incident shock. Usually this is not difficult since the second temperature is approximately double the first. The reflected shock is naturally appropriate for work at very high temperatures. It has the advantage that the twice-heated gas near the end of the tube is stationary and, with proper experimental arrangements, can be maintained in a constant condition for several milliseconds. On the other hand, departures from ideal aerodynamic conditions are more serious in the reflected shock and these cause errors in calculating the temperature and other properties of the shocked gas.

The temperatures and pressures of the gas in the incident and reflected waves and the flow velocity in the former can be calculated from the measured incident shock velocity and knowledge of the enthalpy of the gas over the temperature range involved. The latter includes the enthalpy contribution from the reaction, but this complication is frequently avoided by diluting the reactants sufficiently with inert gas. The shock velocity is determined by measuring the time taken by the shock front to pass between detectors mounted in the wall of the tube; thin film resistance-thermometer gauges or photo-electric detectors are commonly used to send impulses to an electronic counter or oscilloscope.

The shock tube can be applied to a kinetics problem basically in two ways. The first is by continuous monitoring of the concentration of an appropriate species at a fixed station (near the 'end plate' if the reflected shock is used). This operation is often conducted by absorption or emission spectrophotometry, infrared photometry being particularly valuable in the latter case. Windows are fitted in the wall of the shock tube so that the intensity of light received by a photo-detector via a monochromator or interference filter can be recorded by an oscillograph. Response times of less than one microsecond are usually required. Figure 2–7 illustrates such a record of light absorption obtained in a study of the decomposition of nitrosyl chloride in incident shocks. The nitrosyl chloride was the absorber and the arrival of the shock front at the window is shown as an abrupt rise in the absorption trace AA (due to compression) followed by a decline as decomposition occurred. Time markers BB from a frequency generator were recorded simultaneously. The square wave CC is a calibration trace; it was produced by a chopper disc before the nitrosyl chloride was admitted to the tube and shows the deflections for zero and total absorption of the light.

Rate constants are derived from such records by the initial rate method or by fitting the trace to the appropriate integrated rate expression in the usual way. When the incident shock is used, allowance has to be made

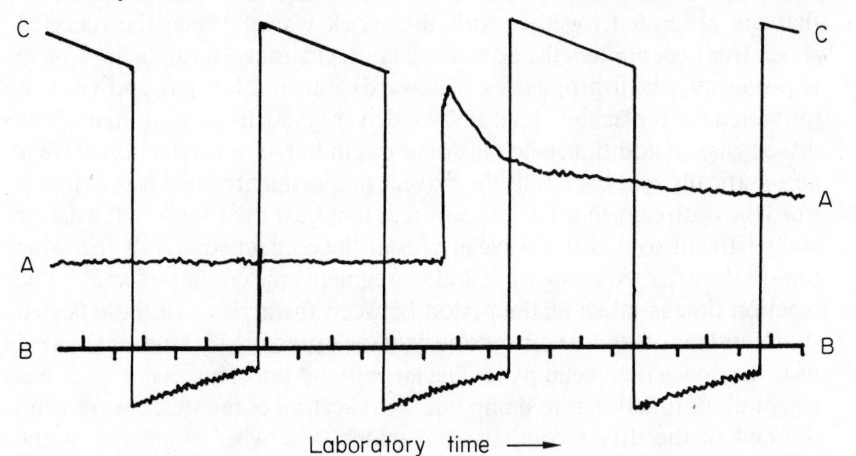

Fig. 2–7 Drawing of an oscilloscope record showing dissociation of nitrosyl chloride in a shock wave; AA, optical absorption trace; BB, time markers; CC, optical calibration trace (Palmer[45]).

for the fact that the compression in the shock wave from the initial pressure p_1 to the final pressure p_2 causes the gas to move bodily forward. Thus, an event occurring t seconds after the initiation of the reaction in the shock front appears at the window $(p_1/p_2)t$ seconds after the appearance of the shock front. Hence the apparent or 'laboratory' time scale has to be expanded by the ratio p_2/p_1. In addition, when absolute concentrations are required from an absorption trace, allowance must be made for the effect of temperature on the extinction coefficient. This is determined from the observed or extrapolated absorption at zero reaction time.

The progress of reaction can also be followed by mass spectrometry. Gas from the reflected shock is sampled directly into the electron beam of a time of flight instrument via a pin-hole in the end plate of the shock tube. Bradley and Kistiakowsky[46] used the technique originally to study the reaction

$$N_2O \rightarrow N_2 + \tfrac{1}{2}O_2 \qquad \text{2–XIII}$$

and obtained complete mass spectra of the reactant, products, and intermediate O atoms at intervals of 50 μs.

The second way in which shock waves are used in gas kinetics is by means of what is described conventionally as the *single pulse* or 'chemical' shock tube. The terms are equivalent—and, from the kinetics viewpoint, equally unfortunate. The procedures that have just been discussed are also concerned with events following single shock waves and are no less 'chemical' than those about to be described. The most prevalent form of the apparatus is shown by Fig. 2–6. The reactant gas is heated in the reflected shock, quenched after a definite time interval, and subsequently analysed. Thus, the procedure corresponds to the 'one-off' static method. Its characteristic feature is summarized by the words 'quenched after a definite time interval'. This is achieved by making use of rarefaction waves that are generated together with the shock waves. When the reflected shock front encounters the advancing contact surface, a rarefaction wave is produced which propagates backwards into the hot gas and cools it (provided the molecular weight of the driver is much less than that of the driven gas). In addition, when the diaphragm bursts, a rarefaction wave is automatically sent back into the driver compartment where it is reflected from the upstream end. The procedure is to adjust the length of the driver compartment so that this wave arrives at the contact surface at the same time as the reflected shock front and so augments the cooling effect.[47] The reaction time is taken as the period between the arrivals of the reflected shock and the combined rarefaction wave at a pressure transducer situated near the downstream end plate. The large dump tank shown in Fig. 2–6 is essential; its function is to damp out the reflection of the shock wave from the end of the driver compartment which otherwise would reheat the reactant gas. Before an experiment, the dump tank is filled with inert gas to the same pressure as that of the reactant gas and it is connected to the shock tube immediately before the diaphragm is burst. Afterwards a sample of the shocked gas near the end plate is withdrawn and analysed. In the particular apparatus illustrated in Fig. 2–6 the reacted gas in this region can be isolated by a valve and so protected from contamination by the driver gas.

The reaction temperature is calculated, as usual, from the shock velocity. The accuracy of the method suffers, however, from uncertainties both in this calculation, which tends to overestimate the temperature in the reflected wave, and in the measured reaction time, which may be longer than it appears because of a finite cooling time. For these reasons, the determination of either the temperature or the reaction time is commonly avoided by adopting a comparative procedure.(48) This hitherto has been applied only to first-order reactions. It consists in adding to the dilute mixture of the reactant in the inert gas a second compound which undergoes a first-order reaction with known Arrhenius parameters. Thus the two reactions, 'unknown' and 'known', are induced by the same shock wave. In one form of the procedure,(48) the rate constants of both reactions are calculated from the measured amounts of reaction and residence time. By repeating the experiment at different shock velocities, it is possible to make a direct comparison of the Arrhenius parameters of the two reactions without calculating the temperatures explicitly. Elimination of T between the Arrhenius equations for the known and unknown reactions yields the following relation:

$$\log k = [\log A - (E/E_s)\log A_s] + (E/E_s)\log k_s \qquad 2\text{–}47$$

where E_s, etc., refer to the known reaction. Hence, a plot of $\log k$ against $\log k_s$ is a straight line from which the values of (E/E_s) and (A/A_s) and therefore E and A can be obtained. A practical example of this is given in Section 2–4–1. The alternative procedure(49) is to use the temperature calculated from the shock velocity to determine the effective reaction time. This is obtained from the known rate constant by applying the first-order rate equation 2–2 to the analytical result from the known reaction. Application of the equation in reverse to the result from the unknown reaction then gives the rate constant required. This method requires simpler instrumentation than the first method and the effect of error in the calculated temperature can be minimized by choosing a standard reaction with Arrhenius parameters similar to those anticipated for the reaction under study. Naturally, both variations of the comparative method are committed to the assumption that the known and unknown reactions occur independently.

2–4–1 Two investigations by shock-tube methods

These will give a general indication of the range of conditions encompassed by shock-tube technique. Although there has been no lack of investigation of more complex reactions, the fact that both examples are studies of unimolecular reactions is symptomatic of the type of reaction to which the shock tube has been most successfully applied.

Olschewski, Troe, and Wagner[50] studied the thermal decomposition of water vapour in a 500- to 5000-fold excess of argon at temperatures between 3500 and 6000 K. Their shock tube was a brass cylinder 8 cm in diameter and 9 m long with a 4 m driver section which could be pressurized to 25 atm. Both incident and reflected shocks were used, and the disappearance of the water vapour was followed by recording the intensity of a narrow band of infrared radiation emitted by the water molecules. A highly sensitive and rapidly responding indium antimonide detector was used. In any one experiment, the disappearance of the water molecules followed first-order kinetics, but the value of the rate constant was found to be directly proportional to the concentration of argon. This indicated that the decomposition was unimolecular in the second-order region. A study of the effect of temperature showed that the second-order rate constant could be expressed by

$$k = 10^{14.7} \exp(-105/RT)\ \mathrm{cm^3\ mole^{-1}\ s^{-1}} \qquad \text{2–48}$$

It was concluded that the rate-determining step of the decomposition is

$$H_2O + Ar \rightarrow H + OH + Ar \qquad \text{2–XIV}$$

Since extrapolation of equation 2–48 to temperatures below about 4500 K gave values of k equal to one-half the observed values, it was further concluded that, at the lower temperatures, reaction 2–XIV is followed almost immediately by the reaction

$$H + H_2O \rightarrow H_2 + OH \qquad \text{2–XV}$$

Observations of the water concentration towards the end of the reaction time indicated that this reaction came to a temporary equilibrium with its reverse reaction (see also Section 7–3–4). Thus the investigation yielded the Arrhenius parameters for the initial step and cast some light on the mechanism of the overall reaction.

Our second example comes from a study by Jeffers and Shaub[51] of the unimolecular *cis-trans*-isomerization of 2-butene at 1040–1325 K. The comparative single-pulse method was used, the unimolecular decomposition of *t*-butanol to *iso*butene and water being adopted as the internal standard. The shock tube was quite small and simply constructed from thick-walled glass tubing of 2 cm internal diameter. The driver and driven compartments were 45 and 112 cm long respectively, and the latter was connected to a 16 litre steel dump tank situated 5 cm downstream of the diaphragm. Helium was used as driver gas and argon as diluent. Bursting pressures of the 0.0005 in 'Mylar' diaphragms were about 4.5 atm absolute. Reaction times (220–300 μs) were determined from the signals transmitted to a raster oscilloscope by a pressure pick-up and the reacted gas was analysed by gas chromatography.

A graph of $\log k$ for the isomerization plotted against $\log k_s$ for the standard reaction was found to be linear in accordance with equation 2–47. From the slope of the graph, its intercept on the ordinate and the known values of A_s and E_s the authors derived the values $A = 10^{13.38}\,s^{-1}$ and $E = 61.6$ kcal mole^{-1} for the isomerization; and, since the pressures in the reflected shocks were about 2000 torr, they concluded that these parameters refer to the reaction at its unimolecular high-pressure limit.

Bibliography

(A) H. W. Melville and B. G. Gowenlock, *Experimental Methods in Gas Reactions*, Macmillan, 1964.

(B) L. Batt, Experimental methods for the study of slow reactions, Chap. 1 of *Comprehensive Chemical Kinetics*, Vol. 1, C. H. Bamford and C. F. H. Tipper (eds), Elsevier, 1969.

(C) A. Maccoll, *Homogeneous Gas Phase Reactions*, Chap. 10 of *Investigation of Rates and Mechanisms of Reactions*, being Vol. 8, Part 1 of *Technique of Organic Chemistry*, 2nd edn, S. L. Friess *et al.* (eds), Interscience, 1961.

(D) E. S. Swinbourne, *Analysis of Kinetic Data*, Nelson, 1971.

(E) A. G. Gaydon and I. R. Hurle, *The Shock Tube in High Temperature Chemical Physics*, Chapman and Hall, 1963.

(F) E. F. Greene and J. P. Toennies, *Chemical Reactions in Shock Waves*, Arnold, 1964.

References

1 J. H. Purnell and C. P. Quinn, *Proc. Roy. Soc.*, **A270,** 267, 1962; *J. Chem. Soc.*, **1961,** 4128.

2 E. S. Swinbourne, *Aust. J. Chem.*, **11,** 314, 1958.

3 E. F. Linhorst and J. H. Hodges, *J. Am. Chem. Soc.*, **56,** 836, 1934.

4 D. R. Warren, *J. Appl. Chem.*, **1,** 40, 1951.

5 R. J. Ellis and H. M. Frey, *J. Chem. Soc.*, **1964,** 5578.

6 A. Maccoll, *J. Chem. Soc.*, **1955,** 965.

7 H. Mouquin and R. L. Garman, *Ind. Eng. Chem.* (*Anal. Edn*), **9,** 287, 1937.

8 S. G. Yorke, *J. Sci. Instr.*, **22,** 196, 1945; **25,** 16, 1948.

9 S. G. Foord, *J. Sci. Instr.*, **11,** 126, 1934.

10 G. L. Pratt and J. H. Purnell, *Chem. & Ind.*, **1960,** 1080.

11 H. C. Berg and D. Klepner, *Rev. Sci. Instr.*, **33,** 248, 1962.

12 H. W. Melville and B. G. Gowenlock, *Experimental Methods in Gas Reactions*, pp. 53–62, Macmillan, 1964.

13 G. L. Pratt and J. H. Purnell, *Anal. Chem.*, **32,** 1213, 1960; R. B. Cundall, K. Hay, and P. W. Lemeunier, *J. Sci. Inst.*, **43,** 652, 1966; E. R. Allen, *Anal. Chem.*, **38,** 527, 1966.

14 F. Kaufman and L. J. Decker, *8th Symposium* (*Internat.*) *on Combustion*, p. 133, Williams and Wilkins, Baltimore, 1962.

15 I. C. Hisatsune, B. Crawford, and R. A. Ogg, *J. Am. Chem. Soc.*, **79,** 4648. 1957; E. J. Dillemuth, D. R. Skidmore, and C. C. Schubert, *J. Phys. Chem.*, **64,** 1496, 1960.

16 P. G. Ashmore, B. P. Levitt, and B. A. Thrush, *Trans. Faraday Soc.*, **52,** 830, 1956.

17 D. M. Golden, R. Walsh, and S. W. Benson, *J. Am. Chem. Soc.*, **87,** 4053, 1965; **88,** 3480, 1966.

18 D. G. H. Marsden, *Rev. Sci. Instr.*, **33,** 288, 1964; R. W. Vreeland and D. F. Swinehart, *J. Am. Chem. Soc.*, **85,** 3349, 1963.
19 J. M. Hay and D. Lyon, *Proc. Roy. Soc.*, **A317,** 1, 1970.
20 I. C. Hisatsune and L. Zafonte, *J. Phys. Chem.*, **73,** 2980, 1969.
21 D. H. Fine, P. Gray, and R. Mackinven, *12th Symposium Internat. on Combustion*, p. 545, The Combustion Institute, 1969.
22 A. O. Allen, *J. Am. Chem. Soc.*, **56,** 2053, 1934.
23 P. J. Robinson, *Trans. Faraday Soc.*, **61,** 1655, 1965; **63,** 2668, 1967.
24 S. W. Benson, *J. Chem. Phys.*, **22,** 46, 1954; T. Boddington and P. Gray, *Proc. Roy. Soc.*, **A320,** 71, 1970.
25 H. M. Frey, *Trans. Faraday Soc.*, **60,** 83, 1964; H. M. Frey and D. C. Marshall, *ibid.*, **61,** 1715, 1965.
26 P. Corbett, A. M. Tarr, and E. Whittle, *Trans. Faraday Soc.*, **59,** 1609, 1963; J. C. Amphlett and E. Whittle, *ibid.*, **64,** 2130, 1968.
27 D. E. Hoare, J. B. Protheroe, and A. D. Walsh, *Trans. Faraday Soc.*, **55,** 548, 1959.
28 H. T. J. Chilton and B. G. Gowenlock, *Trans. Faraday Soc.*, **49,** 1451, 1953; J.-Y. Young and D. C. Conway, *J. Chem. Phys.*, **43,** 1296, 1965.
29 G. M. Harris, *J. Phys. Chem.*, **51,** 505, 1947.
30 M. F. R. Mulcahy and M. R. Pethard, *Aust. J. Chem.*, **16,** 527, 1963.
31 J. J. Batten, *Aust. J. Appl. Sci.*, **12,** 11, 1961; H. W. Melville and B. G. Gowenlock, *Experimental Methods in Gas Reactions*, p. 368, Macmillan, 1964.
32 G. I. Taylor, *Proc. Roy. Soc.*, **A219,** 186, 1953; **A225,** 473, 1954.
33 R. Aris, *Proc. Roy. Soc.*, **A235,** 67, 1956.
34 C-G. Wan and E. N. Ziegler, *Chem. Eng. Sci.*, **25,** 723, 1970.
35 P. V. Danckwerts, *Chem. Eng. Sci.*, **2,** 1, 1953; K. B. Bischoff, *ibid.*, **16,** 131, 1961.
36 S. J. Price, *Canad. J. Chem.*, **40,** 1310, 1962.
37 M. Szwarc, *J. Chem. Phys.*, **16,** 128, 1948.
38 M. H. Blades, A. T. Blades, and E. W. R. Steacie, *Canad. J. Chem.*, **32,** 298, 1954.
39 R. Walsh, D. M. Golden, and S. W. Benson, *J. Am. Chem. Soc.*, **88,** 650, 1966.
40 O. Levinspiel, *Chemical Reaction Engineering*, Chap. 10, Wiley, 1962.
41 M. F. R. Mulcahy and D. J. Williams, *Aust. J. Chem.*, **14,** 534, 1961.
42 J. H. Raley, F. F. Rust, and W. E. Vaughan, *J. Am. Chem. Soc.*, **70,** 88 1948.
43 H. F. Carroll and S. H. Bauer, *J. Am. Chem. Soc.*, **91,** 7727, 1969.
44 A. G. Gaydon and I. R. Hurle, *The Shock Tube in High Temperature Chemical Physics*, p. 37, Chapman and Hall, 1963.
45 H. B. Palmer, *J. Inst. Fuel*, **34,** 359, 1961; B. Deklau and H. B. Palmer, 8*th Symposium* (*Internat.*) *on Combustion*, p. 139, Williams and Wilkins, Baltimore, 1962.
46 J. N. Bradley and G. B. Kistiakowsky, *J. Chem. Phys.*, **35,** 256, 1961; S. C. Barton and J. E. Dove, *Canad. J. Chem.*, **47,** 521, 1969.
47 A. Lifshitz, S. H. Bauer, and E. L. Resler, *J. Chem. Phys.*, **38,** 2056, 1963.
48 W. Tsang, *J. Chem. Phys.*, **40,** 1171, 1964.
49 P. Cadman, M. Day, and A. F. Trotman-Dickenson, *J. Chem. Soc.*, **1970,** 2498.
50 H. A. Olschewski, J. Troe, and H. Gg. Wagner, 11*th Symposium* (*Internat.*) *on Combustion*, p. 155, The Combustion Institute, 1967; K. W. Michel, H. A. Olschewski, H. Richtering, and H. Gg. Wagner, *Z. phys. Chem.*, N.F., **44,** 160, 1965.
51 P. M. Jeffers and W. Shaub, *J. Am. Chem. Soc.*, **91,** 7706, 1969.

Problems

2–1 Show that the average half-life of molecules undergoing the reaction A → Products is given by $t_{0.5} = (2^{n-1}-1)/(n-1)k[A]_0^{n-1}$, where n ($\neq 1$) is the order of reaction, k is the rate constant and $[A]_0$ is the concentration at zero time. Derive the corresponding expression for the second order reaction A+B → Products when $[B]_0 \gg [A]_0$.

2–2 The following problems should be solved graphically. (a) Determine the order of the reaction $n\text{-}C_4H_{10} \rightarrow C_3H_6 + CH_4$ from the following initial rates of formation of propene at various pressures of n-butane at 791 K (Ref. 1)

$p_{C_4H_{10}}$ torr	20	24	29	37	61	77	92	119	156
$(dp_{C_3H_6}/dt)_0$ torr s^{-1}	2.4	3.3	4.3	6.8	12.7	17.2	26	39	53

(b) Use equation 2–4 in appropriate form to determine the rate constant of the second-order reaction $C_2H_4 + H_2 \rightarrow C_2H_6$ from the following pressure-change measurements made at 773 K on a mixture originally of composition 564 H_2:192 C_2H_4 (Pease, *J. Am. Chem. Soc.*, **54**, 876, 1932).

$10^{-2} \times$ Time s	0	3	6	9	15	24	33	42
Total pressure torr	756	749	743	737	725	708	693	680

(c) Di*tert*butyl peroxide decomposes by a first-order reaction substantially with the stoichiometry $((CH_3)_3CO)_2 \rightarrow 2(CH_3)_2CO + C_2H_6$. Determine the rate constant from the following pressure-change measurements with the peroxide vapour at 420 K. (Raley *et al.*, *J. Am. Chem. Soc.*, **70**, 88, 1948.)

Time min	0	6	14	22	30	38	46	∞
Pressure torr	180	199	221	242	262	280	297	517

Supposing the (apparent) departure of the overall stoichiometry from 1 → 3 to have been due to dead-space, what was the dead-space factor?

2–3 The kinetics of pyrolysis of toluene were studied by passing the vapour at 12 torr pressure through a tube 1.7 cm in diameter and 24 cm long heated to 1121 K. First-order rate constants k_a were derived from the yields of gaseous products and residence times t_c assuming plug-flow of the vapour at 1121 K. k_a was found to increase uniformly by about a factor of 2 as t_c was increased over the range 0.1 to 0.6 s. It is suggested that this may have been due to imperfect thermal equilibration. Test this hypothesis by graphing values of k_a/k calculated by equation 2–20 against t_c. ($T_0 = 298$ K, $E = 85$ kcal (356 kJ) mole^{-1} and $\kappa^\circ \simeq D^\circ_{1,1} \simeq 0.25$ cm^2 s^{-1} at 1 atm.)

Supposing similar experiments were to be carried out at 1 torr pressure, what error in k_a could be expected from lack of thermal equilibrium at $t_c = 0.35$? Assuming the same value of D° as above, estimate from Fig. 2–4 the errors in k_a incurred by neglecting diffusion at 1 torr with $t_c = 0.3$ and 3 s. (These residence times give about 2 and 15% reaction respectively.)

2–4 Compare the effects of residence time t_c on the degree of reaction $(1-f)$ for the same first-order reaction carried out under diffusionless plug-flow and perfectly stirred-flow but otherwise identical conditions (including negligible volume change). (This is best done by drawing two graphs of $([A]_0-[A])/[A]_0$ against kt_c on the same axis). What particular features are evident at very small and large values of kt_c? How is t_c for 50% reaction related to k in each case?

2–5 (a) Derive the stirred-flow equation 2–43; and (b) use it in appropriate form to calculate rate constants for the first order reaction $CH_3COOC_2H_5 \rightarrow CH_3COOH + C_2H_4$ from the following experimental data, obtained by passing ethyl acetate vapour (A) in nitrogen carrier-gas through a stirred-flow reactor and measuring the ratio $[CH_3COOH]/[A]$ ($\equiv Y$) in the effluent gas. (de Graaf and Kwart, *J. Phys. Chem.*, **67**, 1458, 1963.)

Temp. (K)	RT/PV (mole^{-1})	$10^6 N_{A,o}$ (mole s^{-1})	$10^6 N_{N_2}$ (mole s^{-1})	Y
649	162	6.80	77.60	0.0283
689	172	2.75	47.55	0.234

2–6 The kinetics of the reaction $C_2H_5Cl \rightarrow C_2H_4 + HCl$ in shock waves were studied by the comparative single pulse method, the equivalent decomposition of *i*-C_3H_7Cl being used as internal standard ($\log(A_s\, s^{-1}) = 13.64$; $E_s = 51.1$ kcal (214 kJ) mol^{-1}). (Tsang, *J. Chem. Phys.*, **41**, 2487, 1964.) The measured reaction time in one experiment was 1.4×10^{-3} s and the amounts of C_2H_4 and C_3H_6 formed corresponded to 0.019 and 1.24% reaction respectively. Estimate the effective temperature of the shock wave and the value of k at that temperature.

The following are some additional values of k and k_s obtained from similar experiments

$10^2 k\, s^{-1}$	32.1	4.58	1.86	147
$k_s\, s^{-1}$	19.4	3.49	1.43	78.5

Determine the Arrhenius parameters for the ethyl chloride decomposition.

2–7 Nitrogen dioxide and SO_2 react together at an appreciable rate above about 150 °C to form NO and SO_3 and perhaps other products. The stoichiometry of the reaction is not known very precisely. Outline methods by which this could be rectified. Assuming the overall reaction is found to be substantially

$$NO_2 + SO_2 \rightarrow NO + SO_3$$

specify appropriate apparatus (with diagrams) and experimental procedures to determine the kinetics over a wide range of temperatures and pressures.

3 Reaction mechanisms and elementary reactions

3–1 Evidence for free radicals in gaseous reactions

That free atoms and radicals participate in a great many gaseous reactions is a generally accepted conclusion that has been taken for granted in the previous two chapters. Nevertheless at this stage it is appropriate to touch lightly on the kind of evidence upon which the conclusion is based. In the first place, numerous radicals have been positively identified by spectroscopic techniques in reactions occurring in flames. This will receive some attention in Chapter 7. Similarly, the presence of a considerable number of alkyl and other radicals in pyrolytic reactions has been specifically established by mass spectroscopy[1] and, in shock-wave studies, by u.v. spectroscopy.[2] In addition, various radicals have been identified in systems subjected to flash photolysis, as will be described in Chapter 4. For the most part, however, radicals have been directly identified in the gas phase only in reactions occurring at high temperatures or light intensities or under otherwise extreme conditions where they are to be expected in high concentrations. With reactions conducted under more usual circumstances it has been possible, in a few cases, to trap out radicals and identify them by their ESR spectra;[3] but in general, the evidence for the role of radicals in such reactions, though very clear, is indirect.

In this category is the early evidence obtained by Paneth and Rice, who were able to extract radicals from the pyrolyses of metal alkyls and the like by passing the decomposing gases over metal films.[4] On a more general plane, both spectroscopy and experiments of the Paneth–Rice type have shown that the primary act in photo-initiated chain reactions is usually the photolysis of a molecule to atoms or radicals. This is strong evidence that similar species are responsible for the chain properties. And since the same reactions when brought about thermally instead of photolytically manifest similar properties, it follows that radicals are likewise operative in the thermal reactions. Furthermore, the kinetics of a number of such reactions—the hydrogen–oxygen reaction for example—are known to be continuous with the kinetics determined under more extreme conditions where radicals can be observed directly. There is also continuity of behaviour between certain gaseous reactions and their counterparts in the liquid phase, where the presence of numerous specific radicals functioning as reaction intermediates has been established by ESR.[5] Finally, studies of reactions conducted with high concentrations of radicals produced by special methods to be discussed in Chapter 4 have proved to be highly relevant to understanding the kinetics of many 'ordinary' gaseous reactions.

3–2 Complex reactions. Chain and non-chain mechanisms

It was noted briefly in Chapter 1 that, although reactions proceeding by free-radical mechanisms are very common, there are nevertheless other reactions which occur as single elementary molecular events. The former are frequently termed *complex* and the latter *molecular*. This section and Sections 3–3 and 3–4 are concerned largely with the general characteristics which distinguish the one from the other. A complex reaction proceeds by a *mechanism*; that is, by a sequence of elementary reactions in which radicals are formed, enter into various transformations, and finally disappear. Stable reaction products may be formed in any or all of these steps. Naturally it is also proper to speak of the mechanism of a molecular reaction. For example, with the molecular isomerization of cyclopropane to propene,

$$(CH_2)_3 \rightarrow CH_2 = CH - CH_3 \qquad \text{3–I}$$

there has been some discussion as to whether reaction occurs simply by transfer of a hydrogen atom from one carbon atom to another or alternatively by fission of a carbon–carbon bond followed by isomerization of the resulting diradical. These are both *intramolecular* mechanisms and obviously any elementary reaction, whether 'molecular' or not, must have a mechanism in this sense. But, in this chapter, 'mechanism' means the sequence, or sequences, of elementary reactions which make up a complex reaction.

Complex reactions can be subdivided into chain and non-chain reactions. The latter are not synonymous with molecular reactions. For example, the mechanism of thermal decomposition of di-*t*-butyl peroxide to acetone and ethane occurs in the following three elementary steps:

$$(CH_3)_3COOC(CH_3)_3 \rightarrow 2(CH_3)_3CO \qquad \text{3–II}$$

$$(CH_3)_3CO \quad (+M) \rightarrow (CH_3)_2CO + CH_3 \quad (+M) \qquad \text{3–III}$$

$$CH_3 + CH_3 \quad (+M) \rightarrow C_2H_6 \quad (+M) \qquad \text{3–IV}$$

But, although it has a free-radical mechanism, it is not a chain reaction. For a chain reaction to take place, the radicals produced initially from the reactants must be continuously regenerated by subsequent reactions with the reactants.

The synthesis of hydrogen bromide was given as an example of a chain reaction in Section 1–1 where, however, the basic features of chain mechanisms were not fully described. These are now illustrated by reference to another pyrolytic reaction, namely the decomposition of biacetyl to (principally) carbon monoxide, ketene, and methane. The mechanism is as follows:[(6)]

Initiation	$CH_3COCOCH_3 \rightarrow 2CH_3CO$	3–V
Propagation	$CH_3CO + M \rightarrow CH_3 + CO + M$	3–VI
	$CH_3 + CH_3COCOCH_3 \rightarrow CH_4 + CH_2COCOCH_3$	3–VII
	$CH_2COCOCH_3 \rightarrow CH_2CO + CH_3CO$	3–VIII
Termination	$CH_3 + CH_3 \quad (+M) \rightarrow C_2H_6 \quad (+M)$	3–IX

The reaction chains, of which the radicals CH_3CO, CH_3, and $CH_2COCOCH_3$ are the *chain carriers*, are *initiated* by reaction 3–V, *propagated* by reactions 3–VI to 3–VIII, and *terminated* by reaction 3–IX. The essential feature of such a mechanism, which distinguishes it from a non-chain mechanism, is the cycle of propagating reactions. Once 'triggered' by the production of a radical by the initiating reaction, the cycle continues to regenerate chain carriers (and to produce reaction products) until two methyl radicals unluckily collide and their free spins are neutralized to form a stable molecule of ethane. In this way, many more molecules of biacetyl undergo reaction in a given time than would be the case if the initial decomposition (reaction 3–V) were the only reaction by which they disappear. Thus the *chain length*, that is, the ratio of the number of reactant molecules disappearing to the number of radicals produced by the initiating reaction, depends on the relative rates of the propagating and terminating reactions. In the present case, these are equivalent to the relative rates at which the methyl radicals react with biacetyl molecules or with each other. Since, in general, the relation between the rates of the propagating and terminating reactions depends on the temperature and pressure and other circumstances, the chain length is not a constant but depends on the reaction conditions.

The type of reaction chain just described is a *straight* or *unbranched chain*: for each chain carrier entering the cycle of propagating reactions, a single chain carrier is produced. There are also reactions with *branched* (or *branching*) *chains* in which each propagating cycle produces more than one chain carrier and thus introduces a net gain in the number of carriers in the system. This indeed corresponds to the popular image of a chain reaction. In fact, however, reactions with branching chains constitute a special class. Since these manifest an entirely new range of kinetic phenomena, they will be considered separately in Chapter 6. Except for the following section, which deals with characteristics common to all chain reactions, our concern in this chapter is with unbranched chains.

3–3 General characteristics of chain reactions

3–3–1 Inhibitors and retarders

The rate of a chain reaction depends on the chain length, which may amount to hundreds or thousands or more. Hence addition of a foreign substance which reacts with any of the chain carriers in such a way as to terminate the chains prematurely is likely to cause a great reduction in the rate. If the added molecules are intrinsically much more reactive towards the chain carriers than the reactant molecules, the effect may be very drastic indeed. When the chains are long, a very small amount of such an *inhibitor* will bring the reaction virtually to a standstill because every

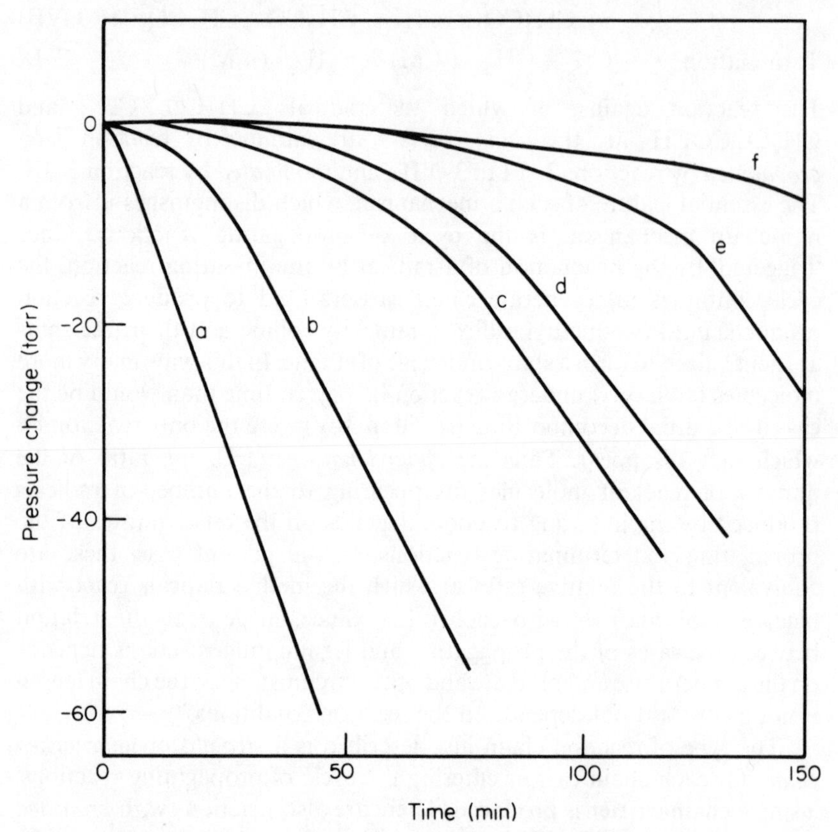

Fig. 3–1 Inhibition of the oxidation of acetaldehyde by various amines.

Pressure–time curves determined (a) with no amine added, (b)–(f) with additions of 1–2 torr methyl-, ethyl-, *n*-butyl-, *t*-butyl- and *iso*propyl-amine. Acetaldehyde pressure = oxygen pressure = 100 torr (Cullis and Waddington[7]).

radical produced by the initiating reaction is captured by an inhibitor molecule before it can initiate the propagating cycle. Since, however, the inhibitor molecules are destroyed by reaction with the chain carriers, the effect is essentially a temporary one. Provided the addition is small, the reaction rate will eventually assume its normal value after an *induction period* during which the inhibitor is consumed. Figure 3–1 shows such behaviour in the reaction between acetaldehyde and oxygen to form peracetic acid:

$$CH_3CHO + O_2 \rightarrow CH_3C(OOH)O \qquad \text{3–X}$$

Curve (a) is a pressure–time curve obtained with the reaction in the absence of inhibitor and the remaining curves were obtained after small quantities of various alkyl amines had been added to similar reaction mixtures. Whereas the 'undoped' reaction began at almost its maximum rate, no reaction could be detected for an hour in the presence of about 1% isopropylamine and about three hours elapsed before the original rate was restored.

Since a stable molecule cannot react with a radical except to form another radical, the function of most inhibitors is to produce a radical that is incapable of attacking the reactant molecules and so continuing the chains. The fate of the inactive radical is eventually to dimerize or react with some other radical. However, 'stable' free radicals like NO and NF_2, which are sometimes used as inhibitors, function by combining directly with the chain carriers.

When the amount of foreign substance added or its reactivity is such that it competes less effectively for the chain carriers, the reaction may be merely *retarded* without the occurrence of an induction period. The addition of an alcohol to acetaldehyde–oxygen mixtures in similar amounts to those of the amines shown in Fig. 3–1 produces such an effect.(8)

The occurrence of inhibition or retardation following the addition of a *small* amount of an appropriate substance is the most straightforward proof of the presence of a chain reaction, since it does not seem possible to see how the effect could be brought about in any other way. On the debit side, as it were, these phenomena can be produced by minute amounts of impurities, and they cause much of the difficulty in obtaining reproducible reaction rates which is frequently a severe trial in studies of chain reactions.

3–3–2 Initiation by light and sensitizers

The occurrence of chain reactions initiated by light was mentioned in Section 3–1. In fact the existence of chain reactions was first discovered in the photosynthesis of hydrogen chloride from elementary hydrogen and chlorine. Absorption by a chlorine molecule of a quantum of sufficient energy is followed by dissociation,

$$Cl_2 + h\nu \rightarrow Cl + Cl \qquad \text{3–XI}$$

and the reaction occurs by the so-called *Nernst chain*:

$$Cl + H_2 \rightarrow HCl + H \qquad \text{3–XII}$$

$$H + Cl_2 \rightarrow HCl + Cl \qquad \text{3–XIII}$$

Photochemical chain reactions are recognized by high values of the *quantum yield*, ϕ, the number of molecules of reactant destroyed or product formed per photon absorbed. ϕ for the photosynthesis of hydrogen chloride can be as high as 10^6. It will be appreciated that, in general, the quantum yield is not the same as the chain length; some of the photo-activated molecules may be deactivated before they can dissociate or otherwise react to produce the radicals which initiate the chains. Apart from the manner of initiation, the general characteristics of photochemical chain reactions are in no way different from those of chain reactions initiated thermally. Indeed the same reaction frequently can be initiated either photochemically or thermally or, of course, by both methods simultaneously.

Since no activation energy is required for photo-initiation, the photochemical reaction often occurs at temperatures that are too low for thermal initiation to be effective. For example there is no difficulty in initiating the oxidation of acetaldehyde (reaction 3–X) photochemically at room temperature[9] whereas, in the absence of light, equivalent reaction rates are obtained only at about 120 °C. Thus, sensitivity to light is often a feature of chain reactions, though, except when there is a large quantum yield, it is not an immediately diagnostic one.

The rate of initiation may also be augmented by the addition of a *sensitizer*, that is, a foreign substance which decomposes thermally or otherwise reacts to produce active radicals at a greater rate than the 'indigenous' reactants. For example, the initial rate of thermal decomposition of propane at 546 °C is increased about tenfold by the addition of 0.5 per cent oxygen.[10] In this case, the initiating radicals are produced by reaction between the oxygen and the reactant molecules, probably by the reaction

$$RH + O_2 \rightarrow R + HO_2 \qquad \text{3–XIV}$$

Sensitization is the reverse of inhibition and, like the latter, is temporary when produced by a small amount of a very active substance; the sensitizer is eventually used up, and the reaction rate then falls to its normal value.

As might be expected, the addition of a less reactive substance can produce a smaller accelerating effect more akin to the contrary of retarder action; the additional rate of initiation is comparable to that which occurs 'indigenously' and the sensitized effect persists during most of the lifetime of the reactants. The pyrolysis of acetaldehyde is accelerated in this way by a few per cent of biacetyl, the initiating radicals being produced by reaction 3–V.[11]

Chain reactions can also be *photo-sensitized* by introducing a foreign substance which produces radicals by absorption of light. This provides a method of causing compounds to react which do not absorb light themselves. For example, the chain decomposition of paraffin hydrocarbons can be brought about at temperatures more than 100° lower than their normal pyrolysis temperatures by 'mercury photo-sensitization'; that is, by adding a little mercury vapour and irradiating with ultraviolet light (which is absorbed by mercury but not by the hydrocarbon).[12] The radicals are generated by reaction of the photo-activated mercury atoms with the hydrocarbon molecules:

$$Hg + h\nu \rightarrow Hg^* \qquad \text{3–XV}$$

$$Hg^* + RH \rightarrow R + H + Hg \qquad \text{3–XVI}$$

In the same way hydrocarbons can be caused to react with oxygen at much lower temperatures than 'normal'[13] or alternatively the effect can be obtained by photo-dissociating a little admixed bromine vapour.[14]

3–3–3 Effects of surface and inert gas

As will appear in more detail in Chapter 4, solid surfaces, including those of glass and silica, can catalyse the recombination of radicals. The surface of the reaction vessel is, therefore, a potential retarder or inhibitor. The degree to which it acts as such depends on the nature of the surface, the size of the reaction vessel, the pressure of the reacting gas, and the kind of radicals involved. In general these factors are interrelated in a complex way, but two limiting conditions can be recognized. The catalytic effect of the surface material may be so strong that the chemisorption of the radicals which leads to recombination occurs after relatively few collisions with the surface. The rate of recombination then becomes entirely determined by the rate of diffusion of the radicals to the surface from the bulk of the gas. The time for a particle to diffuse a given distance x is governed by the Einstein–Schmolukowski law:

$$\tau = x^2/2D \qquad \text{3–1}$$

where D is the diffusion coefficient. Thus the average time taken for a chain carrier to reach the wall from the interior of gas contained in a spherical or cylindrical vessel of radius $\mathbf{R}$ is proportional to $\mathbf{R}^2/D$. If the chains are terminated only at the surface, the chance of the chain carrier reacting to continue the chain is proportional to this time; hence the reaction rate is proportional to $\mathbf{R}^2$, inversely proportional to D, and (so long as the condition of very fast reaction at the surface is fulfilled) independent of the nature of the surface. For vessels of irregular shape, the ratio of the volume to the surface area (V/S) may be taken as qualitatively equivalent to $\mathbf{R}$. Since, for a gas of given composition, D is inversely proportional to the pressure, it follows that the reaction rate becomes directly proportional to the pressure on this account. The effect is not easily recognized, however (with unbranched chains), because the rates of propagation and initiation normally also increase with increased pressure of the reactants. On the other hand, addition of an *inert gas* to a fixed pressure of the reactants will lower the value of D without usually affecting the rates of initiation and propagation. Thus an accelerating effect of inert gas is a good, though not an infallible, indication of heterogeneous termination. (It is not infallible because a unimolecular step in the reaction mechanism may be accelerated by inert gas.)

The second limiting condition applying to heterogeneous chain termination occurs when the recombination of the radicals at the surface is sufficiently slow to control their rate of disappearance from the system. Diffusion does not then influence the rate of termination, consequently the reaction rate becomes proportional to $\mathbf{R}$ rather than $\mathbf{R}^2$, and is unaffected by inert gas. On the other hand, it is directly influenced by the nature of the surface.

It will be seen that, under the limiting conditions just outlined, the effects on the reaction rate of inert gas and the nature of the surface are mutually exclusive; the presence of one implies the absence of the other. It seems, however, that circumstances corresponding to either limiting condition are not often found in practice, the actual situation usually being intermediate between the two. Chain reactions are commonly found to be accelerated by additions of inert gas and, at the same time to be sensitive to changes—sometimes otherwise imperceptible changes—in the condition of the surface. Because of this, and because the chains usually are terminated homogeneously as well as heterogeneously, clear-cut practical demonstrations of either of the relations between reaction rate and $\mathbf{R}$ (or V/S) mentioned above are rare for reactions with unbranched chains. Perhaps the best known comes from the early work of Trifonoff[15] who found the quantum yield of the photosynthesis of hydrogen chloride in glass tubes to be directly proportional to $\mathbf{R}^2$ at pressures in the region of 10 torr. As the pressure was raised, however, the quantum yield became gradually less dependent on $\mathbf{R}$, showing that in these circumstances heterogeneous termination was progressively replaced by homogeneous termination. Quantitative investigations of the effects of heterogeneous termination have more often been made with branching-chain reactions where the phenomena are shown up in greater relief. These matters will receive some attention in Chapter 6.

Naturally the rate of heterogeneous termination can be expected to depend on the kind of radical and the kind of surface present. The effect of a change of radical occurs in a striking way in the pyrolysis of paraffin hydrocarbons. In clean quartz vessels, the rates of decomposition of ethane and isobutane are retarded by increasing the surface-to-volume ratio of the reaction vessel, whereas those of propane and *n*-butane are not. It seems that the difference lies in the fact that hydrogen atoms are chain carriers in the former reactions but not in the latter.[16] The recombination of hydrogen atoms is known to be more than usually susceptible to surface catalysis (see Section 4–3–1). As to the effect of the nature of the surface, an example may be quoted from the kinetics of oxidation of methane at 500 °C ($CH_4 + O_2$). Hoare and Walsh[17] found the reaction rate to be reduced by a factor of 100 when the surface of a quartz vessel was coated with a layer of lead monoxide. Other treatments of the surface, such as 'conditioning' by carrying out a large number of successive oxidations or heating to 950 °C *in vacuo*, produced similar though less drastic effects. (The strong chain-breaking effect of lead monoxide is generally held to be responsible for the 'anti-knock' effect of lead tetraethyl on combustion in the internal combustion engine.)

Initiation at the surface

Since the recombination of radicals to molecules is catalysed by solid surfaces, thermodynamics assures us that the same surfaces must also be catalysts for the fission of the stable molecules into radicals. This phenomenon is a commonplace at high temperatures, as witness the fact that hydrogen is rapidly dissociated to atoms on an incandescent filament. Since a minute rate of generation of radicals may be sufficient to initiate an appreciable chain reaction, the possibility arises that chains may be initiated heterogeneously at lower temperatures where the production of radicals would not otherwise be detected.(18) In practice, it is not uncommon to find acceleration of a reaction by an increase in the surface-to-volume ratio (S/V), but the effect can be difficult to distinguish from that of ordinary heterogeneous catalysis superimposed on a gas-phase reaction. Nevertheless the reality of heterogeneous initiation seems to be established. Thus, according to Chaikin,(18) the rate of the thermal reaction between hydrogen and chlorine at 286 °C and 115 torr is almost independent of S/V. This shows that the chains both begin and end either at the surface or in the gas phase. When, however, oxygen is added to bring about termination in the gas, the rate of the retarded reaction becomes directly proportional to S/V, showing that initiation occurs at the surface. The same result can be obtained in a more pronounced way by packing the reaction vessel with pieces of glass. This has very little effect with pure hydrogen and chlorine but, when oxygen is present, it causes a marked increase in rate.(19)

3–4 Simultaneous chain and molecular reactions

The initiating step of a thermal chain reaction normally involves breaking a chemical bond and hence requires a substantial activation energy. Its infrequency on this account is compensated for by the numerous propagating 'links' in the chain which require relatively little activation energy. The possibility always arises, however, that the same reaction may also proceed by a molecular mechanism. Although, in this case, the reaction observed is the sum of independent molecular events, the rearrangement of chemical bonds occurring in these may require considerably less activation energy than the outright rupture needed to provide radicals for initiation of chains. The net result under any particular conditions may be that either mechanism is the faster, or again, that both proceed at comparable rates.

Simultaneous molecular and chain mechanisms are sometimes found in pyrolytic reactions where they can be recognized by studies of the effects of inhibitors or retarders. An example is demonstrated by Fig. 3–2. This shows that addition of increasing amounts of propene reduces the

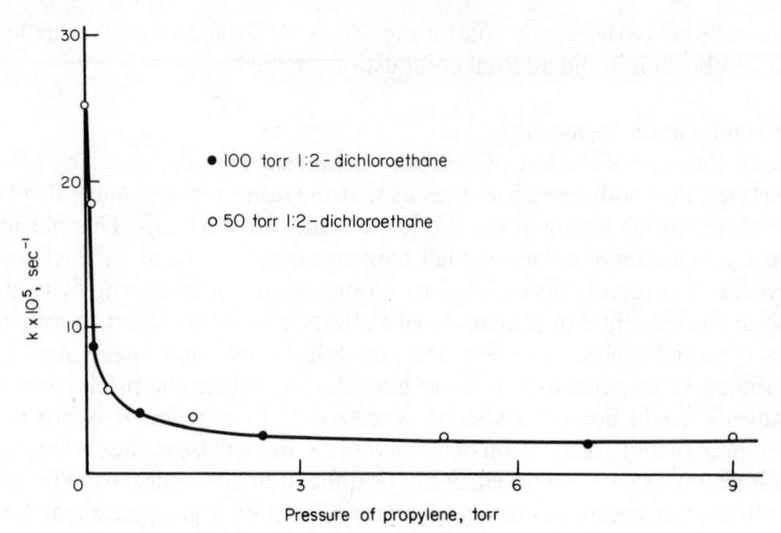

Fig. 3–2 Inhibition of the pyrolysis of 1,2-dichloroethane to a limiting rate by additions of propylene. k is the combined first-order rate constant for the chain and molecular reactions. (Barton and Howlett[20]).

rate of decomposition of 1,2-dichloroethane to a definite limit beyond which continued addition has no further effect. In this, as in other similar reactions, different inhibitors, though initially more or less effective in reducing the rate, nevertheless reduce it ultimately to the same limit.[21] The limiting rate is that of an independent molecular reaction which is accompanied by a chain reaction in the absence of the inhibitor.

Since the possibility of two such simultaneous processes must always be considered, examination of the effects of inhibitors or retarders is an essential preliminary to any kinetic study of reaction mechanism. Retarders commonly used in pyrolytic and other studies are unsaturated compounds, such as propene, cyclohexene, and toluene. These react with the chain carriers to give allyl-type radicals, for example,

$$R\cdot + CH_3{-}CH = CH_2 \rightarrow RH + CH_2{\overset{\cdots}{=}}CH{\overset{\cdots}{=}}CH_2 \qquad \text{3–XVII}$$

The latter are inactive towards the reactant molecules and so interrupt the chains. Because of its free-radical character, nitric oxide was formerly much used as an inhibitor, but it has fallen out of favour because, under some conditions, it will initiate chains as well as terminate them. This property has been the cause of much confusion.[16]

Although, occasionally, it has been possible to develop criteria to predict for which members of a series of similar reactions molecular or chain mechanisms will predominate[22] this usually still needs to be determined by experiment.

It should be noted that a limiting 'unretardable' rate, based merely on the disappearance of the reactants, does not necessarily signify the presence of an independent molecular mechanism. It may simply represent the rate of the process which initiates the chains. Thus, the rate of pyrolysis of biacetyl (see Section 3–2) cannot be reduced to less than that of reaction 3–V. In such cases, however, the products recovered from the fully inhibited reaction necessarily contain stoichiometrically equivalent amounts of products derived from the inhibitor, whereas this is not so when there is a genuine molecular reaction (see also Section 3–9–1).

3–5 Relation between kinetics and mechanism of chain reactions

It might be supposed that a chain mechanism would lead to complicated kinetics for the overall reaction, but this is not necessarily the case. For reasons we are about to develop, the order of reaction is often found to be close to an integral or half-integral value; and the rate constants measured at different temperatures usually conform well to the Arrhenius equation. Thus, empirically, a chain reaction may be characterized over a reasonably wide range of conditions by an 'overall' order, activation energy, and frequency factor. Naturally the kinetics observed, whether simple or complex, are the consequence of the particular mechanism operating and the relation of the one to the other is the subject of the following discussion. This will be based on the *stationary-state approximation*, the basis of which is stated qualitatively in the next few paragraphs.

3–5–1 The stationary-state approximation

Returning to the mechanism of pyrolysis of biacetyl given in Section 3–2, let us consider the rate of change in the concentration of methyl radicals at the beginning of the reaction. We shall assume, for the moment, that the concentration of biacetyl is being held constant by immediately replacing the amount consumed by reaction. It will also be assumed for simplicity (what is, in fact, very nearly the case) that the acetyl radicals formed in reactions 3–V and 3–VIII instantly decompose by reaction 3–VI. The rate of change of the concentration of methyl radicals is the net result of the rates of the elementary reactions by which they are formed and destroyed. Thus, denoting biacetyl by A, the methyl radical by R, and the radical $CH_2COCOCH_3$ by R′, we have

$$\frac{d[R]}{dt} = 2k_5[A] + k_8[R'] - k_7[R][A] - 2k_9[R]^2 \qquad 3\text{–}2$$

The rate constant of the recombination k_9 is defined, as is customary, by the rate of formation of the product, ethane. At the beginning of the reaction the concentration of R rises from zero as the result of reactions 3–V and 3–VI, a process which is soon assisted by reaction 3–VIII. As soon as the concentration becomes finite, however, its increase begins to be counteracted by the consumption of the radicals by reactions 3–VII and 3–IX. Since the rates of the latter increase with increasing [R], they must eventually overtake the rate of formation, and thereafter the radical concentration will remain stationary. That is,

$$\frac{d[R]}{dt} = 0 \qquad 3\text{–}3$$

The intense reactivity of the radicals has two consequences: first, their stationary concentration is reached in a very short time; and second, their concentration always remains very low compared with that of the biacetyl (this because no sooner is a radical formed than it reacts). The result is that, in practice, the condition imposed initially that the concentration of biacetyl should be held constant is superfluous, since the stationary state is attained before a significant amount of the reactant is consumed. Otherwise expressed, the radical concentration always adjusts to a stationary value in a time which is short compared to the time over which a change in reactant concentration normally can be measured. This means that equation 3–3 is valid at any time during the reaction. It does not mean, however, that the actual value of the stationary concentration remains constant throughout the reaction. In fact [R] will decrease progressively as the reactant is consumed. The point is that, whereas $-(d[A]/dt)$ is finite and measurable at any time, $-(d[R]/dt)$ is effectively zero for the period of the rate measurement. Similar considerations apply to the behaviour of the other radicals present. As we shall see, these conclusions, which, of course, are not peculiar to the example we have chosen, are immensely valuable for discussing the kinetics of complex reactions. An example in which the stationary-state approximation is examined quantitatively is given in Section 3–6.

3–5–2 Dependence of overall kinetic parameters on reaction mechanism

In the following we shall study the relation between mechanism and overall kinetics in the course of a discussion of pyrolytic, halogenation, and oxidation reactions. Such reactions are the best-known types of chain reactions occurring in the gaseous phase. For the sake of brevity, our concern will be mostly with reactions in which the chains begin and end homogeneously and which, therefore, are free from surface effects. Assuming the mechanism of a reaction is known, the kinetics to which it will give rise, can be deduced, in principle, by solving a set of equations analogous to equation 3–2, there being such an equation for each radical species. A massive simplification is obtained by adopting the steady-state approximation (equation 3–3) which converts each differential equation into an algebraic one. When several radicals are involved in propagating the chains, however, even the algebraic equations become intractable without recourse to numerical methods. Fortunately, in dealing with a particular radical species, it frequently happens that one of its reactions is so much faster than alternative reactions that this reaction can be regarded to a good approximation as the *only* reaction undergone by the particular species. As already noted, reaction 3–VI in Section 3–2 is of this type for the particular conditions of decomposition of biacetyl under consideration; the acetyl radical decomposes so rapidly that reactions 3–V and 3–VIII can be regarded as producing methyl radicals directly, and the behaviour of the acetyl radical need be considered no further. We are then left with only two radical species as chain carriers and hence only two equations to solve. Indeed, many reactions can be profitably discussed in terms of two-carrier mechanisms, as will be demonstrated in the course of the following discussion.

3–5–3 Pyrolysis

The mechanism of the decomposition of biacetyl given in Section 3–2 is typical of the pyrolyses of many, perhaps most, saturated organic compounds. The overall reaction

$$\mathrm{A} \rightarrow \text{saturated} + \text{unsaturated compounds} \qquad \text{3–XVIII}$$

is initiated by spontaneous fission of a carbon–carbon bond and propagated by reactions whereby complex radicals decompose to unsaturated molecules and simpler radicals, alternating with reactions in which the simpler radicals abstract hydrogen atoms from the reactant molecules. The chains end with mutual neutralization of radicals either by combination, as in equation 3–IX, or disproportionation (see Section 3–7) or both. Such are known as *Rice–Herzfeld mechanisms.* They can be generalized by the following reaction scheme:

$$\text{Initiation} \rightarrow \mathrm{R} \qquad \text{(i)}$$

$$\mathrm{R} + \mathrm{A} \rightarrow \mathrm{R}' + \text{Product} \qquad \text{(a)}$$

$$\mathrm{R}'(+\mathrm{M}) \rightarrow \mathrm{R} + \text{Product(s)} \qquad \text{(b)}$$

$$\left.\begin{array}{r} \mathrm{R}+\mathrm{R}+(+\mathrm{M}) \rightarrow \\ \mathrm{R}+\mathrm{R}'+(+\mathrm{M}) \rightarrow \\ \mathrm{R}'+\mathrm{R}'+(+\mathrm{M}) \rightarrow \end{array}\right\} \text{Termination} \qquad \begin{array}{r}\text{(c)}\\ \text{(d)}\\ \text{(e)}\end{array}$$

The stationary-state equations are

$$\frac{d[\mathrm{R}]}{dt} = 0 = \rho_{\mathrm{i}} - k_{\mathrm{a}}[\mathrm{R}][\mathrm{A}] + k_{\mathrm{b}}[\mathrm{R}'] - 2k_{\mathrm{c}}[\mathrm{R}]^2 - k_{\mathrm{d}}[\mathrm{R}][\mathrm{R}'] \qquad \text{3–4}$$

and

$$\frac{d[\mathrm{R}']}{dt} = 0 = k_{\mathrm{a}}[\mathrm{R}][\mathrm{A}] - k_{\mathrm{b}}[\mathrm{R}'] - k_{\mathrm{d}}[\mathrm{R}][\mathrm{R}'] - k_{\mathrm{e}}[\mathrm{R}']^2 \qquad \text{3–5}$$

Here ρ_{i} is the total rate of initiation by processes which need not be specified for the moment. Similarly, for the time being, equations 3–4 and 3–5 ignore the possibility that reactions (b) to (e) may be influenced by the presence of an 'inert' molecule (M). Addition of equations 3–4 and 3–5 yields

$$\rho_{\mathrm{i}} - 2\{k_{\mathrm{c}}[\mathrm{R}]^2 + k_{\mathrm{d}}[\mathrm{R}][\mathrm{R}'] + k_{\mathrm{e}}[\mathrm{R}']^2\} = 0 \qquad \text{3–6}$$

This simply expresses the fact that, in the stationary state, the rate at which chain carriers are generated by the initiating reactions is exactly balanced by the total rate at which they are destroyed by the terminating reactions.

Assuming the chains are long, almost every R′ radical produced by reaction (a) regenerates an R radical by reaction (b). Hence, to a good approximation,

$$k_a[\mathrm{R}][\mathrm{A}] = k_b[\mathrm{R}'] \qquad 3\text{–}7$$

and thus,

$$[\mathrm{R}'] = (k_a[\mathrm{A}]/k_b)[\mathrm{R}] \qquad 3\text{–}8$$

Substituting equation 3–8 in equation 3–6 and taking the square root yield,

$$[\mathrm{R}] = \left(\frac{\rho_i}{X}\right)^{1/2} \qquad 3\text{–}9$$

where

$$X \equiv 2\{k_c + (k_a[\mathrm{A}]/k_b)k_d + (k_a[\mathrm{A}]/k_b)^2 k_e\} \qquad 3\text{–}10$$

Now the overall reaction rate is given by

$$-\frac{d[\mathrm{A}]}{dt} = k_a[\mathrm{R}][\mathrm{A}] \qquad 3\text{–}11$$

whence, from equation 3–9

$$-\frac{d[\mathrm{A}]}{dt} = k_a\left(\frac{\rho_i}{X}\right)^{1/2}[\mathrm{A}] \qquad 3\text{–}12$$

This expression is not of the form

$$-\frac{d[\mathrm{A}]}{dt} = k[\mathrm{A}]^n \qquad 3\text{–}13$$

Therefore the mechanism, as it stands, does not yield a definite reaction order. However, simplifications are possible in appropriate circumstances. In general the values of the rate constants for the radical–radical reactions k_c, k_d, k_e do not differ greatly one from the other; hence the relative magnitudes of the terms in equation 3–10 depend chiefly upon the value of $(k_a[\mathrm{A}]/k_b)$. It is easily seen that, if $(k_a[\mathrm{A}]/k_b)$ is sufficiently large, the rate equation becomes

$$-\frac{d[\mathrm{A}]}{dt} = \left(\frac{\rho_i}{2k_e}\right)^{1/2} k_b \qquad 3\text{–}14$$

and conversely, if $(k_a[\mathrm{A}]/k_b)$ is small,

$$-\frac{d[\mathrm{A}]}{dt} = \left(\frac{\rho_i}{2k_c}\right)^{1/2} k_a[\mathrm{A}] \qquad 3\text{–}15$$

For an unsensitized pyrolysis, the most likely initiation reaction is unimolecular fission of the weakest bond in the reactant molecule to give two radicals; in which case

$$\rho_{\mathrm{i}} = 2k_{\mathrm{i}}[\mathrm{A}] \qquad 3\text{–}16$$

and the final rate expressions corresponding to equations 3–14 and 3–15 are:

$$-\frac{d[\mathrm{A}]}{dt} = \left(\frac{k_{\mathrm{i}}}{k_{\mathrm{e}}}\right)^{1/2} k_{\mathrm{b}}[\mathrm{A}]^{1/2} \qquad (k_{\mathrm{a}}[\mathrm{A}] \gg k_{\mathrm{b}}) \qquad 3\text{–}17$$

and

$$-\frac{d[\mathrm{A}]}{dt} = \left(\frac{k_{\mathrm{i}}}{k_{\mathrm{c}}}\right)^{1/2} k_{\mathrm{a}}[\mathrm{A}]^{3/2} \qquad (k_{\mathrm{a}}[\mathrm{A}] \ll k_{\mathrm{b}}) \qquad 3\text{–}18$$

In the first case, the reaction has an order of 0.5, an overall frequency factor of $(k_{\mathrm{i}}/k_{\mathrm{e}})^{1/2}k_{\mathrm{b}}$, and an overall activation energy of $\{E_{\mathrm{b}}+\frac{1}{2}(E_{\mathrm{i}}-E_{\mathrm{e}})\}$. The corresponding parameters in the second case are $n = 1.5$, $A = (k_{\mathrm{i}}/k_{\mathrm{c}})^{1/2}k_{\mathrm{a}}$, and $E = \{E_{\mathrm{a}}+\frac{1}{2}(E_{\mathrm{i}}-E_{\mathrm{c}})\}$.

The half integral orders follow directly from the bimolecular nature of the termination reactions, and the appearance of such an order in an experimental rate expression is a safe indication that the chains are mutually terminated. On the other hand, obviously neither equation 3–17 nor equation 3–18 can apply over an indefinitely wide range of [A]. Indeed the order must progressively decrease from 1.5 at low pressures to 0.5 at high pressures of A. Over the pressure range where $(k_{\mathrm{a}}[\mathrm{A}]/k_{\mathrm{b}})$ is of the order of unity the reaction will be approximately first order.

In physical terms, equation 3–17 represents the situation where the R radicals are in such low concentration relative to the R′ radicals that, compared with an R′ radical, each R radical has a negligible chance of colliding with another radical. Equation 3–18 represents the opposite situation. It will be recalled that both equations refer to mechanisms with first-order initiation and second-order termination. This, however, is only one of several possibilities; for example, initiation could occur by a bimolecular process and one or more of the termination reactions could be in its third-order region. Table 3–1 shows the overall reaction orders to which various combinations give rise. Following established nomen-

Table 3–1 Pyrolysis by two-carrier chain reactions

Overall orders of reaction for various types of initiation and termination reactions

First-order initiation		*Second-order initiation*		
Second-order termination	*Third-order termination*	*Second-order termination*	*Third-order termination*	*Overall order*
		$\beta\beta$		2
$\beta\beta$		$\beta\mu$	$\beta\beta$M	$\frac{3}{2}$
$\beta\mu$	$\beta\beta$M	$\mu\mu$	$\beta\mu$M	1
$\mu\mu$	$\beta\mu$M		$\mu\mu$M	$\frac{1}{2}$
	$\mu\mu$M			0

clature, the symbols μ and β signify radicals which undergo first- and second-order propagation reactions respectively. Thus R in the reaction scheme on p. 87 is a β radical; R′ is a μ radical if it decomposes by first-order kinetics or a β radical if a second body (M) is needed. The entries $\beta\beta$, $\beta\mu$, etc., signify the nature of the single terminating reaction which is assumed to predominate; M refers here to a third body which, in the absence of foreign gases and at early stages of reaction, is a molecule of the reactant. The student might care to verify for himself that the orders given in Table 3–1 can be deduced by making the appropriate substitutions and approximations in equation 3–12. It should perhaps be noted that $\beta\mu$- or $\beta\mu$M-type termination can only be dominant when the rate constant k_d is notably greater than either k_c or k_e, a situation which is not commonly found.

We shall now see how the previous considerations can help to elucidate from kinetic data the mechanism of thermal decomposition of *acetaldehyde.*

The pyrolysis of acetaldehyde has a long history and until quite recently not only the mechanism but the experimental facts were in dispute. Much of the confusion was caused by the undetected influence of impurities on the kinetics, a situation which, as we have seen, is always liable to occur in investigations of chain reactions. The principal products are methane and carbon monoxide

$$CH_3CHO \rightarrow CH_4 + CO \qquad \text{3–XIX}$$

together with much smaller amounts of hydrogen, ethane, acetone, and other compounds. Additions of radical-producing substances cause the reaction to accelerate and, in the case of di-*t*-butyl peroxide, the effect is proportional to the square root of the concentration added.[23] The reaction is inhibited by propene[24] but, in quartz vessels, it is not sensitive to the ratio of surface to volume.[25, 26] From these and other facts we have not the space to consider, it seems certain that a homogeneously initiated and terminated chain reaction is involved.

The nature of the products leaves little doubt as to the identity of the propagating reactions and the mechanism is based on the following reactions:

$$\text{Initiation} \rightarrow CH_3 \text{ or } CH_3CO \qquad \text{(i)}$$

$$CH_3 + CH_3CHO \rightarrow CH_4 + CH_3CO \qquad \text{3–XX}$$

$$CH_3CO \quad (+M) \rightarrow CO + CH_3 \quad (+M) \qquad \text{3–VI}$$

$$CH_3 + CH_3 \quad (+M) \rightarrow C_2H_6 \quad (+M) \qquad \text{3–IX}$$

$$CH_3 + CH_3CO \quad (+M) \rightarrow (CH_3)_2CO \quad (+M) \qquad \text{3–XXI}$$

$$CH_3CO + CH_3CO \quad (+M) \rightarrow (CH_3CO)_2 \quad (+M) \qquad \text{3–XXII}$$

This leaves the nature of the initiation and the relative importance of the terminating steps to be determined for the particular reaction conditions. Our discussion refers to aldehyde pressures of about 30–300 torr at about

500–540 °C.[25–27] Under these conditions the reaction order is 1.5. It will be seen from Table 3–1 that this is compatible with three possibilities: first-order initiation combined with second-order termination by reaction 3–IX, or second-order initiation combined with either third-order termination by reaction 3–IX or second-order termination by reaction 3–XXI. Second-order initiation is eliminated, however, by the result of some experiments made by Boyer, Niclause, and Letort.[11] Their approach was to relate the rate of the pyrolysis when sensitized by a trace of biacetyl ($\mathscr{R}_s$) to the rate found with the pure aldehyde ($\mathscr{R}$) under otherwise identical conditions. Equation 3–12 shows that

$$\mathscr{R} = a\rho_i^{1/2} \qquad \text{3–19}$$

and that

$$\mathscr{R}_s = a(\rho_i + \rho_i')^{1/2} \qquad \text{3–20}$$

where ρ_i' is the rate of initiation brought about by the biacetyl and a is constant for both experiments. From these relations it is possible to deduce the rate of that part of the reaction which is initiated by the biacetyl. This is given by

$$\mathscr{R}_B = (\mathscr{R}_s^2 - \mathscr{R}^2)^{1/2} = a\rho_i'^{1/2} \qquad \text{3–21}$$

A study of the influence of the concentrations of biacetyl and acetaldehyde on $\mathscr{R}_B$ produced the relation:

$$\mathscr{R}_B \propto [(CH_3CO)_2]^{1/2}[CH_3CHO] \qquad \text{3–22}$$

Since equations 3–19 and 3–21 are of the same form and

$$\mathscr{R} \propto [CH_3CHO]^{3/2} \qquad \text{3–23}$$

it follows that

$$\rho_i \propto [CH_3CHO] \qquad \text{3–24}$$

That is, initiation of the unsensitized pyrolysis is by a first-order reaction. Presumably it occurs by fission of the carbon–carbon bond

$$CH_3CHO \rightarrow CH_3 + CHO \qquad \text{3–XXIII}$$

The CHO radical probably produces another chain carrier by the reactions:

$$CHO + M \rightarrow CO + H + M$$

$$H + CH_3CHO \rightarrow H_2 + CH_3CO$$

It was noted previously that first-order initiation implies termination by reaction 3–IX to produce ethane. Since the rates of initiation and termination are equal, this implies further that the rate of formation of ethane should be given by

$$\rho_{et} \equiv \frac{d[C_2H_6]}{dt} = k_{23}[CH_3CHO] \qquad \text{3–25}$$

and therefore should be first order with respect to the acetaldehyde pressure. Niclause *et al.*[27] found this to be accurately the case at 520 °C and approximately so at lower temperatures. Lastly, a first-order initiating reaction also requires the terminating reaction (3–IX) to be in its second-order region. This can be checked in the following way. According to the above analysis, the rate of formation of methane is given by the expression

$$\mathscr{R} \equiv \frac{d[\mathrm{CH_4}]}{dt} = k_{20}\left(\frac{k_{23}}{k_9}\right)^{1/2} [\mathrm{CH_3CHO}]^{3/2} \qquad 3\text{–}26$$

(cf. equation 3–18). Equation 3–26 can be combined with equation 3–25 to yield the reaction

$$(k_9/k_{20}^2) = (\rho_{\mathrm{et}}/\mathscr{R}^2)[\mathrm{CH_3CHO}]^2 \qquad 3\text{–}27$$

The rate constant k_{20} is not affected by pressure; consequently, if reaction 3–IX is second order, the combination of the measurable quantities on the right should likewise be independent of pressure. This is confirmed by experiment for pressures above 100 torr. At lower pressures, the values of (k_9/k_{20}^2) so deduced decrease, indicating that the rate of reaction 3–IX is 'falling-off' towards its third-order region.

Summarizing, we can now assert that the mechanism of acetaldehyde pyrolysis at about 520 °C and pressures greater than 100 torr is mainly as follows:

$$\mathrm{CH_3CHO \rightarrow CH_3 + CHO} \qquad 3\text{–XXIII}$$

$$\mathrm{CH_3 + CH_3CHO \rightarrow CH_4 + CH_3CO} \qquad 3\text{–XX}$$

$$\mathrm{CH_3CO \quad (+M) \rightarrow CO + CH_3 \quad (+M)} \qquad 3\text{–VI}$$

$$\mathrm{CH_3 + CH_3 \rightarrow C_2H_6} \qquad 3\text{–XXIV}$$

At 520 °C and 200 torr the chain length is about 6000. It may or may not be an encouragement to a prospective kineticist to learn that this mechanism was proposed by Rice and Herzfeld in 1934,[28] but it was not until over 30 years later that it could be regarded as established.

3–5–4 Oxidation

The great majority of gaseous reactions of molecular oxygen that have been studied experimentally are of the branched-chain type. As such, their characteristics will be discussed in Chapter 6. However, in special circumstances to be mentioned shortly, reactions of the type,

$$\mathrm{R'H + O_2 \rightarrow R'OOH} \qquad 3\text{–XXV}$$

proceed by straight chains, and it will be useful to consider these briefly. Besides taking our examination of the relation between kinetics and mechanism a step further, the discussion will introduce some considerations which will be taken up again in Chapter 6. Reaction 3–XXV signifies the formation of a hydroperoxide from a saturated hydrocarbon or other organic 'substrate', and the special circumstances to which reference was made above are constituted by sensitization or photo-initiation of the reaction at temperatures below about 50 °C.

Granted the chain character of the reaction and the stoichiometry expressed by reaction 3–XXV, when the pressure or the reaction vessel is large enough to prevent the chains from being terminated at the wall, the following mechanism seems almost inevitable:

$$\text{Initiation} \rightarrow R' \text{ or } R'OO \qquad \text{(i)}$$

$$R'OO + R'H \rightarrow R'OOH + R' \qquad \text{(A)}$$

$$R' + O_2 \rightarrow R'OO \qquad \text{(B)}$$

$$\left.\begin{array}{r} R'OO + R'OO \rightarrow \\ R'OO + R' \rightarrow \\ R' + R' \rightarrow \end{array}\right\} \text{Termination} \qquad \begin{array}{r} \text{(C)} \\ \text{(D)} \\ \text{(E)} \end{array}$$

In general, the reacting species are too complicated for reactions B–E to be affected kinetically by the presence of a third body under the usual reaction conditions. If this mechanism is compared with that of the pyrolytic reactions given on p. 87, it will appear that the two mechanisms are quite similar if the radical R′OO is considered to take the place of R in the pyrolytic mechanism. The only difference is that the addition of R′ to the quasi free-radical oxygen molecule in reaction B is substituted for the decomposition reaction (b) in the pyrolytic mechanism. In the terminology of the latter the chain carriers of the oxidation, R′OO and R′, are both β-type radicals.

The oxidation takes place, of course, between two reactants, and it will be seen that, formally, the mechanism deals with them symmetrically. It might therefore be supposed that the rate would be about equally dependent on the concentrations of the oxygen and substrate molecules. Here again, however, the overall kinetics depend on the relative velocities of the elementary reactions and, as will now be explained, the effect of these is to introduce a profound asymmetry into the kinetics.

For long chains the approximation

$$k_A[R'OO][R'H] = k_B[R'][O_2] \qquad 3\text{–}28$$

is valid and solution of the stationary-state equations gives the following relation for the dependence of the reaction rate on the reactant concentrations:

$$\mathscr{R} = -\frac{d[R'H]}{dt} = \frac{d[R'OOH]}{dt} = \frac{k_A \rho_i^{1/2}[R'H]}{2^{1/2}\{k_C + Bk_D + B^2 k_E\}^{1/2}} \qquad 3\text{–}29$$

where

$$B \equiv (k_A[R'H])/(k_B[O_2]) \qquad 3\text{–}30$$

Equation 3–29 is analogous to equation 3–12. For a given mode of initiation, the way in which the rate responds to the concentrations of the reactants is determined by the relative magnitudes of the terms in the denominator. Suppose, for simplicity, what will not be far from the truth, that reactions C, D, and E occur on every collision; then, since $k_D = 2(k_C k_E)^{1/2}$ approximately,§ equation 3–29 reduces to

$$\mathscr{R} = \left(\frac{\rho_i}{2}\right)^{1/2} \left[\frac{k_A[R'H]}{k_C^{1/2} + Bk_E^{1/2}}\right] \qquad 3\text{–}31$$

Since further k_C and k_E are unlikely to be very different, it is easy to see that the kinetics depend almost entirely on the value of B. If B is small, the rate expression approximates to

$$\mathscr{R} = (\rho_i/2k_C)^{1/2} k_A[R'H] \qquad (k_A[R'H] \ll k_B[O_2]) \qquad 3\text{–}32$$

and the rate is independent of $[O_2]$; but if B is large,

$$\mathscr{R} = (\rho_i/2k_E)^{1/2} k_B[O_2] \qquad (k_A[R'H] \gg k_B[O_2]) \qquad 3\text{–}33$$

and the rate is independent of $[R'H]$. Suppose again (what is, in fact, *not* the case) that k_A and k_B were equal. Leaving aside, for the moment, any influence of ρ_i, the dependence of the rate on $[R'H]/[O_2]$—or on $[R'H]$ at constant $[O_2]$—would be as shown by the curve x in Fig. 3–3, which is calculated from equation 3–31 with equal values of k_C and k_E. In any experimental investigation the range over which the reactant concentrations can be varied is necessarily limited. A change in the concentration ratio from about 0.1 to about 10 is not unusual. This corresponds to the distance between the arrows in Fig. 3–3. Thus, in the present instance, the observed influence of $[R'H]$ on the rate would approximate to linearity (equation 3–32) at 'low' concentrations of $[R'H]$ relative to $[O_2]$ and to independence (equation 3–33) at 'high' concentrations. Since, in the assumed circumstances, equation 3–31 is symmetrical, an exactly similar curve would describe the dependence of $\mathscr{R}$ on $[O_2]$ at constant $[R'H]$.

The situation is greatly altered, however, if k_A and k_B are not equal. Let us suppose that $k_B = 200k_A$. With equal frequency factors, this corresponds to a difference in the activation energies of about 6 kcal mole^{-1} at 300 °C or about 3 kcal at 25 °C. The dependence of the rate on $[R'H]$ and $[O_2]$ *over the same concentration range as before* is now given by curves y and z respectively (the vertical scale of z being one fifth that of y). Over the whole accessible range, the rate increases practically linearly with the concentration of R'H and is independent of the concentration of O_2; that is to say, equation 3–32 applies. Obviously equation 3–33 would apply instead, if the magnitudes of k_A and k_B were reversed.

It is now time to turn to the experimental facts. Once again the chemistry of acetaldehyde provides us with a clear example. The photo-oxidation to peracetic acid was studied by McDowell and Sharples[9] at pressures from 50 to 300 torr and temperatures from 20 to 30 °C. The kinetics are summarized by the equation

$$-\frac{d[CH_3CHO]}{dt} = \kappa I_{abs}^{1/2}[CH_3CHO] \qquad 3\text{–}34$$

§ See equation 1–12 and following remark on p. 11.

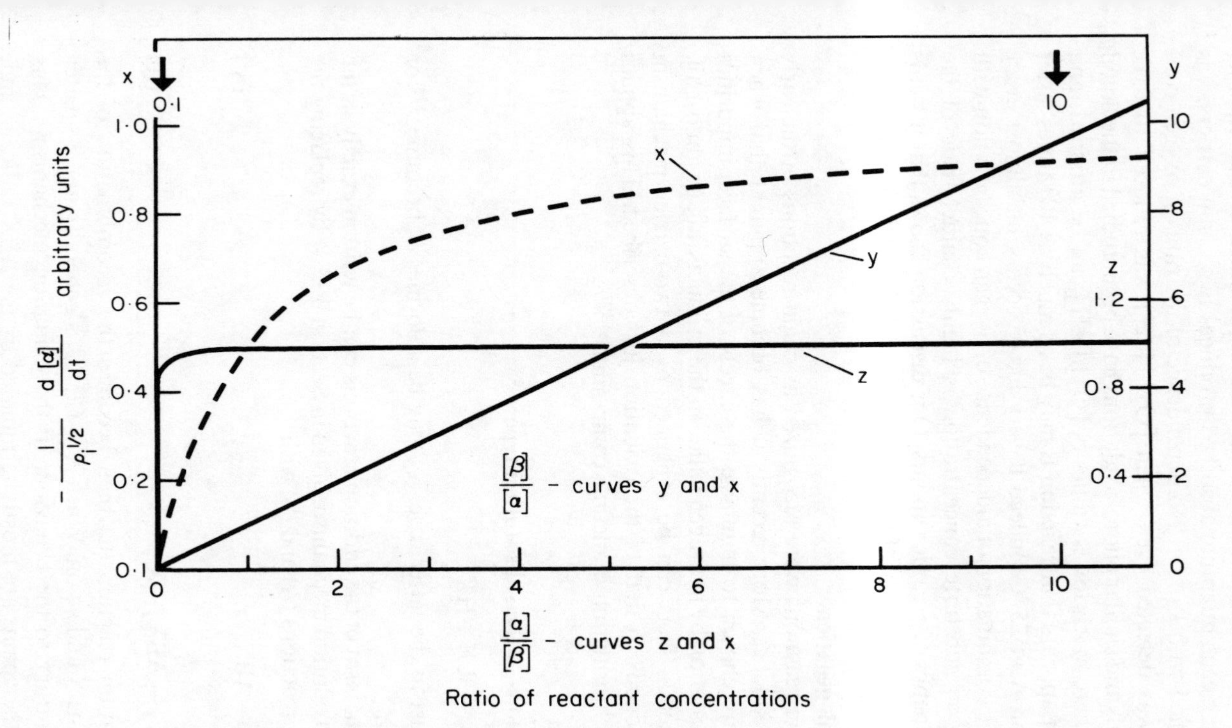

Fig. 3–3 Two-carrier chain reaction $\alpha + \beta \rightarrow$ Products. Dependence of rate on concentrations of reactants with constant rate of initiation (schematic).

Curve x applies when the bimolecular propagation rate constants k_A and k_B are equal; curves y and z apply when $k_B = 200\,k_A$ (see text). Curves y and z apply qualitatively to peroxidation with $\alpha = O_2$ and $\beta = RH$ and to addition halogenation with $\alpha = A{:}B$ and $\beta = Cl_2$ (cf. Fig. 3–4).

where I_{abs} is the number of quanta of light *absorbed* by the aldehyde molecules per second and κ is a constant. This relation is of the same form as equation 3–32. Therefore it appears that k_B is much greater than k_A and, assuming each quantum absorbed produces two chain carriers, κ is to be identified with $k_A/k_C^{1/2}$. No dependence of the rate on oxygen concentration was detected even when $[O_2]/[R'H]$ was reduced to 0.03. However, this and similar photo-oxidations have been studied in the liquid phase,[29] where much lower values of $[O_2]/[RH]$ can be attained; and from these it appears that reaction B may be as much as 10^6 times faster than reaction A at 25 °C. Indeed it very likely occurs on almost every collision. The combination of independence of oxygen concentration with dependence on substrate concentration is highly characteristic of the oxidation kinetics of organic vapours. Our discussion shows that it stems from the intense reactivity of the oxygen molecule towards free radicals.

3–5–5 Halogenation

Halogenation reactions were recognized as chain reactions in the early days of gas kinetics. More recently it has become apparent that when conditions are chosen to minimize the effects of the surface, impurities and self-retardation by hydrogen halide in the products, all, or almost all, the experimental facts can be explained by a two-carrier mechanism closely analogous to that of the oxidation just described. This applies both to the chlorination of olefinic compounds:[30]

$$A{:}B + X_2 \rightarrow ABX_2$$

and to reactions of the substitution type:

$$R'H + X_2 \rightarrow R'X + HX$$

where, at least in the latter case, X_2 may be chlorine,[30] bromine,[31] or iodine.[32]

The mechanism of the addition reactions can be written exactly in the form of the oxidation mechanism given in Section 3–5–4 by replacing the propagating reactions (A) and (B) with

$$X + A{:}B \rightarrow ABX \qquad (A')$$

and

$$ABX + X_2 \rightarrow ABX_2 + X \qquad (B')$$

The termination reaction (C) then becomes the recombination of the halogen atoms X and so on. When X = Cl, $k_{A'}$ is generally much greater than $k_{B'}$ (contrary to the behaviour in the oxidation mechanism). This means that the chains are mostly terminated by the ABX radicals; and, with photo-initiation, the kinetics are described by the equivalent of equation 3–33:

$$\mathscr{R} = \frac{d[ABX_2]}{dt} = (I_{abs}/k_{E'})^{1/2} k_{B'}[Cl_2] \qquad 3\text{–}35$$

(the light being absorbed by the chlorine molecules). The experimental results presented in Fig. 3–4 show that the reaction rate is independent of [A:B] over most of the range. Unlike the situation with the oxygen concentration in photo-oxidation, a decrease in rate can be detected at the lowest concentrations of [A:B]; this is because the magnitudes of $k_{A'}$ and $k_{B'}$ are less different in the chlorination reaction.

Substitution halogenation provides another variation on the theme of equation 3–29. Here the corresponding reactions on p. 93 are replaced by

$$X + R'H \rightarrow HX + R' \qquad (A'')$$

and

$$R' + X_2 \rightarrow R'X + X \qquad (B'')$$

together with the corresponding terminating reactions. The relatively slow abstraction reaction (A″) replaces the fast reaction (A′) of addition halogenation. Thus the kinetics tend to revert to those of oxidation. In the bromination of methane, for example, reaction (A″) is considerably slower

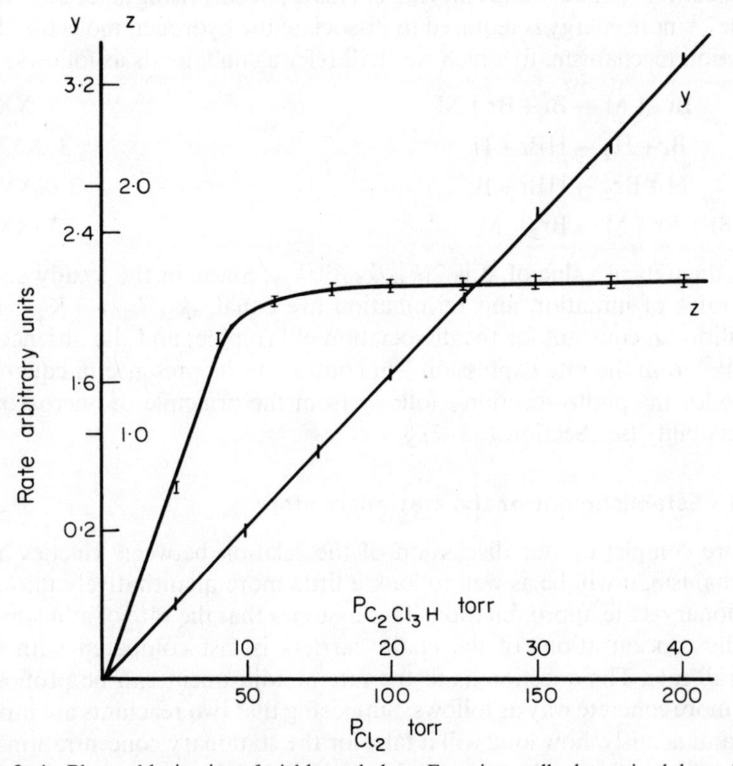

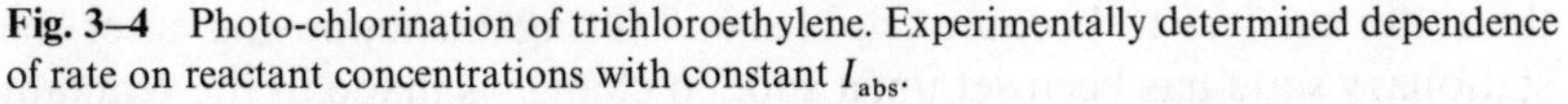

Fig. 3–4 Photo-chlorination of trichloroethylene. Experimentally determined dependence of rate on reactant concentrations with constant I_{abs}.
Curve y: dependence on $[Cl_2]$ with $[C_2Cl_3H] = 20$ torr.
Curve z: dependence on $[C_2Cl_3H]$ with $[Cl_2] = 100$ torr.
(Dainton, Lomax, and Weston[33]).

than reaction (B″), with the result that the initial-rate kinetics of the photo-reaction conform to the equation:[33]

$$\frac{d[CH_3Br]}{dt} = \left(\frac{I_{abs}}{k'_{C'}[M]}\right)^{1/2} k_{A''}[CH_4] \qquad 3\text{–}36$$

which is analogous to equation 3–32 but different from equation 3–35. The total concentration [M] appears in equation 3–36 because reaction (C′) is the recombination of atoms and is third order. A precisely equivalent equation describes the photo-bromination of hydrogen.[34]

When the latter reaction is carried out in the absence of light (at temperature about 250 °C higher) the rate expression found experimentally for the initial rate is of the form:

$$\frac{d[HBr]}{dt} = \kappa'[Br_2]^{1/2}[H_2] \qquad 3\text{–}37$$

The square root dependence on the bromine concentration in place of $I_{abs}^{1/2}$ in equation 3–36 shows that the reaction is initiated by the thermal dissociation of the bromine molecule. This is not surprising since at 58 kcal mole^{-1} more energy is required to dissociate the hydrogen molecule. The reaction mechanism, to which we shall refer again later, is as follows:

$$Br_2 + M \rightarrow Br + Br + M \qquad 3\text{–XXVI}$$

$$Br + H_2 \rightarrow HBr + H \qquad 3\text{–XXVII}$$

$$H + Br_2 \rightarrow HBr + Br \qquad 3\text{–XXVIII}$$

$$Br + Br + M \rightarrow Br_2 + M \qquad 3\text{–XXIX}$$

The theoretical value of κ' is $2(k_{26}/k_{29})^{1/2}k_{27}$. Since, in the steady state, the rates of initiation and termination are equal, $(k_{26}/k_{29}) = K_{26}$, the equilibrium constant for the dissociation of bromine; and the absence of $[M]^{1/2}$ from the rate expression—in contrast to its presence in equation 3–36 for the photo-reaction—follows from the principle of microscopic reversibility (see Section 1–3–2).§

3–6 Establishment of the stationary state

Before completing our discussion of the relation between kinetics and mechanism, it will be as well to look a little more quantitatively into the stationary-state approximation. This assumes that the rate of adjustment of the concentrations of the chain carriers is fast compared with the overall rate. The question as to the rate of adjustment can be proposed in a more concrete way as follows. Supposing that two reactants are mixed instantaneously, how long will it take for the stationary concentrations of the chain carriers to become established? Or again, supposing, after the stationary state has been set up, a sudden change is made in the reactant

§ According to Norrish and Ritchie,[35] $[M]^{1/2}$ is also absent from their experimental expression equivalent to equation 3–36 for the photosynthesis of hydrogen chloride. This result—the enigma variation of equation 3–29—has still to be explained.

concentrations, how long will the reaction take to establish the new stationary concentrations? For the approximation to be valid these transitory periods must be very short compared to the time-scale of the overall reaction.

To answer these questions it is best to proceed by way of an example. For this we shall choose the thermal synthesis of hydrogen bromide which, as we have just seen, takes place by reactions 3–XXX to 3–XXXIII. We shall be concerned with an equimolecular mixture of hydrogen and bromine reacting at 1 atm pressure and 300 °C. It can easily be shown that at any time during the reaction,

$$\frac{d[\mathrm{Br}]}{dt}+\frac{d[\mathrm{H}]}{dt}=2k_{26}[\mathrm{Br}_2][\mathrm{M}]-2k_{29}[\mathrm{M}][\mathrm{Br}]^2 \qquad 3\text{–}38$$

We assume that $[Br_2]$ and $[H_2]$ remain constant for the period under consideration and, since [H] is much smaller than [Br], we neglect $d[\mathrm{H}]/dt$. For convenience let $[\mathrm{Br}] = n$, and replace the other terms in equation 3–38 as follows:

$$\frac{dn}{dt}=\rho_{\mathrm{i}}-\delta n^2 \qquad 3\text{–}39$$

The solution of this equation for the condition that $n = 0$ at zero time is

$$n=(\rho_{\mathrm{i}}/\delta)^{1/2}\tanh\left[t(\rho_{\mathrm{i}}\delta)^{1/2}\right] \qquad 3\text{–}40$$

Since $\tanh x \to 1$ as $x \to \infty$, at the beginning of the reaction the concentration of chain carriers rises asymptotically from zero to the stationary value, $(\rho_{\mathrm{i}}/\delta)^{1/2}$. It readily follows from equation 3–40 that the time taken to reach 99 per cent of the stationary value is given by

$$t_{99}=2.65/(\rho_{\mathrm{i}}\delta)^{1/2} \qquad 3\text{–}41$$

Furthermore,

$$(\rho_{\mathrm{i}}\delta)^{1/2}=2k_{29}[\mathrm{M}](K_{26}[\mathrm{Br}_2])^{1/2} \qquad 3\text{–}42$$

which, on putting in the values[36,37] becomes

$$(\rho_{\mathrm{i}}\delta)^{1/2}=2\times(6\times10^{15})\times[2.1\times10^{-5}](1.4\times10^{-19}\times[1.2\times10^{-5}])^{1/2}$$
$$=2.3\,\mathrm{s}^{-1} \qquad 3\text{–}42(a)$$

and therefore $t_{99} = 1.1$ s.

Suppose now that the first measurement of the consumption of hydrogen is taken 30 s after the beginning of the reaction (at which time somewhat less than 1 per cent would have disappeared). If the stationary concentration of bromine atoms had been established instantaneously at zero time, the fraction of the original hydrogen consumed would have been

$$\frac{-(\Delta[\mathrm{H}_2])_{\mathrm{ss}}}{[\mathrm{H}_2]_0}=30k_{27}\left(\frac{\rho_{\mathrm{i}}}{\delta}\right)^{1/2} \qquad 3\text{–}43$$

However, the true consumption is given by

$$\frac{-(\Delta[H_2])_{tr}}{[H_2]_0} = k_{27} \int_0^{30} n\,dt \qquad 3\text{–}44$$

If equation 3–44 is integrated and the result compared with equation 3–43, using the above value of $(\rho_i\delta)^{1/2}$, it will be found that $(\Delta[H_2])_{ss}/(\Delta[H_2])_{tr}$ is 1.01. That is, the error caused by assuming that the rate measurement refers to stationary concentrations is about 1 per cent.

These calculations show that the chain carriers attain their stationary concentration rapidly at the beginning of the reaction, more rapidly, in fact, than the reactants can normally be mixed (though not more rapidly than a photo-reaction can be initiated). It follows that the carrier concentration should also adjust rapidly to the gradual decrease in reactant concentrations during the course of the reaction. To test this, one may suppose that, after the initial stationary concentration has been set up, the concentration of bromine is instantaneously diminished in some way by 10 per cent. It is left as an exercise for the student to show that the time required for the concentration of bromine atoms to fall to within 1% of its new stationary value is $0.83/(\rho_i\delta)^{1/2}$ and hence, for the present conditions, about 0.4 s.

This example may be considered typical of reactions that can be studied by the ordinary experimental methods described in Chapter 2. But with very fast reactions such as occur in flames and some shock-initiated reactions, where the radical concentrations become commensurate with those of the reactants, the steady-state approximation generally has to be abandoned. The relation between the kinetics and mechanism can then be derived only by solving the differential equations, and this has usually to be done by numerical methods.[38]

3–7 Types of elementary reactions involving free radicals

It has come to be realized from accumulated experience that the great majority of the elementary reactions by which complex reactions are accomplished are themselves mechanistically very simple. A single bond is broken or a single atom is exchanged or some other equally simple rearrangement of binding electrons takes place. Examples will be recognized in the mechanisms already discussed and others will be found in the following pages. In fact, most thermally activated elementary reactions involving free radicals can be classified into six types. At this stage, it will be useful to have a list of these together with a few examples. The latter are drawn mostly from reaction systems we shall not have the opportunity to discuss further and, for this reason, references are provided to assist the student to discover these systems for himself. Excluding primary photochemical processes and reactions of electronically excited species, the most common types of elementary free-radical reaction are:

1. *Molecular fission:* $AB \quad (+M) \rightarrow A\cdot + \cdot B \quad (+M)$
   where A and B may be radicals or atoms, for example

$$C_6H_5CH_2CH_3 \rightarrow C_6H_5CH_2 + CH_3 \qquad \text{3–XXX}^{(39)}$$
$$NOCl + M \rightarrow NO + Cl + M \qquad \text{3–XXXI}^{(40)}$$

2. *Combination of radicals:* $A\cdot + \cdot B \quad (+M) \rightarrow AB \quad (+M)$
   For example

$$C_6H_5CH_2 + CH_3 \rightarrow C_6H_5CH_2CH_3 \qquad \text{3–XXXII}^{(41)}$$
$$NO + Cl + M \rightarrow NOCl + M \qquad \text{3–XXXIII}^{(42)}$$

3. *Atomic transfer:* $A + XB \rightarrow AX + B$
   where X is an atom and either A or XB is a radical (but not both) and similarly with the reverse reaction, for example

$$CF_3 + CH_3OCH_3 \rightarrow CF_3H + CH_2OCH_3 \qquad \text{3–XXXIV}^{(43)}$$
$$CH_3O + CH_4 \rightarrow CH_3OH + CH_3 \qquad \text{3–XXXV}^{(44)}$$
$$CH_3 + N_2O \rightarrow CH_3O + N_2 \qquad \text{3–XXXVI}^{(45)}$$
$$HO_2 + CO \rightarrow CO_2 + OH \qquad \text{3–XXXVII}^{(46)}$$

4. *Radical-to-molecule addition:* $A\cdot + BC \quad (+M) \rightarrow ABC\cdot \quad (+M)$
   For example

$$CH_3 + C_2H_4 \quad (+M) \rightarrow C_3H_7 \quad (+M) \qquad \text{3–XXXVIII}^{(47)}$$
$$CF_3 + (CF_3)_2CO \rightarrow (CF_3)_2COCF_3 \qquad \text{3–XXXIX}^{(48)}$$

5. *Radical decomposition:* $ABC\cdot \quad (+M) \rightarrow BC + A\cdot \quad (+M)$
   For example

$$C_3H_7 \quad (+M) \rightarrow C_2H_4 + CH_3 \quad (+M) \qquad \text{3–XL}^{(49)}$$
$$CH_3COCH_2 \quad (+M) \rightarrow CH_2CO + CH_3 \quad (+M) \qquad \text{3–XLI}^{(50)}$$

6. *Disproportionation of radicals:* $A\cdot + XB\cdot \rightarrow AX + B$
   For example

$$CH_3O + CH_3 \rightarrow CH_2O + CH_4 \qquad \text{3–XLII}^{(51)}$$
$$OH + OH \rightarrow H_2O + O \qquad \text{3–XLIII}^{(52)}$$

Type 6 can be considered a special case of Type 3.

Although the above are the most prevalent, they are not the only types of elementary free-radical reaction possible. Larger radicals, for example, may undergo unimolecular *isomerization.* When the geometry of the radical is favourable, this can occur by a 'tail-biting' atomic transfer, which is an internal version of Type 3. Thus,

$$CH_3CH_2CH_2CH_2CH_2\cdot \rightarrow CH_3\dot{C}H\ CH_2CH_2CH_3 \qquad \text{3–XLIV}$$

In other cases, however, internal rearrangements occur which are much the same as the unimolecular isomerizations undergone by saturated molecules. Cyclopropyl radicals, for example, isomerize to allyl radicals in a similar manner to reaction 3–I (see page 76).[53] In other circumstances such a rearrangement is the prelude to decomposition, as appears to be the case with the reaction

$$C_6H_5OCH_2 \rightarrow C_6H_5CHO + H \qquad \text{3–XLV}^{(54)}$$

Radical transfer reactions sometimes occur that are analogous to the atomic transfers of Type 3. For example, the reaction

$$CH_3 + CH_3COCOCH_3 \rightarrow CH_3CO + CH_3COCH_3 \qquad \text{3–XLVI}$$

takes place in the pyrolysis of biacetyl in parallel with reaction 3–VII in Section 3–2. It seems, however, that for such reactions to take place a multiple bond is required in the molecular partner and the reaction actually occurs in two conventional stages: addition of the radical to the multiple bond (Type 4) followed by decomposition of the new radical (Type 5). In the present instance, this view is supported by the presence of a trace of the compound $(CH_3)_2C(OH)COCH_3$ in the products of the pyrolysis.[6] Another more rare but occasionally important reaction type is the *four-centre reaction* which is well known (though not common) among molecular reactions. For example, the existence of the reaction

$$HS + O_2 \rightarrow OH + SO \qquad \text{3–XLVII}$$

seems necessary to explain the kinetics of oxidation of hydrogen sulphide.[55]

Returning to the six main types, it will be observed that Type 2 is the reverse of Type 1, Type 3 is of the same form in either direction, and Type 5 is the reverse of Type 4. This raises the question as to the prevalence of *the reverse of Type 6*, that is the generation of radicals by atomic exchange between two saturated molecules. Of their nature, reactions of this kind are highly endothermic and therefore the equilibrium at ordinary temperatures is well in favour of the reactants. Nevertheless, such a reaction may generate sufficient radicals to function as the initiating reaction of a chain reaction. The question is whether it generates them faster than an alternative unimolecular bond fission. Conditions for this can be roughly estimated in the following way. The Arrhenius A factor for fission of a reasonably large molecule is likely to be about $10^{15\pm1}$ and that for the atomic exchange not far from the collision rate. Hence, assuming the disproportionating molecules (AX and B) to be present in equivalent concentrations, the two reactions have equal rates when

$$10^{15}\exp(-E_f/RT) \approx 10^{14}[B]\exp(-E_d/RT) \qquad \text{3–45}$$

At, say, 1 atm pressure and 500 °C, it follows that if E_f, the activation energy for fission of the weakest bond in either reactant molecule, is more than about 22 kcal mole^{-1} greater than E_d, the energy required for the disproportionation, the latter reaction will predominate.

Such appears to be the case in the homogeneous hydrogenation of ethylene,

$$H_2 + C_2H_4 \rightarrow C_2H_6 \qquad \text{3–XLVIII}$$

This is probably a two-carrier (H and C_2H_5) chain reaction analogous to halogenation where the chains are initiated by the reaction[56]

$$C_2H_4 + C_2H_4 \rightarrow C_2H_5 + C_2H_3 \qquad \text{3–XLIX}$$

E_d is about 64 kcal mole^{-1}, whereas fission of either reactant into radicals requires E_f to be at least 104 kcal mole^{-1}.

3–8 Relative rates of elementary reactions and rate-constant ratios

When a free radical involved in a complex mechanism participates in two elementary reactions, the two rate constants can be related by determining simultaneously the rates of formation of the respective products. In the pyrolysis of acetaldehyde, for example, the methyl radicals produce methane and ethane by the reactions

$$CH_3 + CH_3CHO \rightarrow CH_4 + CH_3CO \qquad \text{3–XX}$$

and

$$CH_3 + CH_3 \rightarrow C_2H_6 \qquad \text{3–XXIV}$$

Hence,

$$\rho_{me} \equiv \frac{d[CH_4]}{dt} = k_{20}[CH_3][CH_3CHO] \qquad \text{3–46}$$

and

$$\rho_{et} \equiv \frac{d[C_2H_6]}{dt} = k_{24}[CH_3]^2 \qquad \text{3–47}$$

from which, by eliminating $[CH_3]$, the following relation is obtained:

$$\rho_{me}/\rho_{et}^{1/2} = (k_{20}/k_{24}^{1/2})[CH_3CHO] \qquad \text{3–48}$$

Thus the ratio $k_{20}/k_{24}^{1/2}$ can be obtained by determining ρ_{me}, ρ_{et}, and $[CH_3CHO]$; and a study of its variation with temperature yields and the Arrhenius parameters $(A_{20}/A_{24}^{1/2})$ and $(E_{20} - E_{24}/2)$.

It might be thought that of themselves such composite quantities would be of little value; but there are numerous circumstances in which this is far from being the case. In the first place, a test of the constancy of such a ratio as $k_{20}/k_{24}^{1/2}$ over a range of reactant concentrations or other conditions provides an essential check on the validity of the assumed mechanism; that is, in the present case, as to whether the methane and ethane are in fact formed exclusively by the reactions assumed.

This idea has been applied in an elegant way to photo-chlorination reactions.[30] According to the mechanisms discussed in Section 3–5–5, in certain instances the same radical participates in both addition and substitution chlorinations. In the chlorinations of tetrachloroethylene and pentachloroethane, for example, the mechanisms imply that the C_2Cl_5 radical is formed by reactions 3–L and 3–LI respectively:

$$Cl + C_2Cl_4 \rightarrow C_2Cl_5 \qquad \text{3–L}$$

$$Cl + C_2Cl_5H \rightarrow C_2Cl_5 + HCl \qquad \text{3–LI}$$

and in both cases it subsequently undergoes reactions 3–LII and 3–LIII:

$$C_2Cl_5 + Cl_2 \rightarrow C_2Cl_6 + Cl \qquad \text{3–LII}$$

$$2C_2Cl_5 \rightarrow \text{Termination} \qquad \text{3–LIII}$$

The fact that the same value of $k_{52}/k_{53}^{1/2}$ was obtained from studies of the two chlorinations is strong evidence that both mechanisms are correct.

3–8–1 Pressure effects

A second application of the study of rate-constant ratios is to pressure effects. At least one of the elementary reactions in a pyrolytic reaction is inevitably the unimolecular decomposition of a radical. This may be pressure dependent. Indeed, radical decompositions are particularly prone to this effect (see Section 5–2–1). The influence of pressure can be studied by relating the unimolecular rate constant to that of some other elementary reaction known to be independent of total pressure. The latter may conveniently be a true bimolecular reaction. For example, Mulcahy and Williams[57] established the range of pressure dependence of the decomposition of the *t*-butoxy radical by studying the pyrolysis of di-*t*-butyl peroxide in the presence of phenol vapour. Here the butoxy radicals formed by dissociation of the peroxide undergo the alternative reactions:

$$(CH_3)_3CO \quad (+M) \rightarrow (CH_3)_2CO + CH_3 \quad (+M) \qquad \text{3–III}$$

and

$$(CH_3)_3CO + C_6H_5OH \rightarrow (CH_3)_3COH + C_6H_5O \qquad \text{3–LIV}$$

The ratio k_3/k_{54} was determined by measuring the rates of formation of acetone and *t*-butyl alcohol:

$$\rho_{ac}[C_6H_5OH]/\rho_{tba} = k_3/k_{54} \qquad \text{3–49}$$

It was found to vary with total pressure. Since the rate constant of the bimolecular reaction 3–LIV does not depend on pressure, the variation could be attributed to pressure dependence of k_3.

The influence of a third body on the recombination of radicals can be studied in a similar way. Thus in the pyrolysis of biacetyl at 0.5–50 torr the ratio $(\rho_{me}/\rho_{et}^{1/2})$ [biacetyl] increases with decreasing total pressure. This shows that the recombination of methyl radicals under these conditions is in the 'fall-off' region between second- and third-order kinetics (as illustrated in Fig. 1–6).

3–8–2 Cross-combination of radicals

Another rate-constant ratio of some interest concerns the combinations of like and unlike radicals. It is easy to show that the rates of formation of the three combination products AB, AA, and BB of two radicals, A and B are related to the rate constants as follows:

$$\rho_{AB}/(\rho_{AA}\rho_{BB})^{1/2} = k_{AB}/(k_{AA}k_{BB})^{1/2} \qquad 3\text{–}50$$

The value of $k_{AB}/(k_{AA}k_{BB})^{1/2}$ is a measure of any asymmetry in the mutual reactivities of the radicals, an effect which, if pronounced, could strongly affect the kinetics of a reaction with these radicals as chain carriers. The ratio cannot usually be determined by studying the termination products of a single reaction because, as explained previously, the two radicals are

Table 3–2 Cross-combinations involving alkyl radicals (from Kerr and Trotman-Dickenson[59])

Radicals		
A	B	*Experimental value of* $k_{AB}/(k_{AA}k_{BB})^{1/2}$
CH_3	CD_3	1.9
CH_3	C_2H_5	2.0
CH_3	CH_3CO	2.0
CF_2Cl	CF_2ClCF_2	2.0
iso-C_3H_7	*n*-C_4H_9	2.2

seldom present in comparable concentrations. The usual method is to carry out two reactions together. Thus Kerr and Trotman-Dickenson[58] determined the ratio for *n*-propyl and *n*-butyl radicals by photolysing mixtures of *n*-butyraldehyde (C_3H_7CHO) and *n*-valeraldehyde (C_4H_9CHO) and analysing the products for *n*-hexane, heptane, and octane. (The mechanisms of the photolyses are basically the same as the pyrolysis of acetaldehyde given in Section 3–5–3.) Typical values of the cross-combination ratio for alkyl radicals are given in Table 3–2. The value of about 2 obtained in each case is approximately what is to be expected on the basis of simple kinetic theory if the combinations of all three radical pairs are equally likely on collision. Since it is known absolutely that some alkyl radicals combine after only a few collisions, results such as these strongly suggest that this is a common property of alkyl radicals and perhaps of other types of radical as well.

3–8–3 Combination versus disproportionation of radicals

It was noted previously that radicals may not only combine but also disproportionate on collision; for example,

$$C_2H_5+C_2H_5 \rightarrow C_4H_{10} \qquad \text{3–LV}$$

and

$$C_2H_5+C_2H_5 \rightarrow C_2H_6+C_2H_4 \qquad \text{3–LVI}$$

Since both reactions are of the same order in the radical concentrations, their relative rate constants (k_{disp}/k_{comb}) are given directly by the relative rates of formation of the respective products. Table 3–3 shows some values. Like the values of $k_{AB}/(k_{AA}k_{BB})^{1/2}$ in Table 3–2 they are substantially independent of temperature.

Table 3–3 Ratios of rate constants for disproportionation and combination of radicals (from Kerr[60])

Radicals		k_{disp}/k_{comb}
C_2H_5	C_2H_5	0.14
n-C_3H_7	n-C_3H_7	0.15
iso-C_3H_7	*iso*-C_3H_7	0.66
CH_3O	CH_3O	67
CH_3	C_2H_5	0.04
CH_3	n-C_3H_7	0.06
CH_3	*iso*-C_3H_7	0.16
CH_3	CH_3O	1.9

3–8–4 Radical–molecule reactions

The most prevalent and important use of rate-constant ratios is for determining the relative reactivities of a common radical towards different molecules. This can be done when one of the rate constants in a binary ratio is common to a series of reactions. Thus, comparison of $k_{20}/k_{24}^{1/2}$ with $k_7/k_{24}^{1/2}$ determined from the pyrolyses of acetaldehyde (see Section 3–5–3) and biacetyl (see Section 3–2) respectively at, say 500 °C, tells us, after due allowance for the pressure dependence of k_{24}, that the methyl radical abstracts a hydrogen atom four times faster from acetaldehyde than from biacetyl at this temperature. More generally, the composite Arrhenius parameters provide the information that $E_7-E_{20} = 0.7$ kcal mole^{-1} and (less precisely) that $A_7/A_{20} = 0.4$.

Rate-constant ratios for reactions of this type are more commonly determined by using another reaction to generate the appropriate radicals in the presence of the substrate molecules. Photolyses of ketones and azo-compounds

$$RCOR+h\upsilon \rightarrow CO+2R \qquad \text{3–LVII}$$

$$RN{:}NR+h\upsilon \rightarrow N_2+2R \qquad \text{3–LVIII}$$

are much used to generate alkyl radicals, as also are the pyrolyses of the latter compounds and of di-peroxides. When R is methyl, there is least ambiguity as to the reactions giving rise to the relevant products, namely methane and ethane. The result is that the Arrhenius parameters $(A_{trans}/A_{comb}^{1/2})$ and $(E_{trans}-\frac{1}{2}E_{comb})$ have been determined for the transfer of a hydrogen atom to the methyl radical from perhaps a hundred different compounds. Naturally here, as elsewhere, the determination depends on the validity of the assumed mechanism, that is, of the equivalent of equation 3–48 for the particular system. A test of this equation for the

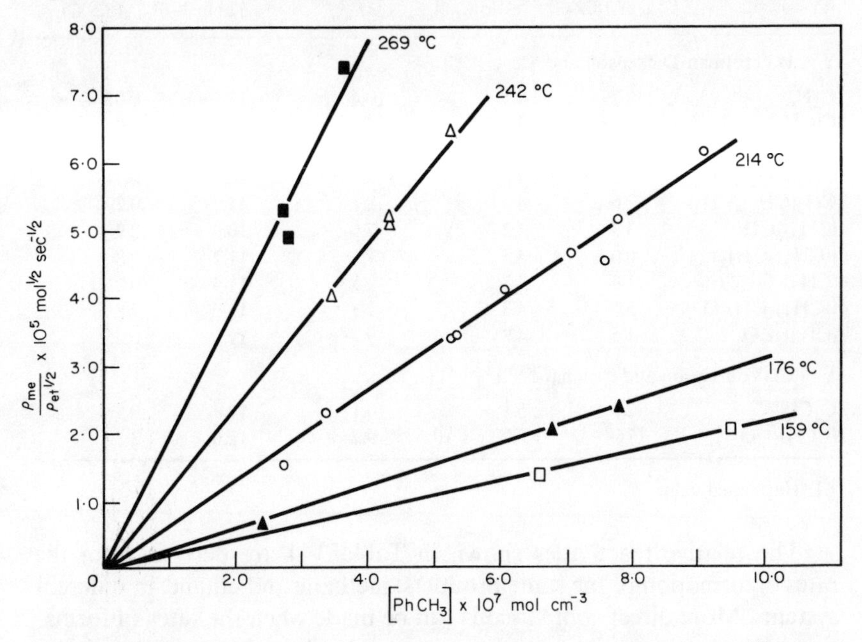

Fig. 3–5 Reaction of methyl radicals with toluene. Variation of the ratio $(\rho_{me}/\rho_{et}^{1/2})$ with $[C_6H_5CH_3]$ (Mulcahy, Williams, and Wilmshurst[61]).

reaction of methyl radicals with toluene is shown in Fig. 3–5. It will be seen that $\rho_{me}/\rho_{et}^{1/2}$ is linearly related to $[C_6H_5CH_3]$ as required by equation 3–48. Lest the reader should be inclined to think that this result is inevitable for such a system, it should be mentioned that equation 3–48 completely fails to describe the situation when toluene is replaced by the iso-electronic molecule phenol. Here the kinetics are complicated by the formation of an intermediate species from which hydrogen atoms are much more rapidly abstracted than from the phenol itself.[62] This shows the importance of establishing the mechanism before ratios of 'rate constants' are calculated from the measured rates.

Values of $k_{me}/k_{et}^{1/2}$ and the corresponding Arrhenius parameters for typical carbon compounds are given in Table 3–4. Some comment on the relative values will be made later in this section. The method of arriving at the absolute values given in the Table is discussed in Section 3–9.

Table 3–4 Relative and absolute rate parameters for atom transfer reactions of methyl radicals

$CH_3 + RY \rightarrow CH_3Y + R \quad k_{trans}$

$CH_3 + CH_3 \rightarrow C_2H_6 \quad k_{comb}$

(k and A in $cm^3\ mole^{-1}\ s^{-1}$; E in kcal $mole^{-1}$)

RY	$k_{trans}/k_{comb}^{1/2}$ at 182 °C	$\log_{10} A_{trans} - \frac{1}{2}\log_{10} A_{comb}$	$E_{trans} - \frac{1}{2}E_{comb} \approx E_{trans}$	$\log_{10} A_{trans}$ (A_{comb} from Table 3–6)	$k_{trans}/k_{comb}^{1/2}$ (per most active Y atom at 182 °C)
Y = H (Trotman-Dickenson and Steacie[63])					
C_2H_6	0.29	4.5	10.4	11.2	0.05
$C(CH_3)_4$	0.49	4.5	10.0	11.2	0.05
$CH_3(CH_2)_2CH_3$	1.6	4.2	8.3	10.9	0.34
$CH_3(CH_2)_3CH_3$	2.1	4.2	8.1	10.9	0.31
$CH_3(CH_2)_4CH_3$	2.6	4.3	8.1	11.0	0.29
$(CH_3)_3CH$	3.3	4.2	7.6	10.9	2.9
$((CH_3)_2CH)_2$	6.0	4.5	7.8	11.2	2.8
$(CH_3)_2O$	1.3	4.7	9.5	11.4	0.21
$((CH_3)_2CH)_2O$	5.6	4.3	7.3	11.0	2.8
$(CH_3)_2CO$	1.5	4.8	9.7	11.5	0.25
Y = Cl (Tomkinson and Pritchard[64])					
C_2Cl_6	2§	5.1	10.1	11.8	0.3
$(CCl_3)_2CO$	17§	5.9	9.7	12.6	3

§ Extrapolated value.

The relative reactivities shown in Table 3–4 are derived from the rates of formation of the same products, methane and ethane, in different systems. More direct comparisons can be made when the rates of formation of products peculiar to each compound can be measured. The relative rates at which halogen atoms abstract hydrogen from many different compounds have been determined in this way.[65] The method consists in halogenating pairs of compounds together,

$$X + RH \rightarrow XH + R \qquad A''$$

$$X + R'H \rightarrow XH + R' \qquad A'''$$

the halogen atoms X being generated either thermally or photochemically. Since long reaction chains occur, almost all the R and R′ radicals abstract halogen atoms from the molecular halogen present to form RX and R′X. Since [X] is the same for both compounds, the relative amounts of RX and R′X formed in a given time are equivalent to the relative rates of reactions A″ and A‴. The method has the advantage that the relative reactivities of structurally different hydrogen atoms in the same molecule can also be determined (assuming the R radicals do not isomerize and adequate analytical resolution is available). Some results of such studies are given in Table 3–5 and will receive some comment presently. The

Table 3–5 Rate parameters for hydrogen atom transfer to chlorine atoms

$Cl+RH \rightarrow HCl+R$

Absolute values are derived from competitive reactions and are based on the values $A = (4.1 \pm 0.3) \times 10^{13}$ cm^3 mole^{-1} s^{-1} and $E = 5.48 \pm 0.14$ kcal mole^{-1} when R = H (from Fettis and Knox[65])

Substrate	$\log_{10} A$ (per H atom)	*E*	$10^{12}k$ (per H atom at 182 °C)
CH_4	12.8	3.8	0.09
C_2H_6	13.2	1.0	4.9
$C(CH_3)_4$	13.1	0.9	5.2
Secondary C—H bonds in			
$CH_3(CH_2)_2CH_3$	13.4	0.3	170
Tertiary C—H bond in			
$(CH_3)_3CH$	13.3	0.1	190
CH_3Cl	13.0	3.3	0.30
CH_2Cl_2	13.1	3.0	0.49
$CHCl_3$	12.8	3.3	0.17
C_2HCl_5	12.7	3.4	0.12

values are referred to the absolute parameters of a standard reaction. These were determined by methods described later.

The competitive method may also be applied to reactions of a common radical with different kinds of atom either in different or in the same molecule. An interesting example of the latter will be mentioned. Timmons and collaborators[66] investigated the relative rates at which bromine atoms abstract hydrogen and deuterium atoms from the side chain of toluene by brominating $C_6H_5CH_2D$ and measuring the HBr and DBr produced. After allowing for the 2:1 ratio of hydrogen to deuterium, their result shows the hydrogen atoms to be the more reactive. Thus,

$$k_H/k_D = (1.08 \pm 0.25) \exp[(1.43 \pm 0.11)/RT] \qquad 3\text{–}51$$

This represents a (large) *primary kinetic isotope effect*. It was assumed, on the basis of other work, that *secondary kinetic isotope effects*, namely the effects of the presence of the deuterium atom on the rate of abstraction of the hydrogen atoms and vice versa, are negligibly small.

We shall conclude this section by noting, without theoretical speculation, some broad features of the relative rates of atom transfer reactions. Table 3–4 shows interesting differences in the reactivities of various molecules towards methyl radicals, first as regards transfer of a hydrogen atom from different molecules, and second as between transfers of hydrogen and chlorine atoms from molecules of similar structure. In the former case, the reaction rate depends on the kind of carbon atom to which the hydrogen atom is attached, increasing in the order primary, secondary, tertiary. For the alkanes, an approximately constant relative rate constant can be assigned (column 6) to each type of hydrogen atom, the sum of the contributions from all the hydrogens being equal to the relative reactivity of the molecule as a whole (column 2). At 182 °C the reactivities of primary, secondary, and tertiary C—H bonds are in the proportions 1:6:60. Qualitatively, the reactivities towards chlorine atoms show the same trend (Table 3–5). This is a general characteristic of C—Y bonds whatever the attacking radical or atom attacked. Similarly there is an increase in reactivity whenever the atom to be abstracted is attached to a carbon atom α to an olefinic bond; the product radical is of the allyl type, as formed in reaction 3–XVII in Section 3–4. Table 3–4 shows that the presence of oxygen, an electronegative atom, in the vicinity of the C—H bond also confers increased reactivity.

Since the values of A within each series of hydrogen-containing compounds in Tables 3–4 and 3–5 are nearly constant, it follows that the different reactivities are to be attributed to differences in activation energy (see also Fig. 5–1). The relative influence of the structure of the 'substrate' molecule decreases with increasing reactivity of the reactant radical: the more reactive is the radical the less is it selective.

Table 3–4 provides a comparison of the rates of abstraction of different atoms by the same radical. Chlorine atoms are transferred more rapidly than hydrogen atoms to methyl radicals and, if the results available are typical, it appears that here the difference is due to the frequency factors rather than the activation energies.

3–9 Absolute rate parameters of elementary reactions from kinetics of complex reactions

For the most part, the rates of formation of products by complex reactions yield no more than ratios of the elementary rate constants. Special procedures have to be adopted to obtain absolute values. Before describing two such methods, however, we shall note briefly some exceptions to the general rule. We saw in Section 3–5–5 that in thermally initiated halogenations of the substitution type, particularly those involving bromine or iodine, the halogen atoms participating in the reaction,

$$\mathrm{X+RH \rightarrow HX+R} \qquad \mathrm{A''}$$

are present in equilibrium with the molecular halogen. Their absolute concentration [X] can therefore be calculated from that of the molecular species by means of the known equilibrium constant. Hence, the absolute value of $k_{A''}$ can be determined from [X] and measured values of [RH] and the overall rate. The Arrhenius parameters of a number of elementary reactions of halogen atoms have been measured in this way.[31,32] A practical example was given in Section 2–1–3.

As we have seen, the elementary reactions by which a chain reaction is initiated and terminated can often be identified by kinetic analysis. In the steady state, the rates of the two processes are equal, and, in principle, the rate of the latter can be determined absolutely by measuring the rate of formation of its products. This does not yield rate constants for the terminating reactions because the absolute concentrations of the radical species are unknown. On the other hand, in general, the concentrations of the molecular species responsible for initiation *are* known, consequently the rate constant of the initiating reaction can be obtained. The Arrhenius parameters for the unimolecular fission of biacetyl to acetyl radicals (see equation 3–V) have been determined in this way.[6] Another version of the procedure consists in measuring the rate of formation of a product formed by the initiation process itself. In the pyrolysis of ethane, a small amount of methane is formed by the reactions,

$$C_2H_6 + M \rightarrow CH_3 + CH_3 + M \qquad \text{3–LIX}$$

$$CH_3 + C_2H_6 \rightarrow C_2H_5 + CH_4 \qquad \text{3–LX}$$

Hence, the absolute rate of reaction 3–LIX can be determined from the rate of formation of methane.[67] Obviously, for these procedures to be practicable, the chains should not be too long and, above all, it has to be proved that the products attributed to initiation or termination do not also form in other ways. The latter condition is seldom easy to establish.

We come now to two special general methods used extensively to determine absolute rate constants. These are the toluene carrier method for molecular fissions and the rotating sector method for radical combinations or disproportionations.

3–9–1 The toluene carrier method

As just mentioned, this is basically a method for determining the kinetic parameters for the thermal fission of a molecule into radicals. The idea is to determine the rate of pyrolysis of the compound in the presence of a large enough excess of inhibitor to prevent any of the radicals resulting from the fission from attacking other molecules of the compound. Because of the high activation energies they require, most fission reactions attain appreciable rates only at relatively high temperatures, and the problem has been to find an inhibitor with the necessary combination of reactivity and thermal stability. Toluene has proved to be suitable and the standard procedure is to determine the rate of decomposition of the compound when 'carried' in a large excess of toluene vapour though a heated tube at low pressures (as described in Section 2–2).

The fission of benzyl bromide will serve as a typical example. This occurs in the following way:

$$C_6H_5CH_2Br \rightarrow C_6H_5CH_2 + Br \qquad \text{3–LXI}$$

$$Br + C_6H_5CH_3 \rightarrow C_6H_5CH_2 + HBr \qquad \text{3–LXII}$$

$$2C_6H_5CH_2 \rightarrow (C_6H_5CH_2)_2 \qquad \text{3–LXIII}$$

The toluene, though intrinsically less reactive than benzyl bromide towards bromine atoms, is present in vast excess and effectively removes them all by reaction 3–LXII. Hence the rate of reaction 3–LXI can be determined from the rate of formation of hydrogen bromide. In this way Szwarc and collaborators[68] obtained the relation:§

$$k_{61} = 10^{13.0} \exp[-50.5 \pm 2/RT]\ \mathrm{s}^{-1} \qquad 3\text{–}52$$

The method has also been used a good deal to determine the kinetics of fission of metal alkyls, for example the reaction

$$Zn(CH_3)_2 \rightarrow ZnCH_3 + CH_3 \qquad 3\text{–LXIV}^{(69)}$$

There are several potential sources of inaccuracy, and for a time the method fell somewhat into disrepute. But with adequate attention to heat transfer in the flow reactor (Section 2–2–3), to the reaction stoichiometry and to the effect of pressure on the unimolecular rate constant, accurate results can undoubtedly be obtained. An important modification is the use of aniline instead of toluene as carrier.[39] This is particularly useful when one of the fission products is a methyl radical. Methane is formed exclusively by the reaction

$$CH_3 + C_6H_5NH_2 \rightarrow C_6H_5NH + CH_4 \qquad 3\text{–LXV}$$

whereas with toluene at high temperatures it can also arise from pyrolysis of the toluene, as was discussed in Section 2–2–3. For this reason investigations with aniline can be carried to higher temperatures without obfuscating the kinetics.

3–9–2 The rotating sector method

This technique occupies a central position in experimental kinetics. It is by far the most practical method for determining the absolute rate constants of radical combination reactions when the radicals contain more than a few atoms. We have already seen that rate-constant ratios involving combination reactions are often determined fairly easily; consequently measurement of relatively few absolute values for combination reactions opens the door to a great many more absolute rate parameters. The method is based on the kinetic behaviour of photochemical reactions involving radical combinations when the activating radiation is intermittent. Although such radiation can be provided in various ways, the most convenient procedure is to interrupt a light beam periodically by interposing a rotating disc incorporating a transparent sector. The reaction rate then observed depends on the duration of each light flash and therefore, with a constant sector angle, on the speed of rotation of the disc. For chain reactions with second-order termination, the range of speeds over which the effect is observed is related to the *average chain lifetime* $\bar{\tau}$; that is, to the average time elapsing between the initiation of a chain by the radiation and its termination by mutual interaction with

§ They subsequently repeated the operation with a dozen different nuclear-substituted benzyl bromides, finding very little effect of the substituents on the activation energy.

another chain. The object is to determine $\bar{\tau}$, which can be related to the rate constant for combination of the chain carriers in the following way. (The method does not distinguish between combination and disproportionation and, in the following, the former is taken to include the latter.)

As we have seen In Section 3–5, it is usually possible to arrange conditions so that a chain reaction is terminated by combination of radicals of the same kind. In these circumstances, the reaction can be represented schematically as follows:

$$\rho_i \rightarrow R \qquad \text{(i)}$$

$$R + Y \rightarrow \text{Products} + R \qquad \text{(p)}$$

$$R + R \rightarrow \text{Termination} \qquad \text{(comb)}$$

where Y is a reactant molecule. Assuming there is no thermal initiation, ρ_i is proportional to I_{abs}, the number of dissociating quanta absorbed in unit time. The number of times R radicals enter into the propagation reaction in unit time is $k_p[R][Y]$. Hence, the average time taken for a *single* R radical to react is

$$\bar{t} = 1/k_p[Y] \qquad 3\text{–}53$$

The average chain lifetime is given by

$$\bar{\tau} = \bar{t} \times \upsilon = (1/k_p[Y])(k_p[R][Y]/\rho_i) = [R]/\rho_i \qquad 3\text{–}54$$

υ being the chain length. But

$$[R] = (\rho_i/k_{comb})^{1/2} \qquad 3\text{–}55$$

where, in this case, k_{comb} is defined in terms of the rate of disappearance of a *single* R radical. Therefore

$$\bar{\tau} = (\rho_i k_{comb})^{-1/2} \qquad 3\text{–}56$$

Hence, if ρ_i is known and $\bar{\tau}$ can be determined, k_{comb} can be obtained.

The determination of $\bar{\tau}$ depends on the fact that, when the initiating light is turned on, the steady-state concentration of radicals is not achieved instantly; nor does it decay instantly to zero when the light is cut off (cf. Section 3–6). We shall assume, for simplicity, that the sector angle is 90°. Irradiation therefore occurs during a quarter of each cycle. Suppose the disc is rotating sufficiently slowly for each period of illumination to be much longer than $\bar{\tau}$. This means that, during almost the whole light period, the concentration of radicals is at the same steady-state level as if the disc were absent altogether; and similarly over almost the whole dark period it is zero. For long chains, the rate at any time is proportional to the radical concentration; hence, the average rate observed with slow rotation of the disc, $\mathcal{R}_{slow}$, is a quarter of that observed without the disc, that is, with continuous irradiation ($\mathcal{R}_c$). Thus,

$$\mathcal{R}_{slow} = \tfrac{1}{4}\mathcal{R}_c = (\tfrac{1}{4})(\rho_i/k_{comb})^{1/2}k_p[Y] \qquad 3\text{–}57$$

Suppose, on the other hand, the light flashes and dark periods are short compared to $\bar{\tau}$. The radical concentration then changes very little between

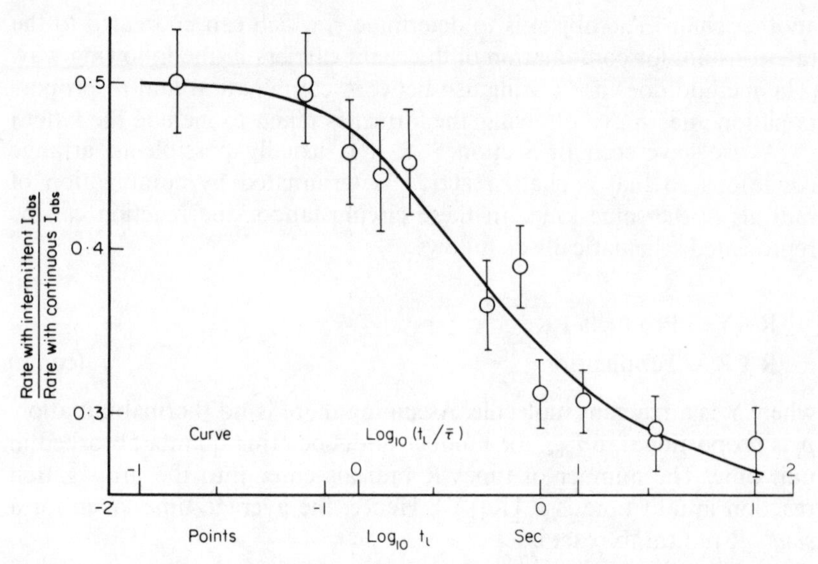

Fig. 3–6 Rotating sector experiments with the photochemical oxidation of propionaldehyde. The average chain lifetime $\bar{\tau} = 0.14$ s is obtained by fitting the theoretical curve to the experimental points. t_1 is the duration of flash. Sector angle = 90° (McDowell and Sharples[9]).

them and, in the limit, it takes up a quasi-stationary value corresponding to irradiation by light of a quarter the intensity of full irradiation. In these circumstance, the rate observed is

$$\mathscr{R}_{\text{fast}} = (\rho_i/4k_{\text{comb}})^{1/2}k_p[\text{Y}] = \tfrac{1}{2}\mathscr{R}_c \qquad 3\text{–}58$$

In other words, the rate observed at high disc speeds is double that observed at low speeds. (The ratio naturally depends on the angle of the transparent sector.) As the speed is reduced the rate falls in a manner which, for a given sector angle, depends uniquely on the ratio of $\bar{\tau}$ to the length of the light flash t_1. The theoretical relation between $\mathscr{R}/\mathscr{R}_c$ and $t_1/\bar{\tau}$ is of the form shown by the curve in Fig. 3–6. Its derivation is given in reference (A) of the Bibliography; reference 70 tabulates values from which curves appropriate to various sector angles can be drawn. The value of $\bar{\tau}$ is determined by fitting the theoretical curve to the experimental values of $\mathscr{R}/\mathscr{R}_c$ plotted against $\log t_1$. The experimental value of t_1 which corresponds to that of $t_1/\bar{\tau} = 1$, that is, $\log_{10} t_1/\bar{\tau} = 0$ when the theoretical curve coincides with the experimental points, is equal to $\bar{\tau}$. In the example shown in Fig. 3–6, $\bar{\tau} = 0.14 \pm 0.05$ s.

To obtain k_{comb} from equation 3–56, it is necessary to know ρ_i. When every quantum of radiation absorbed yields two chain carriers, as is the case, for example, when the chains are initiated by photolysis of chlorine,[30] ρ_i can be obtained directly by measuring I_{abs}. In other cases, it may be derived from the rate of formation of the product of the combination reaction[71] as exemplified by equation 3–47, or from a study of the overall reaction in the presence of inhibitors or retarders.[9]

Some typical Arrhenius parameters for combination or disproportionation reactions are given in Table 3–6.

Table 3–6 Arrhenius parameters for some combination or disproportionation reactions of radicals determined by the rotating sector method

	A (cm^3 $mole^{-1}$ s^{-1})	E (kcal $mole^{-1}$)
$2CH_3$	13.35	0
$2C_2H_5$	13.4	0
$2C_2H_3Cl_2$	12.9	0.3
$2C_2Cl_5$	12.5	0.5
$2CF_3$	13.4	0
$Cl+COCl$	14.6	0.8

3–10 Bond dissociation energies from kinetic measurements

Since chemical reaction involves making and breaking bonds, the starting-point for most theories of chemical reactivity is information on the energies that are required to break particular bonds, or, of course, are the energies that are required to break particular bonds, or, of course, are recovered when the bonds are formed from the radical fragments. The most accessible quantity is the *bond dissociation energy*. This is defined as the energy—strictly the enthalpy—absorbed when the bond is totally severed, the two fragments being removed to an infinite distance apart. The values are usually referred to 25 °C but, in many cases—though not in all—the errors incurred by measuring at other temperatures are commensurate with the errors of measurement.

Bond dissociation energies can be determined in various ways. Except for spectroscopic measurements on diatomic molecules, these generally still depend on incompletely validated assumptions or are not intrinsically very accurate. The acquisition of reliable bond dissociation energies has been a major preoccupation of kineticists for several decades, particularly, perhaps, because the required data can frequently be obtained by kinetic methods. The principles of these methods will now be outlined very briefly. For a comprehensive account of kinetic and other methods, including their accuracy and underlying assumptions, the reader should consult the monograph by Cottrell.[72]

The kinetic methods are based on the fact, illustrated by Fig. 1–5, that the difference between the activation energies of a reaction and its reverse reaction is equal to the equilibrium internal energy absorbed by the reaction. Thus, for a bond fission

$$AB \rightarrow A+B$$

the bond dissociation energy is given by the relation

$$D(\text{A—B}) = E_f - E_{comb} + RT \qquad 3\text{–}59$$

The term RT is present because D is defined as the change in enthalpy rather than internal energy; it is about 0.6 kcal mole^{-1} at 25 °C and is commonly neglected. E_{comb} is usually assumed to be zero (see Table 3–6) and equation 3–59 becomes

$$D(\mathrm{A{-}B}) \approx E_f \qquad 3\text{–}60$$

Thus determination of the activation energy for bond fission by the toluene carrier or some other technique yields the bond dissociation energy directly. For example, the Arrhenius equation 3–52 for the fission of benzyl bromide yields the value 50.5 kcal for $D(C_6H_5CH_2{-}Br)$.

A second standard method, applicable to carbon–hydrogen bonds, is via studies of halogen substitution. For the reversible reaction

$$\mathrm{X+R-H \rightleftarrows HX+R} \qquad \mathrm{A'', -A''}$$

the following relation applies

$$E_{A''} - E_{-A''} = D(\mathrm{R{-}H}) - D(\mathrm{H{-}X}) \qquad 3\text{–}61$$

Since $D(\mathrm{H{-}X})$ is known from thermochemical data, determination of $E_{A''}$ and $E_{-A''}$ suffices to obtain $D(\mathrm{R{-}H})$. It was shown in Section 3–9 that, when X is a bromine or iodine atom, the Arrhenius parameters of reaction A″ can be obtained by determining the initial rates of thermal halogenation of RH over a range of temperature. Alternatively, $E_{A''}$ can be derived from the temperature coefficient of the photo-bromination.[31] $E_{-A''}$ is determined by making use of the fact that reaction A″ is reversible under the usual experimental conditions and the reverse reaction, which causes retardation, comes into play as the HX builds up. By studying the reaction rate over a period of time, with or without the addition of excess HX, it is possible to obtain the value of $E_{-A''}$. This usually amounts to only 1 or 2 kcal mole^{-1}. Alternatively, $E_{-A''}$ may be determined from a separate study of the competition for R radicals between HX and a compound (iodine) with which they react without activation energy. Thus, $E_{A''}$ and $E_{-A''}$ having been determined, $D(\mathrm{R{-}H})$ follows from equation 3–61. The halogenation method was invented by Kistiakowsky and Van Artsdalen[31] using bromine; more recently, it has been developed in various ways and applied to a considerable number of compounds by Golden and Benson using iodine.[74]

Bond dissociation energies do not have to be determined separately for every bond in every molecule. When the necessary information is available, they can be calculated from other bond dissociation energies and standard thermochemical heats of formation. Suppose, for example, the value of $D(CH_3{-}Br)$ is required. From the dissociation energy of the bond $CH_3{-}CH_3$ and the heat of formation of ethane, it is easy to calculate the heat of formation of the free methyl radical. Likewise, the heat of formation of the bromine atom is equal to $\frac{1}{2}D(\mathrm{Br{-}Br})$. Using these data, $D(CH_3{-}Br)$ can be calculated from the heat of formation of methyl bromide. This type of calculation, besides yielding essential information about molecular properties, provides a useful check on the consistency of dissociation energies determined by different experimental methods. In this way, many of the values now accepted have been arrived at by the convergence of evidence rather than by a single definitive measurement. A short list of bond dissociation energies derived from various sources is

Table 3–7 Bond dissociation energies

(to nearest kcal)

H—H	104	$H_3C—CH_3$	88
		$H_3C—C_2H_5$	85
$H—CH_3$	103	$H_3C—C_2H_4$	26
$H—C_2H_5$	98	$H_3C—COCH_3$	81
$H—C_2H_4$	39	$H_3C—CO$	11
$H—CH(CH_3)_2$	94.5		
$H—C(CH_3)_3$	91		
$H—CH{:}CH_2$	106	C=O	257
$H—CH_2CH{:}CH_2$	87.5	OC=O	127
$H—C_6H_5$	110	$H_3C—OH$	90
$H—CH_2C_6H_5$	85	$H_3C—F$	108
$H—COCH_3$	86.5	$H_3C—Cl$	81
$H—NH_2$	104	N≡N	226
H—O	102	$H_2N—NH_2$	59
$H—O_2$	47	N=O	151
H—OH	119	O=O	119
$H—OCH_3$	100	HO—OH	51
H—F	136	O=SO	132
H—SH	91	F—F	39
H—Cl	103	Cl—Cl	58
H—Br	87.5	Br—Br	46
H—I	71	I—I	36

given in Table 3–7. Most of the values are accurate to about ± 2 kcal mole^{-1} or better.

Bibliography

(A) F. S. Dainton, *Chain Reactions*, 2nd edn, Methuen, 1966.

(B) N. N. Semenov, *Some Problems in Chemical Kinetics and Reactivity*, Princeton U.P., 1959, and Pergamon Press, 1958.

(C) P. G. Ashmore, F. S. Dainton, and T. M. Sugden (eds), *Photochemistry and Reaction Kinetics*, C.U.P., 1966.

(D) A. F. Trotman-Dickenson, *Free Radicals*, Methuen, 1959.

References

1 F. P. Lossing, Chap. 11 of *Mass Spectrometry*, C. A. McDowell (ed.), McGraw-Hill, 1963.

2 H. E. Avery and J. N. Bradley, *Trans. Faraday Soc.*, **60,** 850, 857, 1964; A. G. Gaydon *et al.*, *Proc. Roy. Soc.*, **A279,** 313, 1964; D. Schofield *et al.*, *J. Chem. Phys.*, **42,** 2132, 1965.

3 P. J. Zandstra and J. D. Michaelsen, *J. Chem. Phys.*, **39,** 933, 1963; E. G. Janzen and J. L. Gerlock, *Nature*, **222,** 867, 1969.

4 F. O. Rice and K. K. Rice, *The Aliphatic Free Radicals*, Baltimore, 1935.
5 J. R. Thomas, *J. Am. Chem. Soc.*, **85,** 591, 1963; R. Livingston and H. Zeldes, *ibid.*, **88,** 4333, 1966; J. Q. Adams, *ibid.*, **90,** 5363, 1968; J. M. Rivers and S. Shih, *J. Chem. Phys.*, **50,** 3132, 1969.
6 K. J. Hole and M. F. R. Mulcahy, *J. Phys. Chem.*, **73,** 177, 1969.
7 C. F. Cullis and D. J. Waddington, *Trans. Faraday Soc.*, **53,** 1317, 1957.
8 J. B. Farmer and C. A. McDowell, *Trans. Faraday Soc.*, **48,** 624, 1952.
9 C. A. McDowell and L. K. Sharples, *Canad. J. Chem.*, **36,** 251, 268, 1958.
10 M. Niclause, R. Martin, A. Combes, and M. Dzierzynski, *Canad. J. Chem.*, **43,** 1120, 1965.
11 A. Boyer, M. Niclause, and M. Letort, *J. Chim. Phys.*, **49,** 345, 1952.
12 S. Bywater and E. W. R. Steacie, *J. Chem. Phys.*, **19,** 319, 1951.
13 A. Gray, *J. Chem. Soc.*, **1952,** 3150; **1953,** 1300.
14 G. B. Serbeev and V. Ya. Shtern, *Dok. Akad. Nauk. SSR*, **91,** 1357, 1953.
15 A. Trifonoff, *Z. phys. Chem.*, **B3,** 195, 1929.
16 D. A. Leathard and J. H. Purnell, *Ann. Rev. Phys. Chem.*, **21,** 197, 1970.
17 D. E. Hoare and A. D. Walsh, *5th Symposium* (*Internat.*) *on Combustion*, p. 467, Reinhold, 1955.
18 N. N. Semenov, *Some Problems in Chemical Kinetics and Reactivity*, vol. 1, pp. 213ff, Princeton University Press, 1959.
19 R. N. Pease, *J. Am. Chem. Soc.*, **56,** 2388, 1934.
20 D. H. R. Barton and K. E. Howlett, *J. Chem. Soc.*, **1949,** 155.
21 A. Maccoll and P. J. Thomas, *J. Chem. Soc.*, **1957,** 5033.
22 D. H. R. Barton and F. P. Onyon, *Trans. Faraday Soc.*, **45,** 725, 1949.
23 A. Boyer, M. Niclause, and M. Letort, *J. Chim. Phys.*, **49,** 337, 1952.
24 F. O. Rice and O. L. Polly, *J. Chem. Phys.*, **6,** 273, 1938.
25 K. J. Laidler and M. T. H. Liu, *Proc. Roy. Soc.*, **A297,** 365, 1967.
26 M. T. H. Liu and K. J. Laidler, *Canad. J. Chem.*, **46,** 479, 1968.
27 G.-M. Côme, M. Dzierzynski, R. Martin, and M. Niclause, *Rev. Inst. Franc. Pétrole*, **23,** 1365, 1968.
28 F. O. Rice and K. F. Herzfeld, *J. Am. Chem. Soc.*, **56,** 284, 1934.
29 M. Niclause, J. Lemaire, and M. Letort, *Adv. Photochem.*, **4,** 25, 1966.
30 G. Chiltz, P. Goldfinger *et al.*, *Chem. Rev.*, **63,** 355, 1963.
31 G. B. Kistiakowsky and E. R. Van Artsdalen, *J. Chem. Phys.*, **12,** 469, 1944.
32 R. Walsh and S. W. Benson, *J. Am. Chem. Soc.*, **88,** 4570, 1966.
33 F. S. Dainton, D. A. Lomax, and M. Weston, *Trans. Faraday Soc.*, **53,** 460, 1957.
34 S. W. Benson, *The Foundations of Chemical Kinetics*, pp. 320 *et seq.*, McGraw-Hill, 1960.
35 R. G. W. Norrish, *Proc. Roy. Soc.*, **A301,** 1, 1967.
36 J. K. K. Ip and G. Burns, *J. Chem. Phys.*, **51,** 3414, 1969.
37 E. A. Moelwyn-Hughes, *Physical Chemistry*, p. 550, C.U.P., 1940.
38 T. R. Crossley and M. A. Slifkin, *Progress in Reaction Kinetics*, Vol. 5, p. 409, G. Porter (ed.), Pergamon Press, 1970; R. H. Snow, *J. Phys. Chem.*, **70,** 2780, 1966.
39 G. L. Esteban, J. A. Kerr, and A. F. Trotman-Dickenson, *J. Chem. Soc.*, **1963,** 3873.
40 P. G. Ashmore and M. G. Burnett, *Trans. Faraday Soc.*, **58,** 1801, 1962.
41 R. J. Kominar, M. G. Jacko, and S. J. Price, *Canad. J. Chem.*, **45,** 575, 1967.
42 T. C. Clark, M. A. A. Clyne, and D. H. Stedman, *Trans. Faraday Soc.*, **62,** 3354, 1966.

43 N. L. Arthur, P. Gray, and A. A. Herod, *Canad. J. Chem.*, **47,** 1347, 1969.
44 R. Shaw and A. F. Trotman-Dickenson, *J. Chem. Soc.*, **1960,** 3210.
45 T. N. Bell and K. O. Kutschke, *Canad. J. Chem.*, **42,** 2032, 1964.
46 R. R. Baldwin, R. W. Walker, and S. J. Webster, *Combustion and Flame*, **15,** 167, 1970.
47 R. J. Cvetanovic and R. S. Irwin, *J. Chem. Phys.*, **46,** 1694, 1967.
48 A. S. Gordon, *J. Chem. Phys.*, **36,** 1330, 1962.
49 M. C. Lin and K. J. Laidler, *Canad. J. Chem.*, **44,** 2927, 1966.
50 R. K. Brinton, *J. Am. Chem. Soc.*, **83,** 1541, 1961.
51 J. C. J. Thynne and P. Gray, *Trans. Faraday Soc.*, **59,** 1149, 1963.
52 L. F. Phillips and H. I. Schiff, *J. Chem. Phys.*, **37,** 1233, 1962.
53 G. Greig and J. C. J. Thynne, *Trans. Faraday Soc.*, **62,** 3338, 1966.
54 M. F. R. Mulcahy, B. G. Tucker, D. J. Williams, and J. R. Wilmshurst, *Aust. J. Chem.*, **20,** 1155, 1967.
55 C. F. Cullis and M. F. R. Mulcahy, *Combustion and Flame*, **18,** 225, 1972.
56 M. C. Lin, *Canad. J. Chem.*, **44,** 1237, 1966.
57 M. F. R. Mulcahy and D. J. Williams, *Aust. J. Chem.*, **17,** 1291, 1964.
58 J. A. Kerr and A. F. Trotman-Dickenson, *J. Chem. Soc.*, **1960,** 1602.
59 J. A. Kerr and A. F. Trotman-Dickenson, *Progress in Reaction Kinetics*, G. Porter (ed.), vol. 1, p. 4, Pergamon Press, 1961.
60 J. A. Kerr, *Ann. Reports Prog. Chem.*, **64A,** 73, 1967.
61 M. F. R. Mulcahy, D. J. Williams, and J. R. Wilmshurst, *Aust. J. Chem.*, **17,** 1329, 1964.
62 M. F. R. Mulcahy and D. J. Williams, *Aust. J. Chem.*, **18,** 20, 1965.
63 A. F. Trotman-Dickenson and E. W. R. Steacie, *J. Chem. Phys.*, **19,** 329, 1951.
64 D. M. Tomkinson and H. O. Pritchard, *J. Phys. Chem.*, **70,** 1579, 1966.
65 G. C. Fettis and J. H. Knox, *Progress in Reaction Kinetics*, G. Porter (ed.), vol. 2, p. 1, Pergamon Press, 1964.
66 R. B. Timmons, J. de Guzman, and R. E. Vamerin, *J. Am. Chem. Soc.*, **90,** 5996, 1968.
67 A. B. Trenwith, *Trans. Faraday Soc.*, **62,** 1538, 1966.
68 M. Szwarc, B. W. Ghosh, and A. H. Sehon, *J. Chem. Phys.*, **18,** 1142, 1950; M. Szwarc, C. H. Leigh, and A. H. Sehon, *ibid.*, **19,** 657, 1951.
69 A. N. Dunlop and S. J. W. Price, *Canad. J. Chem.*, **48,** 3205, 1970.
70 G. M. Burnett and H. W. Melville, *Technique of Organic Chemistry*, Chap. 20, A. Weissberger (ed.), 2nd edn, vol. 8, Part 2, Wiley, 1963.
71 R. Gomer and G. B. Kistiakowsky, *J. Chem. Phys.*, **19,** 85, 1951; A. Shepp, *ibid.*, **24,** 939, 1956.
72 T. L. Cottrell, *The Strengths of Chemical Bonds*, 2nd edn, Butterworths, 1958.
73 G. C. Fettis and A. F. Trotman-Dickenson, *J. Chem. Soc.*, **1961,** 3037.
74 D. M. Golden and S. W. Benson, *Chem. Rev.*, **69,** 125, 1969.

Problems

3–1 Discuss the significance of the following experimental results:

(a) The reaction $n\text{-}C_3H_7Br \rightarrow C_3H_6 + HBr$ at 380 °C was found to be little affected by packing the reaction vessel with glass tubing but addition of sufficient *cyclo*hexene reduced the rate by a factor of 9, subsequent additions having no further effect. There was no change in the nature of the main products. Additions of insufficient *cyclo*hexene to produce the limiting rate resulted in S-shaped pressure-time curves. Other olefinic compounds produced the same reduction in rate but different amounts were required. The rate of the residual reaction was first-order in $n\text{-}C_3H_7Br$.

(b) The rate of the reaction $C_3H_8 \rightarrow C_3H_6 + H_2$ at 550 °C was found to be unaffected either by the size of the Pyrex reaction vessel or by coating the surface with PbO. Addition of 0.5% O_2 to propane in a large uncoated vessel produced a temporary 6-fold increase in rate, but the same procedure with a smaller vessel caused the rate to be retarded. When the PbO-coated vessel was used, addition of O_2 produced an induction period during which no reaction could be detected.
(c) The rate of reaction of CH_3CHO with O_2 in Pyrex vessels is substantially independent of the vessel size. It was found to increase by an order of magnitude when the surface was coated with KCl and, in spherical vessels, it then became inversely proportional to $(\text{radius})^{1/2}$.

3–2 The oxidation of tetrachloroethylene ($C_2Cl_4 + \frac{1}{2}O_2 \rightarrow CCl_3COCl$) can be brought about at 25 °C by photosensitization with chlorine (Dickenson and Carrico, *J. Am. Chem. Soc.*, **56**, 1473, 1934). In an experiment carried out with light of wavelength $\lambda = 436 \times 10^{-9}$ m, the total radiant energy flux incident as a parallel beam upon the surface of the reaction mixture was 2.76×10^{-4} J s^{-1} (measured with a thermopile) and the rate of consumption of C_2Cl_4 was 7.1×10^{-8} mole s^{-1}. The length of the reaction vessel was 12 cm and $[Cl_2]$ was 3.17×10^{-6} mole cm^{-3}. The (decadic) absorption coefficient of Cl_2 is 1.64×10^3 cm^2 mole^{-1}. What was the quantum yield for the consumption of C_2Cl_4? ($Nh\nu = 0.12/\lambda$ J mole^{-1}).

3–3 The reaction $CH_3OCH_3 \rightarrow CH_4 + CH_2O$ occurs above 500 °C. It is accelerated by sources of free radicals and retarded by propene, but not affected by addition of inert gas at ether pressures above 100 torr. Its order is close to 1.5 at these pressures. Discuss possible reaction mechanisms, taking cognizance of alternative reaction orders for the elementary steps. What do you consider the most probable mechanism? What minor products would you expect to find and how might studies of the kinetics of their formation help to elucidate the mechanism. In what way, if at all, would you expect the overall order to change if the reaction were to be conducted at substantially lower pressures?

3–4 (a) The photolysis of diethyl ketone below 200 °C occurs exclusively by a free radical mechanism to produce carbon monoxide and the hydrocarbon products: ethane, ethene, and *n*-butane. Quantum yields for CO are $\leqslant 1$. Suggest a reaction mechanism which accounts for the rate ratios:

$$(\mathscr{R}_{C_2H_6} + \mathscr{R}_{C_4H_{10}})/\mathscr{R}_{CO} = 1.0 \quad \text{and} \quad \mathscr{R}_{C_2H_4}/\mathscr{R}_{C_4H_{10}} = 0.14$$

both of which are found to be independent of temperature and pressure between 10 and 60 torr. What kinetic significance do you attach to the second ratio?
(b) The quantum yield for the mercury-photosensitized decomposition of ethane at 470 °C is typically about 4. The main products are ethane and hydrogen together with smaller amounts of *n*-butane. What is the most probable reaction mechanism? The following initial rates were measured by Loucks and Laidler. What can be deduced quantitatively from them concerning the effect of pressure on the rate of decomposition of the C_2H_5 radical?

Pressure (torr)	4.0	27.7	98.8	463
$10^{12}\, d[H_2]/dt$ (mole cm^{-3} s^{-1})	1.34	9.76	22.5	41.2
$10^{12}\, d[C_4H_{10}]/dt$ (mole cm^{-3} s^{-1})	0.129	0.935	1.82	1.54

3–5 The kinetics of the atom-transfer reaction $CH_3 + C_6H_5OCH_3 \rightarrow C_6H_5OCH_2 + CH_4$ at 243 °C were studied by pyrolysing di*tert*butyl peroxide in the presence of a large excess of anisole vapour (An) and measuring the rates of formation of methane and ethane (Mulcahy *et al.*, *Aust. J. Chem.*, **20**, 1155, 1967). The rate constant for the formation of ethane from methyl radicals is 2.2×10^{13} cm^3 mole^{-1} s^{-1}. Determine the rate constant of the transfer reaction as accurately as you can from the following data.

$10^7[An]$ mole cm^{-3}	1.46	3.41	5.69	6.55
$10^{11}\, d[CH_4]/dt$ mole cm^{-3} s^{-1}	1.41	8.77	6.88	22.1
$10^{10}\, d[C_2H_6]/dt$ mole cm^{-3} s^{-1}	5.98	48.6	12.7	95.3

3–6 (a) The kinetics of the chain reaction $F_2+CCl_3F \rightarrow ClF+CCl_2F_2$ in a *Teflon*-lined reaction vessel at about 250 °C are described by the relation

$$-d[CCl_3F]/dt = \kappa[F_2]^{1/2}[CCl_3F]$$

(Foon and Tait, *J. C. S. Faraday I*, **68,** 104, 1972.) Suggest a reasonable mechanism. How do you explain the fact that κ is independent of the pressure of inert gas present?
(b) Initial rates of formation of HI from the homogeneous reaction

$$I_2+CH_2O \rightarrow HI+(ICHO) \xrightarrow{\text{fast}} 2HI+CO$$

are well described by the relation $\mathscr{R}_0 = \kappa[I_2]^{1/2}[CH_2O]$ (Ref. 32). Measurements over the temperature range 180–270 °C gave the value 35.5 kcal mole^{-1} for the activation energy associated with κ. Use this value and the following standard thermochemical data to derive the bond dissociation energies D(H—CHO) and D(H—CO), assuming the value E = 1.5 kcal mole^{-1} for the reaction $CHO+HI \rightarrow CH_2O+I$.

	I—I	H—I	H—H	CH_2O	CO
D(A—B) kcal mole^{-1}	36.0	71.4	104.2	—	—
$-\Delta H_f^\circ$ kcal mole^{-1}	—	—	—	27.7	26.4

(Neglect the effect of temperature on these quantities)

4 Experimental methods II: direct investigation of elementary reactions of free atoms and radicals

This chapter is concerned with methods of investigating directly the kinetics of elementary reactions in which one or both of the reactants is a free radical, a term which here as elsewhere in this book includes free atom. Rate parameters of such reactions are needed basically for two purposes. First, as we have seen in Chapter 3, most gaseous reactions of ordinary laboratory experience are free-radical reactions; and, if their mechanisms are to be understood in a quantitative sense, the absolute rates of the constituent elementary reactions must be known. Secondly, to understand chemical reactivity at a more fundamental level is to understand what determines the rates of elementary reactions and, of these, radical reactions are generally the simplest as regards internal mechanism. In short, absolute rates of elementary radical reactions are required in order to understand the kinetics of complex reactions and to provide basic data for testing fundamental theories of chemical reactivity.

Reference was made in Chapter 3 to indirect methods of determining absolute reaction rates, the most general being the rotating sector method for reactions between two radicals. With methods of this kind the identity of the radicals involved and the nature of the particular reactions they undergo have to be inferred more or less intuitively from the molecular structure and kinetics of formation of the products. Obviously more direct methods are needed to identify elementary reactions directly and to study them in isolation as single processes. In any particular case, this involves possessing an unequivocal source of the reactant radicals and being able to follow their disappearance during reaction by specific analysis. It is also desirable to study the rates of the formation of product radicals, though this is more difficult to achieve with accuracy. The experimental methods outlined in the present chapter represent a variety of current ways of achieving these objectives.

Static and flow methods are both used and the general principles of treatment of the kinetic measurements are identical with those given in Chapter 2. The apparatus and instrumentation are necessarily rather more elaborate since they must cope, among other things, with very fast reaction rates. As with the procedures for slower reactions described earlier, the static method, on the whole, lends itself better to identifying intermediate species in complex systems—in this case radicals—and characterizing their reactions. Flow methods are generally more suitable for determining accurate kinetic parameters of reactions that are already at least reasonably well defined. The following two sections deal with static and flow procedures respectively, after which, by way of example, some typical investigations relevant to the main types of radical reactions will be described briefly. Following this, mention is made of experimental studies of reactions involving radicals and molecules in specific energy states; and the chapter concludes with a short account of procedures whereby reactive molecular collisions can be studied in detail by producing crossed molecular beams of the reactants.

4–1 Static methods (flash photolysis—kinetic spectroscopy)

The two principal ways of studying elementary reactions in a stationary gas are by means of flash photolysis and shock heating. Both methods operate by transmitting a short, massive pulse of energy to the mixture of appropriate molecules. In this way radicals are produced in high enough concentration to be measured by time-resolved ('kinetic') spectroscopy. Shock-tube methods were described in Chapter 2. Here we are concerned with flash photolysis, a technique which gives a remarkable variety of insights into elementary chemical processes. Not only does it enable elementary reactions to be identified directly by observing the rise and fall in radical concentrations but we shall see later that in many cases the quantum states of the reactants and products can also be identified and their history recorded.

A diagram of the apparatus for flash photolysis and kinetic spectroscopy is given in Fig. 4–1. The principle is to photolyse a substantial proportion of a few torr of the potentially reactive gas by an extremely powerful but very brief flash of light and to follow this flash immediately with a second, much weaker, flash directed through the gas to a u.v.-visible spectrometer. The second flash is triggered electronically by the first so as to occur after a specified time interval which can be varied from one experiment to another. Very high concentrations of radicals are produced by the photo-flash. By carrying out a series of otherwise identical experiments with different time intervals between flashes, the absorption spectra of the various radicals and other constituents of the gas are recorded photographically as a function of time.

Each of the flashes is produced by discharging a condenser through an inert gas; a capacity of about 50 μF charged to about 10 kV is used for the photo-flash, so transmitting an energy of about 25 kJ to the inert gas. The duration of the photo-flash can be limited to about 10×10^{-6} s (i.e. 10 μs); and that of the second flash, which is generated by about 0.1 kJ, to about 1–2 μs. In this way events in the reaction vessel can be followed over a period from 0 to 10^4 μs after the radicals are first produced. More recently, with the use of laser pulses to produce the flashes, resolution to 10^{-8} s has become possible.[(1)]

Figure 4–1 shows the photo-flash tube simply laid parallel with the reaction tube, but the two tubes may be made concentric or the one

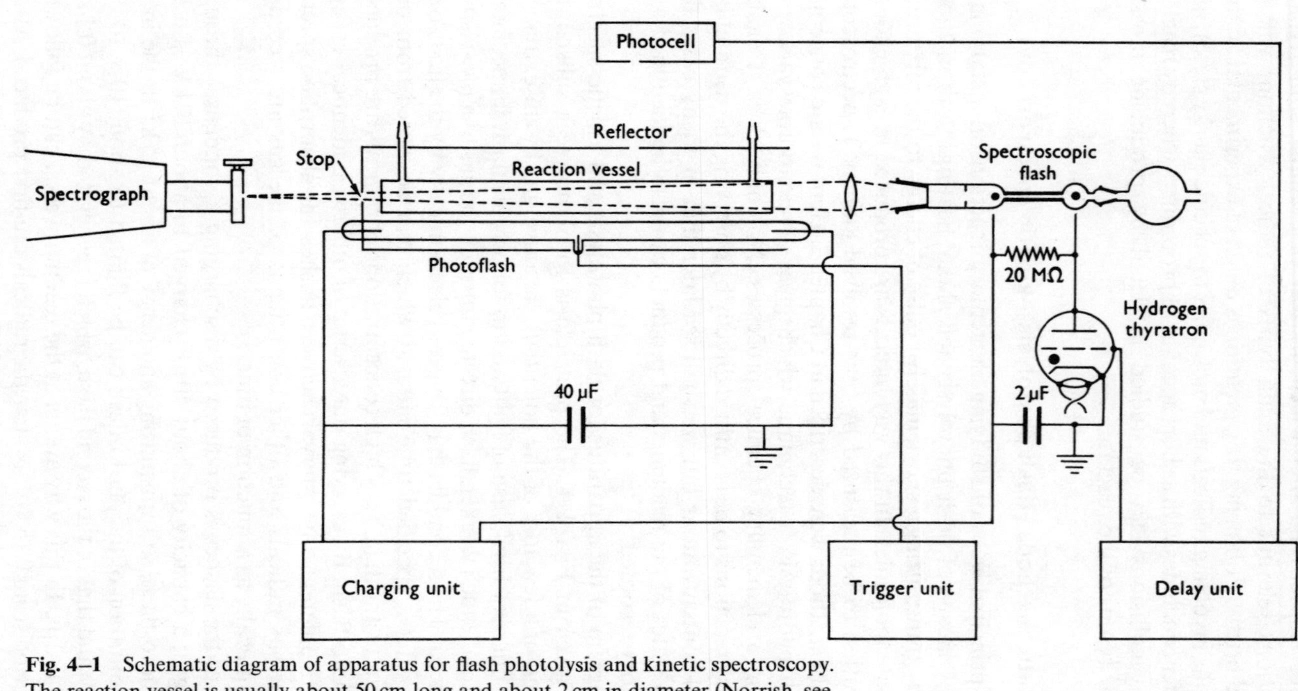

Fig. 4–1 Schematic diagram of apparatus for flash photolysis and kinetic spectroscopy. The reaction vessel is usually about 50 cm long and about 2 cm in diameter (Norrish, see Bibliography, ref. (A)).

coiled round the other for more uniform illumination. The usual method of determining the changes in concentration of the absorbing species by photometry is carried out on photographic plates, the spectra obtained from a number of successive experiments usually being recorded on the same plate. When the spectrum of a particular species is sufficiently discrete, continuous photoelectric recording can be combined with oscillographic display; and this further lays open the way to improved sensitivity and precision by the use of repetitive flashes and electronic integration, for example by a 'computer of average transients'. A recent development is the use of resonance fluorescence from the reacting species, instead of absorption, to monitor their concentration. For an authoritative account of the achievements, potentialities, and limitations of flash photolytic technique reference may be made to the article by Porter.[2] Here it will suffice to mention that over 60 different radical species containing one to four atoms have been detected by the technique, as well as an even greater number of a special class of large aromatic radicals such as C_6H_5, C_6H_5O, etc.[3] From the present point of view, the basic feature of greatest value probably lies in the ability to observe high radical concentrations, in which circumstances the radicals frequently react faster with each other than with the molecular species present. Radical–radical reactions are particularly difficult to observe directly by other means. A second advantage of the technique derives from the fact that the time over which observations are made is generally short compared with the time required for radicals to diffuse to the wall; and so the influence of heterogeneous reactions is eliminated.

Flash photolysis can be carried out either 'adiabatically' or 'isothermally'. During the period of the photo-flash the absorbing molecules become highly excited and some may undergo many dissociations and recombinations. The result is that a large amount of energy from the light appears almost instantly in the gas as heat and the temperature rises immediately to 2000–3000 °C, sometimes higher. The reactions observed are then largely pyrolytic rather than photolytic in character. This is the basis of the 'adiabatic' method, from which much valuable qualitative information about radical reactions in high-temperature pyrolyses and combustion has been obtained. This form of the technique is analogous to that of the shock tube (without, however, the advantage of a precise definition of the reaction temperature). To observe reactions under 'isothermal' conditions the pressure of the absorbing molecules is kept low—usually $\leqslant 1$ torr—and a several hundredfold excess of inert gas is added to absorb the heat. In this way the temperature rise is reduced to a few degrees or less. Naturally the isothermal method is adopted for quantitative measurements of reaction rates. In such experiments the reaction temperature can be varied by placing the apparatus in a thermostat.

The type of basic qualitative information on elementary reactions provided by the technique can be illustrated by investigations carried out by Norrish and others on the decomposition and oxidation of hydrogen sulphide.[4] When hydrogen sulphide was flashed under isothermal conditions, the spectrum of the SH radical appeared immediately. It then decreased in intensity over a period of a few milliseconds, during which time the spectrum of S_2 appeared, increased to maximum intensity, and then declined. The SH radicals are formed by the reaction

$$H_2S + h\nu \rightarrow H + SH \qquad \text{4–I}$$

and these observations show that they disappear by mutual reaction to form S_2:

$$SH + SH \rightarrow H_2S + S \qquad \text{4–II}$$

$$S + SH \rightarrow S_2 + H \qquad \text{4–III}$$

(The alternative reaction

$$SH + SH \rightarrow S_2 + H_2 \qquad \text{4–IV}$$

is known on other grounds to be much slower than reaction 4–II.) The S_2 formed in reaction 4–III subsequently polymerizes to solid sulphur, which in Norrish's experiments was found on the walls of the tube. When the photolysis was carried out adiabatically, the spectra were more intense, but the sequence of events was the same. This shows that the pyrolysis of hydrogen sulphide occurs by the same mechanism as the photolysis.

When oxygen was present, flashing under adiabatic conditions again produced the spectrum of SH; S_2, however, was not seen and its spectrum was replaced by those of OH and SO which appeared simultaneously with the decline of SH. These radicals were evidently formed by the reaction

$$SH + O_2 \rightarrow SO + OH \qquad \text{4–V}$$

Their spectra subsequently gave way to the spectrum of sulphur dioxide, a final product. Thus:

$$SO + O_2 \rightarrow SO_2 + O \qquad \text{4–VI}$$

this reaction being preferred to the alternative reaction

$$SO + OH \rightarrow SO_2 + H \qquad \text{4–VII}$$

because the mutual behaviour of the SO and SO_2 spectra was found to be the same irrespective of whether H_2S, COS, or CS_2 was flashed with oxygen.

These examples are typical of the kind of evidence supplied by flash photolysis regarding the participation of elementary reactions in overall processes. They show that reactions 4–II and 4–III are important steps in the pyrolysis of hydrogen sulphide and that reactions 4–V and 4–VI are important in its combustion. The implication is that, if the overall reactions are to be understood quantitatively, attention should be focused on determining the rates of these particular elementary reactions. For such determinations, the flow methods described in the next section are frequently the most appropriate, but flash photolysis can often be used to advantage particularly, as mentioned previously, for reactions between radicals.

An example of the latter can be found in the work of Porter and Norrish[5] on the kinetics of disproportionation of ClO radicals:

$$ClO + ClO \rightarrow ClOO + Cl \qquad \text{4–VIII}$$

When a mixture of chlorine and oxygen is subjected to flash photolysis, subsequent analysis of the gas reveals little or no chemical change. Nevertheless reactions occur during and immediately after the flash which are manifested by the vivid appearance of the spectrum of ClO on the plate of the spectrograph. From evidence too complex to be given here[6] the radicals are known to disappear by reaction 4–VIII, and the gas reverts to its original composition by the very fast reaction

$$Cl + ClOO \rightarrow Cl_2 + O_2 \qquad \text{4–IX}$$

This reaction is not visible to the spectrograph. Porter and Wright[5] studied the decay of the radicals under various conditions over a period of 10 ms and determined the second-order rate constant k_8 in the usual way by plotting 1/[ClO] against time (equation 2–3). They also found the rate to be independent both of the total pressure (above 55 torr) and temperature.

To obtain the absolute value of k_8 it was necessary to know the absolute concentration of the ClO radicals. This is frequently the least accessible information in investigations of this kind. Here it was obtained from experiments on the flash photolysis of chlorine dioxide. This compound is almost totally decomposed by the flash, and the initial reaction is

$$ClO_2 + h\nu \rightarrow ClO + O \qquad \text{4–X}$$

The intensity of absorption by the ClO radicals was measured before they had time to decay appreciably. This was related to their concentration found from the difference between the concentrations of the chlorine dioxide measured spectrometrically before and immediately after the flash.

Another example of the determination of an elementary rate constant by flash photolytic technique will be found in Section 4–3–3.

4–2 Fast flow methods

As noted previously, quantitative study of an elementary reaction as a single process is frequently best carried out by a flow method. Fundamentally, the same principle is applied as in the flow methods for slower reactions described in Chapter 2. The reaction is conducted in a rapid flow of gas, usually an inert gas; and the changes in concentration of the relevant species with time are thereby converted to changes with reference to the position along a length of tube or with respect to some other spatial arrangement. Reaction times that can conveniently be accommodated are from about 10^{-3} to about 1 s. The aim of investigations of this kind, at all events in the first instance, is generally to determine rate constants or, more generally, Arrhenius parameters. Ideally, no reaction mechanisms in the sense of the term used in Chapter 2 are involved. Nevertheless, in practice, it is seldom possible to study an elementary reaction in complete isolation, and the influence of concurrent or consecutive reactions may have to be determined and allowed for.

The methods to be described mostly depend upon special techniques for identifying radicals and measuring their concentration. Strictly, the concentrations both of the reactant and product radicals as well as those of stable reactants and products should be determined as a function of time. Meaningful concentration measurements of product radicals are frequently made difficult, however, by the subsequent rapid reactions the radicals inevitably undergo; and hitherto many investigators have been content to measure the change in concentration of the reactant radicals and, to varying degrees, those of the molecular species involved. When the reaction is clearly defined, serious error is unlikely to be incurred by this procedure, but cases are not unknown where failure to establish the complete stoichiometry has led to incorrect results.

Since the same techniques for producing fast flows, generating the radicals, and measuring their concentrations are applicable, by and large, to any kind of flow system, they will be discussed first. But before this, we shall anticipate a little of the subsequent discussion to consider briefly a typical experimental arrangement. This is illustrated by Fig. 4–2, a diagram of a flow-tube apparatus used to study reactions of oxygen atoms. It incorporates four basic elements: (i) a device for generating the atoms—in this case a microwave cavity; (ii) pumps to bring the atoms rapidly into contact with the other reactant; (iii) a 'reactor'—here a length of tubing; and (iv) a 'detector' for determining the concentration of the atoms. In this case the detector is a photocell for recording the intensity of chemiluminescence which is proportional to the concentration of the oxygen atoms. This and other features of such apparatus will be enlarged upon later; at this stage we are concerned simply to introduce the following account of the common features of fast-flow technique.

4–2–1 Fast-flow equipment

This incorporates the usual components of high-vacuum apparatus. The pressures at which reactions can be studied largely depend on the methods by which the radicals are generated (see below); the common range is from about 0.1 to 10 torr, though this is seldom covered in a single investigation. Occasionally pressures up to 100 torr have been used. The fast flow is achieved by a mechanical vacuum pump. Many reactions can be studied conveniently in flow tubes with linear gas velocities of a few hundred centimetres per second. With a tube about 2 cm in diameter and 1 m long this requires a pump of about 2 litre s^{-1} nominal speed. To

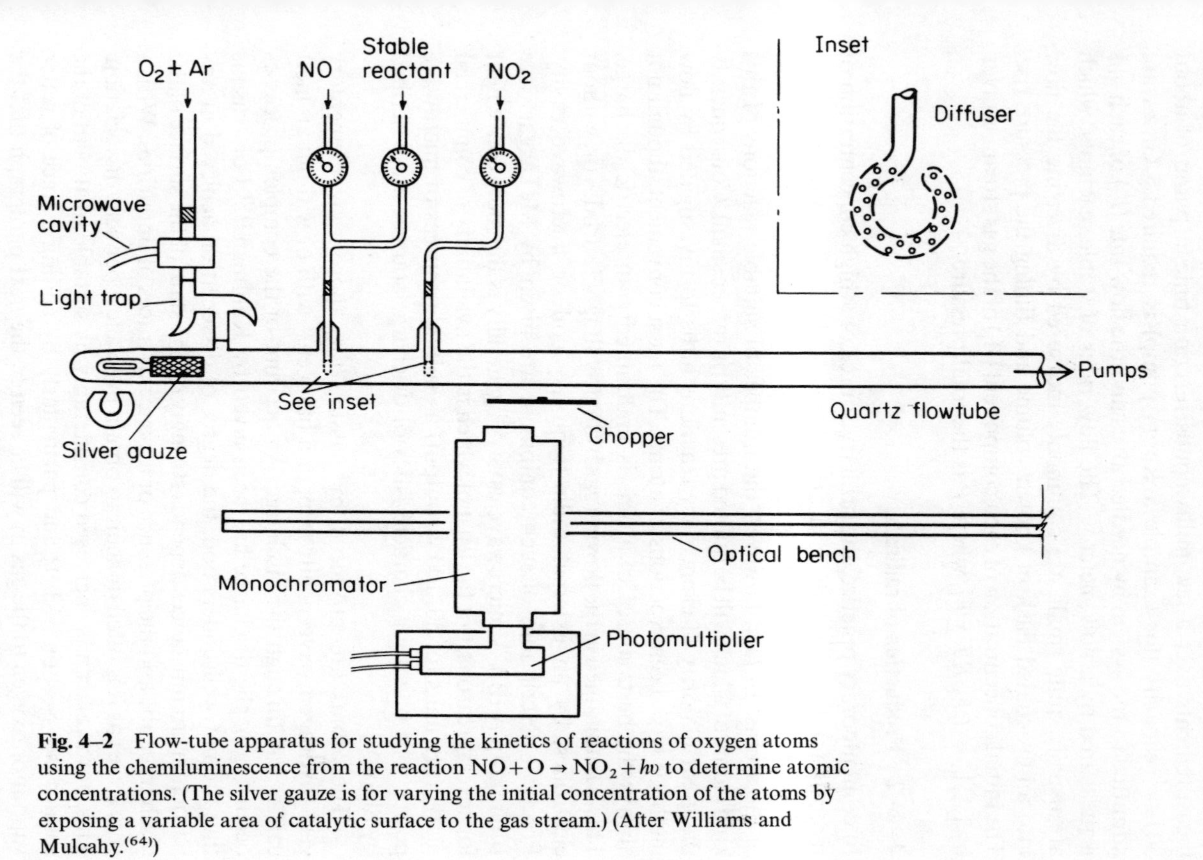

Fig. 4–2 Flow-tube apparatus for studying the kinetics of reactions of oxygen atoms using the chemiluminescence from the reaction $NO + O \rightarrow NO_2 + h\nu$ to determine atomic concentrations. (The silver gauze is for varying the initial concentration of the atoms by exposing a variable area of catalytic surface to the gas stream.) (After Williams and Mulcahy.[64])

achieve 2000–3000 cm s^{-1} in the same apparatus, and thus be able to make measurements over a few milliseconds reaction time, a pump of about 30 l s^{-1} capacity (for example a Roots pump) is required.§ Gases are admitted to the system by needle valves and the flow rate (F_i) of each gas is measured by a flow meter.[7] The flow rates of stable reactants, which are usually quite small, are commonly measured by diverting the flow into an evacuated flask of known volume and timing the pressure rise. The initial concentration of each component (A_i) of the gas stream is given by $[A_i]_0 = PF_i/RT \sum F_i$, where P is the total pressure.

4–2–2 Production of radicals

In contemporary practice, the primary source of a high concentration of radicals for kinetic study is nearly always a device for producing free atoms. More complex radicals are produced by subsequent fast reactions of the atoms. In fact, however, the number of suitable reactions of this kind (some of which will be quoted presently) is rather small. Consequently the great majority of elementary reactions hitherto investigated by flow methods have been reactions of atoms. The most convenient, though not always the best, source of atoms is undoubtedly an electric discharge. This is maintained in the flowing gas by high-voltage electrodes or, without electrodes, by an external radio-frequency coil or microwave cavity. Practical descriptions of such equipment are given by McTaggart (see Bibliography (B)). A microwave cavity[8] generally is the most convenient for several reasons, not the least of which is the availability of commercial diathermy units (of 50–100 W output) as suitable sources of microwave power. The operating characteristics of discharge sources are described by Shaw.[9]

The atoms are produced by passing the appropriate elementary gas—hydrogen, oxygen, nitrogen, or a halogen—with or without an inert carrier gas through the discharge. An account of the complex processes occurring in the discharge has been given by Kaufman.[10] For reasons that are not well understood, the degree of dissociation achieved in any particular circumstances depends strongly on the state of the surface of the discharge tube and the presence or absence of traces of impurities. With a very pure gas it is often difficult to obtain even a few per cent dissociation. This, however, can be increased considerably by suitable treatment of the surface (see Section 4–3–1) and particularly by adding a trace of water vapour or oxygen to the gas. It will be seen in the next paragraph that the latter procedure may not always be legitimate for kinetic work.

Electric discharges are capable of generating many labile species other than free atoms and indeed are a favourite means of producing complex

§ It is usual to prevent corrosive reaction products from reaching the pump by condensing them in a liquid nitrogen trap. This should always be preceded by another trap (preferably heated) filled with silver gauze to destroy radical species, which otherwise are likely to form explosive compounds at low temperatures. In all circumstances, the condensed products should be handled with the greatest respect.

radicals from organic and other compounds for spectroscopic study or preparative chemistry. The reason discharges are seldom suitable for kinetic work with such radicals is because the breakdown of a compound generally produces more than a single radical species and this complicates the subsequent kinetics. For example, a discharge through water vapour was formerly used as a source of hydroxyl until it was realized that oxygen atoms are also produced. Because of the very fast reaction

$$O+OH \rightarrow O_2+H \qquad \text{4–XI}$$

relatively few hydroxyl radicals emerge from the discharge and rather more may be generated downstream by the following reactions:

$$H+O_2+M \rightarrow HO_2+M \qquad \text{4–XII}$$

$$H+HO_2 \rightarrow OH+OH \qquad \text{4–XIII}$$

Even with elementary gases, difficulties can occur. A discharge in oxygen produces electronically excited molecular oxygen (in $^1\Delta_g$ and $^1\Sigma_g^+$ states) as well as atoms. For some years the kinetics of the reaction

$$O+O_2+M \rightarrow O_3+M \qquad \text{4–XIV}$$

were obscured by the unsuspected occurrence of reactions of the type:

$$O_2(^1\Delta_g)+O_3 \rightarrow 2O_2+O \qquad \text{4–XV}$$

Likewise nitrogen from a discharge, known for over half a century as 'active nitrogen', contains excited molecules in addition to atoms.[11]

Free atoms can also be produced by thermal decomposition. Atomic hydrogen is commonly generated by passing the gas over an electrically heated tungsten filament at about 1700 °C.[12] Oxygen atoms can be produced by passing ozone at low pressure through a quartz tube at about 1000 °C[13] or nitrous oxide over a Nernst glower, such as is used in an infra-red spectrometer.[14] Such methods are superior to the use of a discharge in that they do not produce ionized or other unwanted species and are capable of being applied at much higher total pressures.[15] On the other hand, in general, they cannot match the high concentrations of atoms that are readily produced by a discharge.

A notorious difficulty associated with flow technique is caused by recombination of the atoms at the walls. With the size of equipment and pressures usually employed, this can easily cause most of the atoms, particularly halogen or hydrogen atoms, to disappear before they reach the reaction zone or, as noted previously, even emerge from the source. Traces of impurities can drastically affect the catalytic activity of clean quartz or borosilicate glass surfaces and so give rise to erratic behaviour. On the other hand coating the surface with phosphoric or boric acid produces a marked improvement in the level and stability of the atomic concentration (see also Section 4–3–1).

To produce higher radicals in a satisfactory way from atoms, reactions are required which are fast enough to be complete before the product radicals have time to recombine or otherwise react. Such are known as *titration reactions* and they will be enlarged upon later. The best-known examples for this purpose are the very fast reactions

$$H + NO_2 \rightarrow NO + OH \qquad \text{4–XVI}$$

and

$$O + COS \rightarrow CO + SO \qquad \text{4–XVII}$$

used to generate OH and SO radicals respectively. Another such reaction, though appropriate to a different purpose, is the reaction

$$N + NO \rightarrow N_2 + O \qquad \text{4–XVIII}$$

This provides a useful source of oxygen atoms free from molecular oxygen.

4–2–3 Determination of radical concentrations

The special techniques developed for directly determining radical concentrations constitute the basic *raison d'être* of the flow methods. They are of two kinds, absolute and relative. *Absolute methods* in standard use are based on calorimetry, chemical titration, and ESR spectrometry. Application of ultraviolet and mass spectrometry to absolute determination of radical species is possible,[16,17] but restricted at present by lack of accurate extinction coefficients and absolute instrumental sensitivities respectively. These techniques are more readily applicable as relative methods, where such information is not required.

The *calorimetric method*, which hitherto has been applied only to atoms, uses an isothermal catalytic detector.[18] This consists of a coil of wire§ immersed in the gas stream. It is heated to a constant temperature 100° or so above the gas temperature by an electric current and forms one arm of a Wheatstone bridge. The atoms recombine on the surface of the wire and transfer to it the heat of recombination. The intense catalytic activity sets up a diffusion gradient in the gas which is steep enough to ensure that all the atoms arriving in the vicinity of the coil reach its surface. This, however, can be checked with a second coil located downstream. In the presence of the atoms, less power is required to maintain the temperature—that is, the resistance—of the coil at the preselected value, and the difference is related by the known heat of recombination to the number of atoms recombining per second. The constant temperature ensures that heat losses are constant whether or not the atoms are present (provided, of course, the atomic concentrations are low enough not to affect the thermal conductivity of the gas).

The technique can be made both sensitive and accurate and it is most valuable for calibrating relative methods. However, it is not specific and therefore cannot be used directly to follow reactions where active species are formed as well as destroyed. On the other hand the method is well

§ Platinum for H,[19] silver-coated platinum for O,[20] nickel for halogens.[21]

suited to studies of recombination reactions, in which only one kind of active species is involved.[20, 22]

As its name implies, the *titration method* is basically similar to the familiar procedure used in volumetric analysis. An appropriate reactant is added to the gas containing the atoms to be determined and its flow is gradually increased until they are all consumed. The added molecules react so rapidly with the atoms as to exclude any other reaction. Hence the number that must be added to reach the end-point is equivalent to the number of atoms originally present. Application to any particular atomic species depends upon finding a reaction that occurs within fewer than a hundred or so molecular collisions. Some such reactions together with their rate constants at 25 °C are listed in Table 4–1. Techniques for

Table 4–1 Atomic titration reactions

Reaction	k (at 25 °C) ($cm^3\ mole^{-1}\ s^{-1}$)	*Reference* §
(a) $O + NO_2 \rightarrow O_2 + NO$	4×10^{12}	(23)
(b) $N + NO \rightarrow N_2 + O$	2×10^{13}	(24)
(c) $H + NOCl \rightarrow HCl + NO$	2×10^{13}	(25)
(d) $Cl + NOCl \rightarrow Cl_2 + NO$	2×10^{12}	(26)
(e) $H + NO_2 \rightarrow OH + NO$	3×10^{13}	(27)
(f) $O + Br_2 \rightarrow OBr + Br$	5×10^{12}	(28)

§ The references are to determinations of the rate constants.

determining the end-point depend to some extent on the particular reaction. Oxygen and nitrogen atoms can conveniently be titrated visually by means of reactions (a) and (b) of Table 4–1. On addition of nitrogen dioxide to oxygen atoms,

$$O + NO_2 \rightarrow NO + O_2 \qquad \text{4–XIX}$$

a greenish-yellow chemiluminescence appears in the flowing gas. This arises from excited nitrogen peroxide molecules formed in the course of the reaction

$$O + NO \rightarrow NO_2 + h\nu \qquad \text{4–XX}$$

The radiation is intense enough to be visible in a darkened room, but the reaction consumes a negligible proportion of the oxygen atoms. The intensity of the glow at any stage is proportional to $[O][NO]$. With progressive addition of the nitrogen dioxide, it passes through a maximum and, when precisely sufficient has been added to consume all the atoms, it is sharply extinguished.[29] At the end-point, the flow rate of the nitrogen dioxide is equal to that of the oxygen atoms. Thus the method is absolute, precise, convenient, cheap—and spectacular.

Titration of nitrogen atoms with nitric oxide is carried out similarly (Table 4–1, reaction (b)). Gas containing nitrogen atoms emits a yellow glow—'the nitrogen afterglow'—originating from a small concentration of metastable nitrogen molecules formed by recombination of the atoms. When nitric oxide is added, the yellow is replaced by a blue glow emitted by excited nitric oxide molecules formed in a similar way:

$$N+NO \rightarrow N_2+O \qquad \text{4–XVIII}$$

$$N+O \rightarrow NO^* \rightarrow NO+h\nu \qquad \text{4–XXI}$$

When no nitrogen atoms remain, the blue glow is extinguished and is replaced, on further addition of nitric oxide, by the yellow-green glow from reaction 4–XX. Again, measurement of the flow of nitric oxide at the end-point gives the flow of nitrogen atoms.

Titrations are frequently carried out by chemical analysis, usually with the aid of a mass spectrometer (see Fig. 4–3). The pin-hole leak to the ion

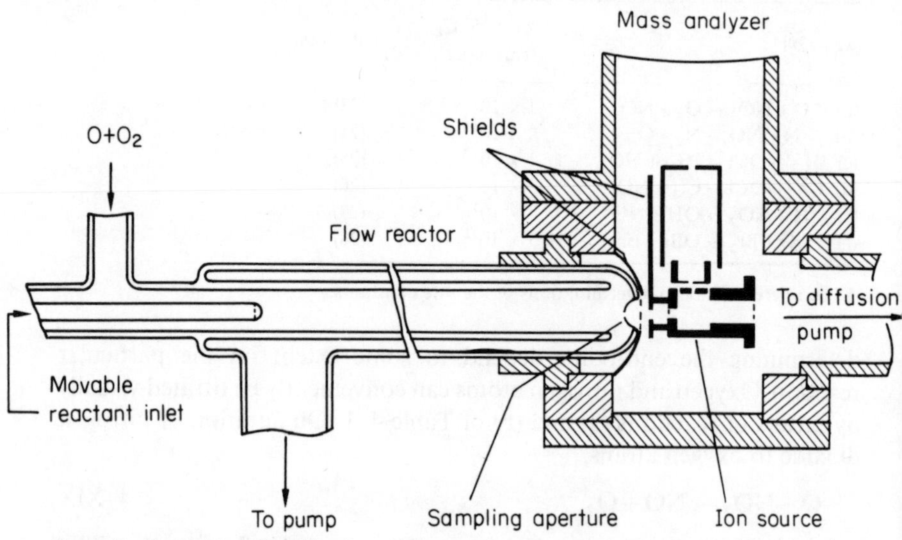

Fig. 4–3 Typical fast-flow kinetics apparatus using a mass spectrometer to determine the concentrations of radicals and stable species (Sullivan and Warneck[30]).

chamber is located sufficiently far downstream from the point of addition of the titrant for the titration reactions to be complete before the products enter the instrument. The titrant is added in increasing amount until the first record of its signal by the mass spectrometer shows that it is in excess or, alternatively, until the signal of a product of the titration reaction does not increase on further addition of the titrant. Hydrogen atoms are frequently determined mass spectrometrically via reaction (e) of Table 4–1.[27] It should be mentioned, however, that the overall stoichiometry of this reaction, as of reaction (c),[31] may be affected by radical recombination at the surface; this must be suitably 'poisoned' if reliable results are to be obtained.[32]

Because of interference by other reactive species, the procedures just described cannot often be used to measure absolute radical concentrations in the midst of reaction. In this and other respects *ESR spectrometry* offers great advantages. The ESR spectrum of a radical is, of course, specific to it and, under appropriate conditions, the intensity of the spectrum is proportional to the radical concentration. Detection is external to the reaction space and does not perturb the reaction; and since, in the present context, magnetic resonance occurs only with species with unpaired electrons and a very few other species, there is basically no interference from the saturated molecules present. Furthermore, interference between different spectra is not usually a problem and the concentrations of several different atoms or radicals can be determined almost simultaneously. Finally, absolute calibration is relatively simple.

These facts combine to make the ESR method very powerful. Its main disadvantages are low spatial resolution and inability to detect complex radicals—in addition, of course, to the necessity to build or buy an ESR spectrometer which, in either case, is an expensive instrument. As shown typically by Fig. 4–4, the detector is a resonant microwave cavity which contains the reacting gas and, with the usual X-band instrument, averages the radical concentration over a few cubic centimetres. For satisfactory accuracy, concentration gradients must therefore be small over this volume. All the atoms of interest and a number of diatomic radicals (OH, ClO, SH, etc.) can be handled but, in the present state of ESR technology, it is not possible to measure, or even detect in the great majority of instances, realistic concentrations of more complex radicals in the gas phase. Concentrations of about 10^{-11} mole cm^{-3} (i.e. about 10^{-4} torr at 25 °C) of the simple species can be measured absolutely to an accuracy of ± 10 per cent but relative measurements can be made much more precisely.

It is not possible to elaborate on ESR technique here. The basic principles as applied to free radicals will be found in the article by Foner[33] and the monograph by Ingram.[34] Krongelb and Strandberg[35] give, among other valuable information, the fundamental theory of concentration measurements; and this is also summarized by Westenberg and de Haas.[36] Practical details of instrumentation are given by Alger[37] and the construction of a cavity suitable for gas-phase work with both atoms and radicals is described by Carrington *et al.*[38] Such cavities are also available commercially (at some expense).

The basis of absolute concentration measurements lies in the use of a stable molecular species with unpaired spins as calibrating gas. Because of the nature of the ESR transitions involved, oxygen is appropriate for atomic species and nitric oxide for unsymmetrical diatomic radicals such as OH. The reference gas fills the cavity of the spectrometer in precisely the same way as the gas containing the radicals. The concentration of the radicals is obtained by comparing the (doubly) integrated intensities of suitable lines in the spectra of the radicals and the reference gas; the latter being admitted to the cavity at a known pressure. The calibration factors

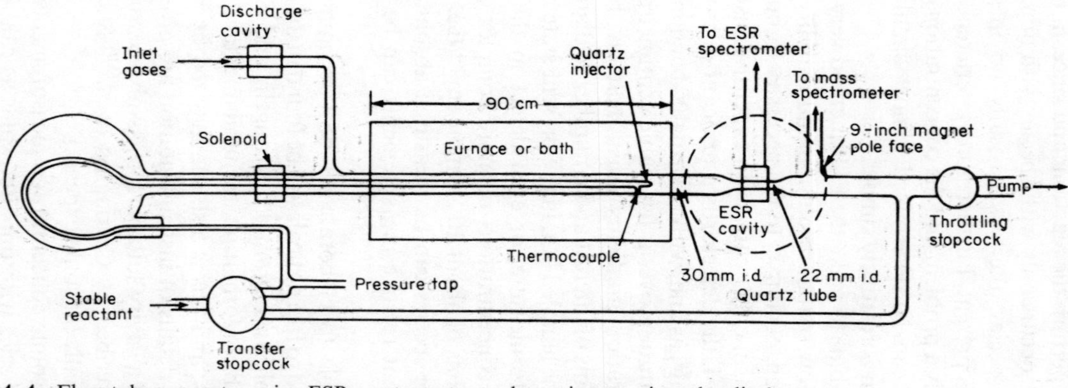

Fig. 4–4 Flow-tube apparatus using ESR spectrometry to determine atomic and radical concentrations. The reaction time is varied by changing the position of the reactant inlet by means of the movable solenoid (Westenberg and de Haas[23]).

for the usual atomic species and OH are given in indispensable papers by Westenberg.[36,39]

Comparison of absolute methods

It is of some moment to know whether the concentrations found by different methods agree. The ESR and titration methods for determining nitrogen and oxygen atoms were compared directly by Westenberg and de Haas.[40] Thus for nitrogen atoms, nitric oxide was used as titrant,

$$N + NO \rightarrow N_2 + O \qquad \text{4–XVIII}$$

It was added progressively to a constant stream of the atoms, and the concentrations of both nitrogen and oxygen atoms were determined by situating the cavity of the ESR spectrometer a short distance downstream of the mixing point. The distance corresponded to a time long enough for reaction to be completed but too short for appreciable recombination of the atoms to occur. Figure 4–5 shows in a particularly graphic way the

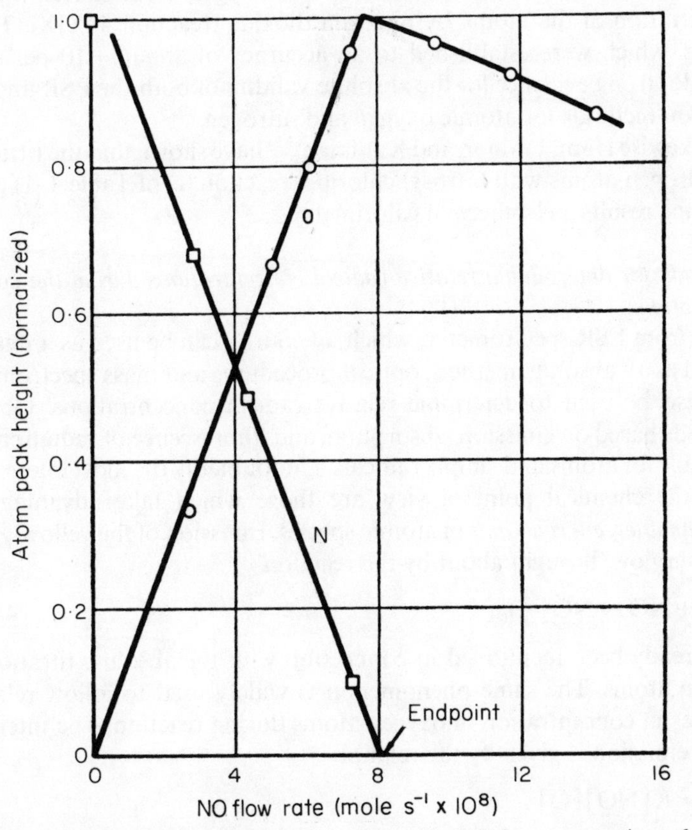

Fig. 4–5 Titration of nitrogen atoms with nitric oxide. Relative concentrations of N atoms and O atoms measured by ESR as a function of the flow rate of NO. Pressure = 1 torr; carrier gas (N_2) flow = 1.5×10^{-5} mole s^{-1}; initial flow of N atoms = 8×10^{-8} mole atoms s^{-1} (Westenberg and de Haas[36]).

disappearance of nitrogen atoms and the simultaneous appearance of oxygen atoms during the progress of such an 'ESR titration'. (The decrease in the concentration of oxygen atoms observed with excess nitric oxide is due to the reaction

$$O+NO+M \rightarrow NO_2+M \qquad \text{4–XXII}$$

which is followed immediately by

$$O+NO_2 \rightarrow NO+O_2 \qquad \text{4–XIX}$$

Since reaction 4–XXII is much slower than reaction 4–XVIII, its effect on the oxygen atom concentration is negligible when an appreciable concentration of nitrogen atoms is present.) Using their absolute ESR calibrations, Westenberg and de Haas were able to show first, that the flow rate of nitric oxide at the end-point is equal to the flow rate of the nitrogen atoms and second, that, up to the end-point, one atom of oxygen is produced for every nitrogen atom consumed. They found similar agreement between ESR determinations of oxygen atom concentrations and titration of the atoms by nitrogen dioxide (reaction 4–XIX). These results, which were established to an accuracy of about ± 10 per cent, provide strong evidence for the absolute validity of both the ESR and the titration methods for atomic oxygen and nitrogen.

Likewise Ham, Trainor, and Kaufman(22) have shown that the titration of hydrogen atoms with nitrosyl chloride (reaction (c) of Table 4–1) gives the same results as isothermal calorimetry.

Methods for determining relative radical concentrations during the course of reaction

Apart from ESR spectrometry, which, of course, can be used as a relative as well as an absolute method, optical procedures and mass spectrometry can also be used to determine relative radical concentrations. Optical methods based on emission, absorption, and fluorescence of radiation are available for atoms and simple radicals. Undoubtedly the most interesting from the chemical point of view are those which take advantage of *chemiluminescent reactions* of atomic species. Emission of the yellow-green 'air afterglow' brought about by the reaction

$$O+NO \rightarrow NO_2+h\nu \qquad \text{4–XX}$$

has already been mentioned in connection with the absolute titration of oxygen atoms. The same phenomenon is widely used to follow relative changes in concentration of oxygen atoms during reaction. The intensity of the emission is given by the relation

$$I = K[NO][O] \qquad \text{4–1}$$

where K is independent of the total pressure (above 0.1 torr(41)). A minute concentration of nitric oxide, much smaller than that of the oxygen atoms, is added to the flowing gas. In these circumstances, the amount of nitric oxide consumed by reaction 4–XX, as also by the non-chemiluminescent reaction

$$O + NO + M \rightarrow NO_2 + M \qquad \text{4–XXII}$$

is immediately regenerated by the fast reaction

$$O + NO_2 \rightarrow O_2 + NO \qquad \text{4–XIX}$$

Thus the concentration of nitric oxide remains constant and equation 4–1 shows that the intensity emitted at any stage of the reaction under investigation is precisely proportional to the instantaneous concentration of the oxygen atoms. The consumption of oxygen atoms by reactions 4–XXII and 4–XIX is negligible. In the experimental arrangement shown in Fig. 4–1 the emission is recorded by a photomultiplier via a monochromator (or appropriate optical filters[42]), and the relative emission intensities at different reaction times are measured by moving the assembly along the flow tube.

Figure 4–6 shows a comparison of simultaneous measurements of relative oxygen atom concentrations by the chemiluminescent and ESR methods made by the author's colleagues, Mr J. R. Steven and Mr D. J. Williams. (The resonant cavity of the ESR spectrometer was fitted with a window and the chemiluminescence was passed through a light-pipe to the photomultiplier.) The constant K in equation 4–1, though independent of the total pressure, is not independent of the nature of the gaseous *milieu*. This is reflected by the different lines for the different carrier gases shown in the figure. Of the two methods the chemiluminescent is the more sensitive, a factor of 10 in sensitivity over the ESR method being easily obtained.

Hydrogen atoms also undergo a chemiluminescent reaction with nitric oxide analogous to reaction 4–XX and the red glow emitted can be used to determine their relative concentration.[43] Similarly, chlorine atoms can be determined by means of the reddish 'chlorine afterglow' emitted from electronically excited chlorine molecules:

$$M + Cl + Cl \rightarrow M + Cl_2^* \rightarrow Cl_2 + M + h\nu \qquad \text{4–XXIII}$$

The intensity is proportional to $[Cl]^2$ under certain specified conditions, but the kinetics are more complicated than the sole occurrence of the above reaction sequence would imply.[44,45]

Absorption spectrometry

This method can be used to determine atomic species notably H,[46] N, O,[47] and halogens.[48] The required sensitivity is obtained by basically the same method as is used in the technique of atomic absorption photometry familiar to analytical chemists; that is, by measuring the absorption of resonance radiation emitted from an appropriate source. However, the apparatus must be designed to cope with the fact that the resonance lines of the atoms just mentioned lie in the vacuum ultraviolet.

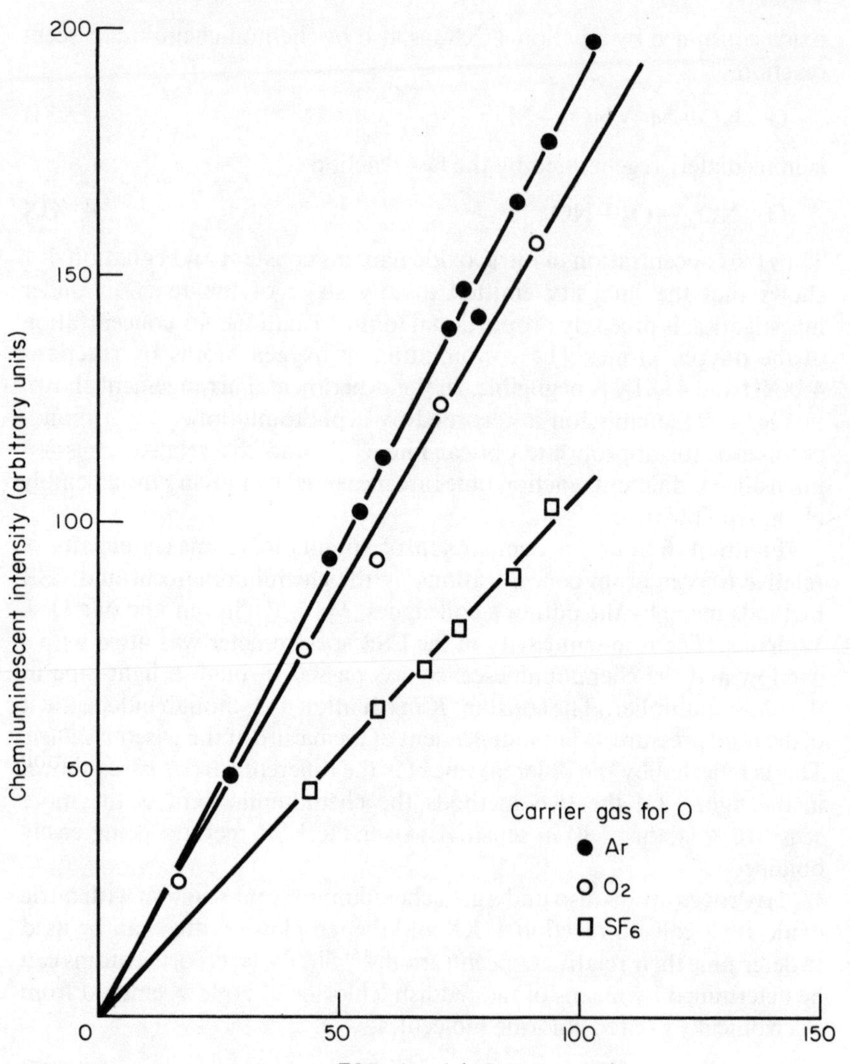

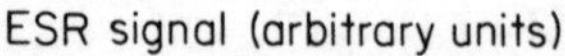

Fig. 4–6 Measurement of oxygen atom concentration. Simultaneous comparison of the intensity of emission of chemiluminescence from the reaction $O + NO \rightarrow NO_2 + h\nu$ with the intensity of the ESR absorption signal from the O atoms (Steven and Williams 1970).

The potential of the method, which has not yet been fully exploited, lies in its specificity and high sensitivity. Hydrogen atoms, for example,[46] can be determined in concentrations as low as 2×10^{-13} mole cm^{-3}. Hydroxyl radicals have been determined in a similar way using a single rotational line in the quartz ultraviolet.[49]

It appears that even greater sensitivity can be obtained by measuring the intensity of *fluorescence* emitted at right angles to an incident beam of resonance radiation,[50] and it is likely that this method will be used more in the future. A particular application is to air-borne investigations of reactions in the upper atmosphere where the radical concentrations are very low.

Mass spectrometry

This is the only technique so far devised for flow systems that is capable of measuring realistic concentrations of radicals containing more than one or two atoms. The review by Lossing in the Bibliography (E) lists over 70 different radical species detected by mass spectrometry, ranging from single atoms to radicals containing more than 20 atoms. The potential scope of the technique is therefore wider than those of the methods just discussed. As ever, this advantage has to be set against inherent difficulties, not the least of which is the fact that the species must be physically extracted from the reaction system in order to be detected. The principles and practice by which a radical can be discriminated from related molecules present in the same reaction system, for example CH_3 from CH_4, are discussed by Foner[33] and Lossing.[51] In general, any of the various conventional types of mass spectrometer as described, for example, by Kiser[52] may be suitable and there is no intrinsic reason to prevent relative concentrations of almost any radical from being determined provided it forms a stable positive ion under electron impact. In addition, with the aid of modern methods of measuring minute electric currents, the conversion of a neutral to a charged species brings with it an enormous potential sensitivity, a fact which indeed is regularly exploited in determining extremely low concentrations of molecular species. The 'record' as regards detection of radicals seems to be held (1971) by Foner and Hudson[53] who observed N_2H_3 radicals at the level of 1 part in 10^7. With a reacting gas at 1 torr this corresponds to an absolute radical concentration of 5×10^{-15} mole cm^{-3}.

Figure 4–3 shows the manner of coupling a mass spectrometer to a flow system. A pin-hole about 0.005 cm in diameter in a thin septum communicates directly between the end of a flow tube and the ionization chamber of the spectrometer. The dimensions of the pin-hole are of the same order as the mean free path of the molecules in the flow tube. The pressure in the spectrometer is maintained at about 10^{-5} torr by a fast pump and the geometry is arranged so that an appreciable proportion of the molecules emerging from the pin-hole as a diffuse molecular beam enter the ion source before they can collide with the walls. The low pressure also ensures that reaction ceases as soon as the gas enters the spectrometer. More advanced apparatus designed for maximum sensitivity[54,55] is shown schematically in Fig. 4–7. Here the pin-hole is replaced by an orifice about 1 mm in diameter through which supersonic continuum flow occurs. The expanding jet which develops inside the orifice crosses a space evacuated by a high-speed diffusion pump and the molecules diverging from the axis of the jet are 'skimmed' off by a second orifice which, like the first, is at the apex of a cone. The angle of the cone is calculated to minimize collisions of the molecules with its inside wall, and its distance from the sampling cone is adjusted to maximize the intensity of the molecular beam entering the ion source.

The ions generated from the beam are discriminated from those arising from the molecular background in the ion source by modulating the beam with a vibrating reed and coherently detecting the modulated ion signal The electronics are arranged so that the detector signal when the beam is cut off is subtracted from the signal generated when it is on, thus eliminating the background. Needless to say, the construction and successful operation of such apparatus are not achieved without a good deal of skill and experience. We shall see later in this chapter that, quite apart from its application to mass spectrometry, the development of molecular beam technique is highly significant for the progress of gas kinetics.

4–2–4 Flow systems

Having discussed ways of generating and detecting radicals in fast-moving gas, we can return to a few essential details of the types of flow system used and to the methods adopted to derive elementary kinetic parameters from the experimental measurements. Most of the discussion of flow systems in Chapter 2 is relevant here, and for the most part it will be necessary only to amplify some points which relate particularly to the present context.

The flow-tube method

The majority of investigations by flow methods hitherto have been conducted with flow tubes, for which typical experimental arrangements have been already encountered in Figs. 4–2, 4–3, and 4–4. In general, when the reaction of a radical with a stable molecule is to be studied, the radicals enter the flow tube in the main bulk of the flowing gas and are mixed with a much smaller flow of the stable molecules by means of a 'diffuser' inlet as shown in Fig. 4–2. At the usual pressures of about 1 torr, mixing can be considered complete within 1 ms.[49] Concentration measurements are made at various stages of reaction either by moving the detector along the flow tube as in Fig. 4–2, or by moving the inlet relative to the detector, as in Figs. 4–3 and 4–4.

Because of the high thermal diffusivity of the gas at low pressures, thermal equilibration between the gas flow and the thermostat is rapid in the absence of reaction, but it can be seriously affected by internal heating when a highly exothermic reaction is involved. In general activation energies are low and some departure from thermal equilibrium can often be tolerated; nevertheless the effect can impose a limit on the radical concentrations which can be used.

When it is legitimate to assume simple plug flow in the flow tube (as discussed below), the rate constant sought is usually derived by plotting the appropriate function of the radical concentration [R] against the distance ΔL between detector positions for a fixed inlet or between inlet positions for a fixed detector; thus, log [R] vs. ΔL or 1/[R] vs. ΔL for a

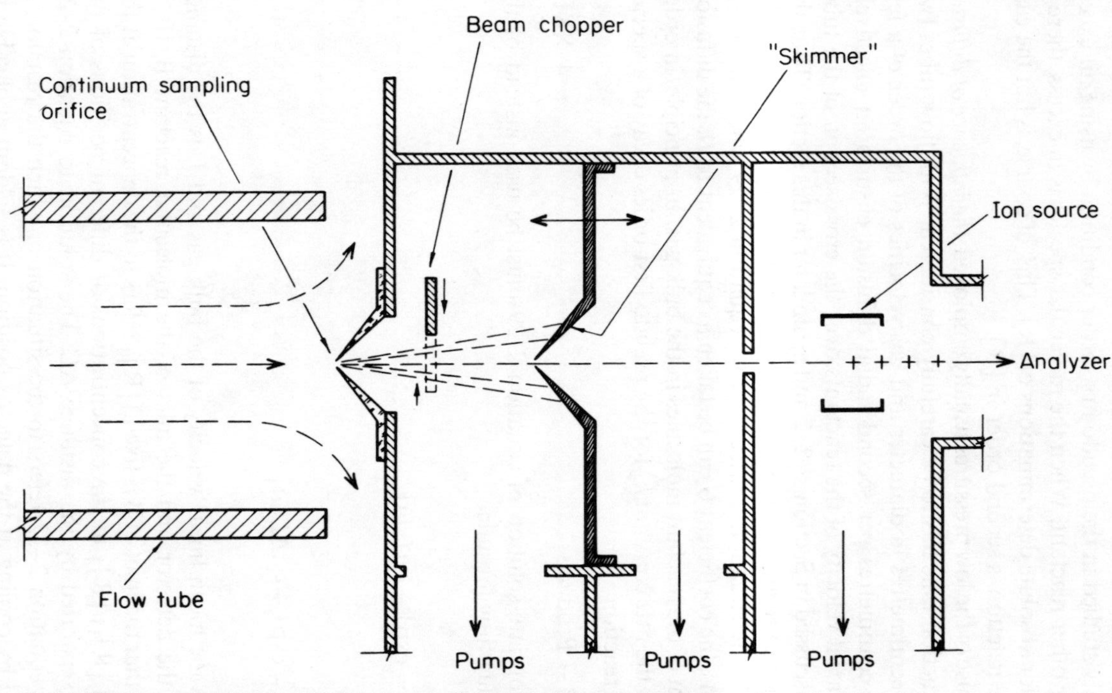

Fig. 4–7 Mass spectrometry of labile species. Schematic diagram of apparatus for molecular beam sampling by supersonic continuum flow ('nozzle' beam).

reaction first- or second-order in [R] respectively. The reaction time t_c required for substitution in the corresponding rate expression 2–2 or 2–3 is linearly related to L by equation 2–15. Radical–molecule reactions are commonly studied in the pseudo first-order condition by using an excess of one or other reactant. When the molecular species is in excess, there is no need for absolute determinations of [R]. This, of course, is not the case when the reaction is second order in [R].

Because of the low pressures usually employed, *the influence of diffusion* on the reaction time has to be carefully considered. With flow tubes two or three centimetres in diameter and gas velocities of the order of a few hundred centimetres per second radial diffusion ensures an effectively uniform axial velocity of the reactants over the cross-section of the tube. This is discussed in Section 2–2–2 and again later in this section. Thus the mass-transfer regime can be considered to be 'plug flow plus axial diffusion'. The situation is described by equation 2–29 where, for these conditions, the coefficient G can be taken as equivalent to D, the diffusion coefficient of the reactant molecules in the bulk gas. In principle, in order to obtain the rate constant k_ψ for the pseudo first-order decay of a species R by the reaction

$$R + B \rightarrow \text{Products} \qquad \text{4–XXIV}$$

the appropriate solution of equation 2–29 must be used instead of the simple plug-flow formulae:

$$\frac{[R]_1}{[R]_0} = \exp[-k_\psi(L/\bar{u})] \qquad \text{4–2}$$

or

$$\frac{[R]_2}{[R]_1} = \exp[-k_\psi(\Delta L/\bar{u})] \qquad \text{4–3}$$

Here $\bar{u}$ is the mean linear velocity of the bulk gas and L is the distance between the detector and the inlet of the molecular reactant B. (It is assumed that the inlet of B is fixed.) $[R]_0$ refers to the concentration at the inlet and $[R]_1$, $[R]_2$ to the concentrations at different positions of the detector separated by the distance ΔL. The solution to equation 2–29 given by equation 2–32 refers to the situation in which the reaction is quenched by cooling at the detector position. It is seldom applicable in the present context since the reaction usually passes the detector undisturbed. Thus the boundary condition 2–31 is inappropriate. The present situation depends on the fate of the radicals after passing the detector, but with most apparatus the boundary condition $[R] = 0$ at $z = \infty$ is probably a reasonable approximation. The initial boundary condition 2–30—which conserves the flux of the reactant across the hypothetical discontinuity at the inlet—applies in the present case, since the reaction begins abruptly at the inlet of B. (This assumes there is no reaction without B.) In these circumstances, equation 4–2 becomes

$$\frac{[\mathrm{R}]_1}{[\mathrm{R}]_0} = \frac{2}{(1+\alpha)} \exp\left[(1-\alpha)\bar{u}L/2D\right] \qquad 4\text{–}4$$

where

$$\alpha \equiv (1+4Dk_\psi/\bar{u}^2)^{1/2} \qquad 4\text{–}5\S$$

Similarly it follows from equation 4–4 that equation 4–3 should be replaced by

$$\frac{[\mathrm{R}]_2}{[\mathrm{R}]_1} = \exp\left[(1-\alpha)\bar{u}(\Delta L)/2D\right] \qquad 4\text{–}6$$

Equations 4–4 and 4–6 reduce to equations 4–2 and 4–3 as the dimensionless quantity α defined by equation 4–5 approaches unity, that is, as $4Dk_\psi/\bar{u}^2 \to 0$. It will be seen that, for a given reaction, the effect of axial diffusion decreases as the gas velocity increases.

When $4Dk_\psi/\bar{u}^2$ is small, a useful approximation can be derived from equation 4–6, namely

$$k_\psi \approx (k_\psi)_{\mathrm{app}}\left[1+(D/\bar{u}^2)(k_\psi)_{\mathrm{app}}\right] \qquad 4\text{–}7^{(56)}$$

where $(k_\psi)_{\mathrm{app}}$ is the apparent value of k_ψ which would be obtained by using the plug-flow equation 4–3. This relation provides a criterion for the applicability of equation 4–3 (or 4–2); thus, the condition for the error in the rate constant due to axial diffusion to be less than, say, 1 per cent is

$$D(k_\psi)_{\mathrm{app}}/\bar{u}^2 < 0.01 \qquad 4\text{–}8$$

With extremely fast reactions, diffusion may not be fast enough to produce the flat velocity profile *across* the flow tube implied by the assumption of 'plug flow plus axial diffusion'. The sufficient condition for a flat profile is given by equation 2–28 which, for a pseudo first-order reaction, becomes

$$k_\psi R^2/D \leqslant 2 \qquad 4\text{–}9$$

This relation is for a homogeneous reaction. A reaction at the surface of the flow tube tends to produce greater non-uniformity in the velocity profile and it seems that the rate constant for such a reaction should be about trebled before being inserted in equation 4–9.(56) With a flow tube 2 or 3 cm in diameter and pressures of about 1 torr, the effect of the velocity profile becomes apparent with reactions occurring in a few milliseconds.(57) To summarize, the conditions under which diffusive effects can be neglected in deriving rate parameters from flow-tube experiments are bounded on the one hand by equation 4–8 for axial diffusion, and on the other by equation 4–9 for radial diffusion.

The effect of the pressure drop along the flow tube must also be considered when flow velocities are high. This was discussed in Section 2–2–2.

§ In terms of the quantities used in equation 2–32, $\alpha \equiv (1+2\mu s)^{1/2}$.

'Diffusion-cloud' method

Radical recombination and other reactions at the walls frequently limit the effectiveness of the flow-tube method, particularly at elevated temperatures. This has led to a number of attempts to develop 'wall-less' flow systems. In general the procedure has been to allow parallel streams of the reactants to interpenetrate by diffusion and to determine the resulting concentration changes before the reactants or products can reach the wall.[58,59] We have the space only to refer the interested reader to publications by Tal'rose[60] for details of the method in its most successful form.

The stirred-flow method

The principles of the stirred-flow method were stated in Section 2–3, and Fig. 2–5 is a diagram of a stirred-flow reactor. The method has been used relatively little for studies of elementary radical reactions. This is somewhat surprising since, for some conditions, it has distinct advantages over the flow-tube method. In particular, it enables the surface-to-volume ratio of the reaction vessel to be reduced and with it the influence of surface reactions. Furthermore, the diffusive effects which tend to complicate the kinetics of the conventional flow tube are turned to advantage in helping to provide the required uniformity of composition in the reactor. Figure 4–8 shows a stirred-flow apparatus for determining the kinetics of atomic reactions by ESR spectrometry. The atoms enter the reactor in the main stream of carrier gas by the central diffuser and can be rapidly mixed with a second reactant either in or just before the reactor. The atomic concentrations entering and leaving the reactor are measured by switching the modulation power from one half of the double ESR cavity to the other. Neglecting the small amount of reaction occurring during the rapid passage through the connecting tubes, the steady state is given by

$$v_{in}[X]_{in} - v_{out}[X]_{out} - Vk_{\psi}[X]_{out} = 0 \qquad 4\text{–}10$$

where v_{in} is the volumetric flow velocity of the entering gas containing the atoms X, and v_{out} is that of the effluent gas containing the products; V is the volume of the reactor and k_{ψ} relates to the sum total of all chemical reactions consuming X, including, of course, reaction with an added reactant. Normally there is a large excess of carrier gas and the difference between v_{in} and v_{out} can be neglected. In this case equation 4–10 yields

$$k_{\psi} = \left[\frac{[X]_{in}}{[X]_{out}} - 1\right]\frac{v}{V} \qquad 4\text{–}11$$

Thus k_{ψ} can be derived very simply from the measurements.

Determination of the manner in which k_{ψ} depends on the concentrations of the various species present, including X, enables the kinetics of the particular reaction under study to be clearly delineated. Suppose, for example, a reaction $X + B \rightarrow$ Products is to be investigated in the un-

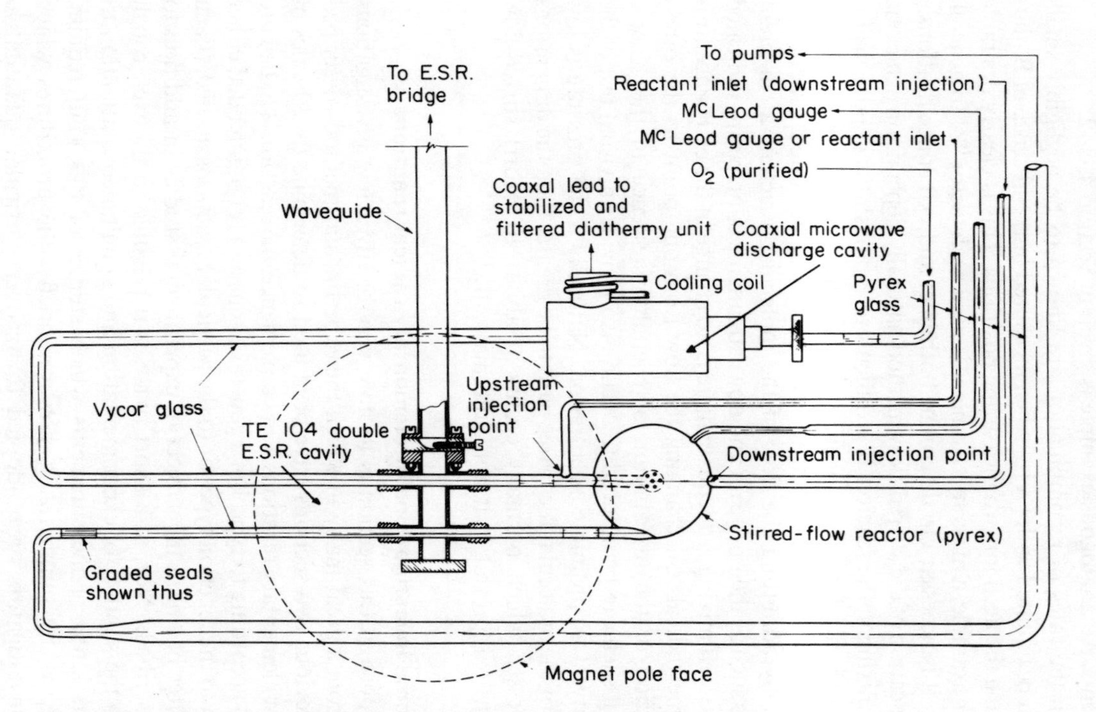

Fig. 4–8 Flow system with ESR double cavity for measuring oxygen atom concentrations at the inlet and outlet of a stirred-flow reactor (Mulcahy, Steven, and Ward[61]).

avoidable presence of gas-phase and first-order wall recombination of X. The behaviour of k_ψ will be according to the relation

$$k_\psi = k_{comb}[M][X]_{out} + k_w + k[B]_{out} \qquad 4\text{–}12$$

from which the value of k can be extracted by means of the appropriate experiments. An analogous example is given in Section 4–3–2.

When the rates of recombination are known to be negligibly small, $[X]_{in}$ can be equated with $[X]_{out}$ when measured in the absence of B, thus eliminating the need for two measuring positions. This is clearly desirable when, for example, a mass spectrometer is used as detector.[62] In general, however, it is necessary to determine both inlet and exit concentrations.

Application of the stirred-flow method naturally depends on achieving 'perfect' mixing in the reactor. It is difficult to establish the conditions for this theoretically for a reactor with complicated geometry such as that shown in Fig. 4–8. The efficiency of mixing can be tested practically, however, by admitting the flow of the molecular reactant at different locations such as the upstream and downstream injection points indicated in the figure. If mixing is adequate there is no change in the exit concentration of the radicals on changing the flow from one point to the other. Observation of the distribution of chemiluminescence in the reactor volume, if such can be induced to occur, also provides a visual test of the uniformity of an atomic concentration. Naturally if the pressure is low enough for the molecular mean free path to be greater than the dimensions of the vessel (which normally occurs at about 10^{-3} torr) stirred-flow conditions will prevail, as it were, automatically.[63]

4–3 Some investigations of elementary radical reactions

Our purpose in this section is to note very briefly a few investigations chosen from a great many in which the experimental methods described in the previous two sections have been used to determine the kinetics of particular elementary reactions. For convenience, the reactions studied are classified according to type, but a general discussion is not intended either in regard to the reaction types or to the particular reactions quoted. (Each of the latter has been the subject of experimental studies in addition to those described.) Nor, in a short space, is it possible to do even rough justice to the subtlety of experimentation and simultaneous attention to numerous kinds of measurements which studies of this kind require. Nevertheless, by summarizing, however briefly, the procedures which certain investigators have adopted to uncover particular quantitative facts, it is hoped to bring the reader a little closer to laboratory realities. At the same time the opportunity will be taken to note in passing some matters of general interest which should find a place in this chapter.

4–3–1 Surface recombination

The strong influence exerted on chain reactions by the recombination of radicals at the walls, together with the more practical fact that the same phenomenon frequently sets a limit to investigations of elementary gas-phase reactions, has led to a number of studies of heterogeneous recombination. Most of these have been concerned with atomic species. Williams and Mulcahy,[64] for example, examined the decay of oxygen atoms over various materials using flow apparatus similar to that shown in Fig. 4–2. At the total pressure used (0.75 torr) recombination occurred almost entirely at the wall of the flow tube which was was coated with the material under study. Relative oxygen concentrations along the tube were determined from the intensity I of the 'air afterglow' in the presence of a trace of nitric oxide. Graphs of $\log I$ against distance showed the decay to be first order and, from their slopes, the values of the rate constants k_w for the several materials were obtained.

Heterogeneous recombinations of atoms are commonly, though not universally, found to be first order. It is usual to express the catalytic activity of the surface in terms of the recombination coefficient γ, defined as the fraction of the total number of collisions of the atoms with the surface leading to recombination. When the rate is sufficiently slow for the concentration gradient from the interior of the vessel to the wall to be negligible, the relation between γ and k_w can be derived in the following way. The number of collisions made by the atoms with the total area of surface S in unit time is given by the kinetic theory formula $\upsilon_c = S(RT/2\pi M)^{1/2}[n_A]$, where M is the molecular weight of the atoms and $[n_A]$ their concentration. Similarly the total number of atoms disappearing from the volume of the vessel V in unit time is given by $\upsilon = Vk_w[n_A]$. Hence it follows that

$$\gamma = \frac{\upsilon}{\upsilon_c} = \left(\frac{2\pi M}{RT}\right)^{1/2}\left(\frac{V}{S}\right)k_w \qquad 4\text{–}13$$

For a cylinder of radius $\mathbf{R}$, $V/S = \mathbf{R}/2$ and, for a sphere, $\mathbf{R}/3$. Since for a given material γ is constant, k_w is inversely proportional to $\mathbf{R}$. It should be noted, however, that with the vessel dimensions and gas pressures commonly adopted, diffusion becomes rate controlling at values of γ notably less than 1 (10^{-1} to 10^{-2} for flow-tube work). In these circumstances, the *apparent* value of k_w becomes inversely proportional to $\mathbf{R}^2$.

Values of γ obtained in a number of investigations are shown in Table 4–2. Their variability reflects the sensitivity of γ to the previous history of the surfaces—particularly those of glass and quartz—as well as to the effects of impurities. In general γ increases with increasing temperature, though in a complicated manner.[65,66] When deactivation of the walls is required, most laboratories have their 'favourite poison'—phosphoric acid, Teflon, and the like. For hydrogen and oxygen and possibly other atoms there is an interesting but quite unexplained correlation between the activity of the surface material and its alkalinity, acidic surfaces being much less active than alkaline.[67]

Table 4–2 Recombination coefficients (γ) for hydrogen and oxygen atoms on various surfaces at 25 °C

Material§	$\gamma \times 10^3$	
	H atoms	O atoms
Pyrex glass	0.5–5	0.02–0.5
Fused quartz	0.5–5	0.05–0.2
Teflon	0.001–0.01	0.01–0.04
Sulphuric acid	—	0.02
Phosphoric acid	0.001–0.01	0.04
Boric oxide	low	0.07
Lead monoxide	—	6
Silver	50	200

§Coated on a glass or quartz surface.

4–3–2 Radical–molecule reactions

As our first example of an investigation of a reaction of this type we refer to the work of Mulcahy and Williams[66] on the reaction

$$O+O_2+M \rightarrow O_3+M \qquad \text{4–XIV}$$

This elementary reaction is responsible for the formation of ozone in the upper atmosphere and its kinetics have been studied on several occasions. Mulcahy and Williams used a stirred-flow system similar to that shown in Fig. 4–8 except that the chemiluminescent method was used to determine the relative concentration of oxygen atoms at the entrance and exit of the reactor.[68] The oxygen atoms were generated upstream by pyrolysing ozone and entered the reactor together with a larger concentration of molecular oxygen in a still larger excess of inert gas M. Pseudo first-order rate constants (k_ψ) were determined by means of equation 4–11, corrections being made to allow for reaction in the connecting tubes. The kinetics were elucidated by studying the influence of various factors on k_ψ, as indicated by equation 4–12. In agreement with previous work, the rate of recombination of oxygen atoms to form molecular oxygen in the gas phase was found to be negligible under the chosen conditions. Thus the values of k_ψ were genuinely first order in [O]. Several features of the kinetics are shown by Fig. 4–9. It will be seen that k_ψ is related to the total concentration (i.e. pressure) by an expression of the form:

$$k_\psi = k_w + k([O_2]+[M])^2 \qquad \text{4–14}$$

The first term, corresponding to the value of k_ψ at zero pressure, relates to recombination at the surface and the quadratic term shows that the gas-phase reaction is, in fact, third order, as indicated by reaction 4–XIV. The mole fraction (x) of oxygen in the gas was small and for these conditions it can be shown that $k = x(1-x)k_{14}$. Thus the slope of each

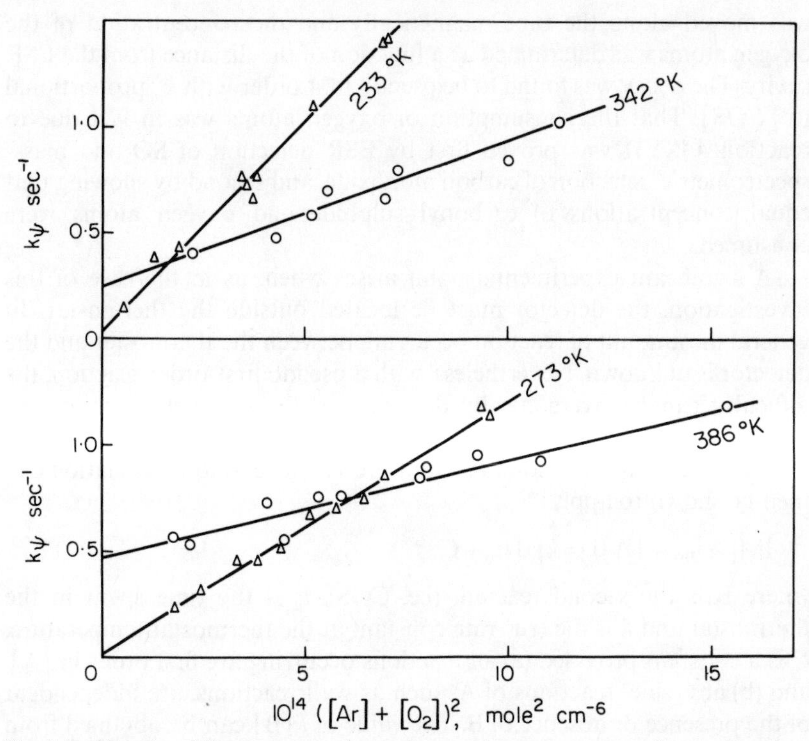

Fig. 4–9 Kinetics of disappearance of oxygen atoms in argon–oxygen mixtures. Pseudo-first-order rate constants graphed as function of total molecular concentration at different temperatures (Mulcahy and Williams[66]).

graph in Fig. 4–9 is directly proportional to k_{14} and its intercept is equal to k_w. The results show that k_{14} decreases with increasing temperature as is commonly the case with third-order reactions. With argon as third body the dependence of k_{14} on temperature can equally well be expressed by the relations

$$k_{14}^{\mathrm{Ar}} = 10^{13.1} \exp(+1.7/RT) \text{ cm}^6 \text{ mole}^{-2} \text{ s}^{-1} \qquad 4\text{–}15$$

or

$$k_{14}^{\mathrm{Ar}} = 10^{14.4}(T/298)^{-3.0} \text{ cm}^6 \text{ mole}^{-2} \text{ s}^{-1} \qquad 4\text{–}16$$

On the other hand, k_w—which refers to a Teflon surface—increases with temperature.

Turning to an investigation of a radical–molecule reaction of a different type, the flow tube–ESR apparatus illustrated in Fig. 4–4 was used by Westenberg and de Haas[23] to study the fast atomic exchange reaction

$$\mathrm{O + COS \rightarrow CO + SO} \qquad 4\text{–XVII}$$

Helium or argon containing a little oxygen was passed through a discharge and the oxygen atoms formed were mixed with an excess of carbonyl sulphide at the central diffuser-injector shown in the diagram. The injector was moved along the tube magnetically and the concentration of the oxygen atoms was determined as a function of the distance from the ESR cavity. The decay was found to be pseudo first order with k_ψ proportional to [COS]. That the consumption of oxygen atoms was in fact due to reaction 4–XVII was proved first by ESR detection of SO and mass-spectrometric detection of carbon monoxide, and second by showing that equal concentrations of carbonyl sulphide and oxygen atoms were consumed.

A significant experimental point arises when, as in the case of this investigation, the detector must be located outside the thermostat. In general the amount of reaction occurring between the thermostat and the detector is unknown. Nevertheless, with a pseudo first-order reaction, the difficulty can be overcome by determining the concentrations of the active species ([A])—that is, [O] in the present example—at the detector with and without the second reactant present. The following relation can then be shown to apply:

$$\ln([A]_{B=0}/[A]) = k[B]t_c + C \qquad 4\text{–}17^{(69)}$$

where B is the second reactant (i.e. COS), t_c is the time spent in the thermostat and k is the true rate constant at the thermostat temperature. C is a constant provided (a) all reactions occurring are first order in [A], and (b) any 'side' reactions of A, such as wall reactions, are independent of the presence or absence of B. The value of $k[B]$ can be obtained from the slope of a graph of the left-hand side of equation 4–17 against t_c. In this way the effects both of the main reactions downstream of the thermostat and side reactions everywhere are eliminated. Condition (a) is normally easily met, but condition (b) in regard to wall reactions is something of a skeleton in the gas kinetics cupboard. Neither in this nor in most other ways of determining gas-phase rate constants is it easy to justify strictly except *a posteriori*.§

The Arrhenius plot of the values of k_{17} obtained by Westenberg and de Haas is shown in Fig. 4–10. The values at the lower temperatures included in the graph from an investigation by Wagner and collaborators were obtained using essentially the same experimental technique.[54] The higher temperature values, however, were obtained using molecular-beam mass spectrometry.[71] At the highest temperature common to the investigations in the two laboratories (535 °C), the absolute values of k_{17} differ by a factor of 1.7. Bearing in mind the difficulties occasioned by the great speed of the reaction and the high temperature, this is to be considered good agreement.

4–3–3 Radical–radical reactions

Recombination reactions of *atoms* are well suited to study by flow methods since they require the presence of a third body and are therefore relatively

§ A case where it definitely is not justified will be found in reference 70.

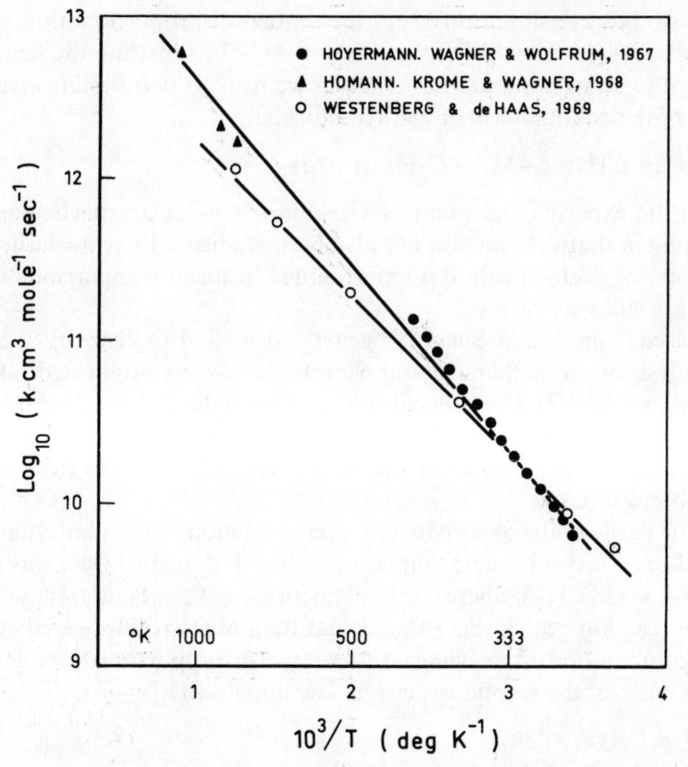

Fig. 4–10 Arrhenius plot of rate constants for the reaction $O+COS \rightarrow CO+SO$. ○ and ● determined by ESR; ▲ determined by molecular-beam mass spectrometry. The Arrhenius parameters are $A = 10^{13.28}$ $cm^3 \, mole^{-1} \, s^{-1}$, $E = 4.5$ kcal $mole^{-1}$ (Westenberg and de Haas[23]); $A = 10^{13.8}$, $E = 5.5$ (Homann *et al.*[71]).

slow. Campbell and Thrush,[72] for example, used flow-tube apparatus of the general form shown in Fig. 4–2 to determine rate constants for the reaction

$$N+N+M \rightarrow N_2+M \qquad \text{4–XXV}$$

with nitrogen, argon, and helium as third-body molecules. The nitrogen atoms were generated by a discharge and their decay along the flow tube was observed by recording the relative intensity of emission of the yellow 'nitrogen afterglow'. The chemiluminescence is emitted concomitantly with the recombination of the atoms and its intensity is proportional to the square of their concentration.

For our final example we return to flash photolysis–kinetic spectroscopy (strictly speaking, kinetic spectrometry). As we observed previously, this technique is particularly suitable for fast radical–radical reactions; that is, for disproportionations and two-body recombinations of complex radicals. The limitation imposed at present by difficulties in obtaining absolute extinction coefficients for radicals can be expected to be progressively reduced in the future. In general, the kinetics of reactions of this type are inaccessible or only just accessible by flow methods.

A study by flash photolysis of the mutual disproportionation of ClO radicals was described briefly in Section 4–1. To illustrate the technique applied to a fast combination reaction we refer to two similar investigations of the combination of methyl radicals:[(73)]

$$CH_3+CH_3 \quad (+M)\rightarrow C_2H_6 \quad (+M) \qquad \text{4–XXVI}$$

From the experimental point of view these studies are particularly interesting in that the reaction has also been studied—15 years earlier—by the rotating sector method. Opportunities to make comparisons of this kind are still very rare.

Basco, James, and Suart[(73a)] generated methyl radicals by the flash photolysis of azomethane (about 5 torr) in an excess of purified nitrogen (reaction 3–LVIII). The concentrations of the radicals were determined at intervals up to 100 μs after the flash by measuring the optical absorbance photographically. A band in the methyl radical spectrum near 216 nm was found suitable.

The decay of the absorbance A after extinction of the photo-flash was found to be second order; that is, a graph of $1/A$ against time was linear. Since $A = \varepsilon L[CH_3]$, where ε is the extinction coefficient and L is the length of the reaction vessel, this showed that the radicals disappeared entirely by recombination (there being no alternative disproportionation reaction). Since, further, the second-order rate law implies that

$$\varepsilon L\, d(1/A)/dt = 2k_{26} \qquad \text{4–18}$$

the measured slope of the graph, $d(1/A)/dt$, gave the value of $2k_{26}/\varepsilon L$.

The critical determination of ε was accomplished in the same investigation by measuring the absorbance as a function of time as it rose to a maximum and began to decline *during* the photo-flash. The extra absorbance which would have been observed at the end of the flash if no radicals had recombined during the flash was calculated from the measured decay after the flash was extinct; and this was added to the observed absorbance to give the absorbance of the total number of radicals generated. This number was taken to be twice the number of molecules of ethane in the reaction vessel at the end of the experiment. Thus the 'total' absorbance of the radicals A_T was related to their concentration to give $\varepsilon = A_T/2[C_2H_6]L$. The value of k_{26} obtained by substituting ε and L in equation 4–18 is given in Table 4–3. Within the assigned limits of error it was found to be independent of the total pressure from 25 to 175 torr, showing that reaction 4–XXVI is substantially second order at these pressures. The second value of k_{26} given in Table 4–3 was obtained independently by van den Berg, Callear, and Norstrom using a similar experimental technique.[(73b)] These investigators used mercury dimethyl as an alternative source of methyl radicals and found the same result as with azomethane. The values of k_{26} obtained by the two groups agree well with each other but, what is much more significant, they also agree with the value obtained by the rotating sector method. The latter is an average

Table 4–3 Determinations of the rate constant for the reaction $CH_3 + CH_3 \rightarrow C_2H_6$ in the second-order region

Method	*Source of* CH_3	*Temp.* (°C)	$10^{13}k$ ($cm^3\ mole^{-1}\ s^{-1}$)	*Reference*
Flash photolysis	$(CH_3)_2N_2$	25	2.6 ± 0.3	(73a)
Flash photolysis	$(CH_3)_2N_2$; $(CH_3)_2Hg$	20	2.43 ± 0.24	(73b)
Rotating sector	$(CH_3)_2CO$	125–175	2.2	(74)

value for the temperature range 125–175 °C but, since there is evidence that the activation energy is zero, the differences between the temperatures at which the three investigations were conducted are probably not significant. The precision of the rotating sector measurements is not known, but it is surely difficult not to be impressed by the concurrence of such remarkably different techniques.

4–4 Experimental studies of quantum states of products and reactants of elementary reactions by kinetic-spectroscopic methods

So far in this chapter it has been assumed that the subject of any particular kinetic study of an elementary reaction is the rate of the complete process by which the reactants in a state of Boltzmann equilibrium pass over to products in the same condition. The spectroscopic techniques described have been used first to identify reactive species and afterwards to indicate regions of their spectra suitable for photometric determination of their concentration. In flash-photolysis experiments time-resolved photographs of the complete absorption spectra are frequently obtained; nevertheless, in the applications considered so far, the purpose of recording the spectra has been primarily photometric. In principle, however, a spectrum recorded during the progress of reaction is capable of yielding far more information about the particular species concerned than its concentration. In general, the radical or molecule absorbs or emits a photon in a much shorter time than it takes to undergo any other event. Hence its spectrum, assuming it can be adequately resolved, provides a virtually instantaneous record of its quantum state. And, if the recording can be made faster than the occurrence of the 'other events'—reaction and collisional deactivation of various kinds to be specified later—it becomes possible to observe the quantum states assumed by the species as it enters into or emerges from reaction.

Although reactions begin and end in a state of Boltzmann equilibrium, for the greater number the existence of an activation energy signifies the importance of the more highly energized reactant molecules. And, as regards bimolecular reactions, the existence of steric factors different from unity suggests that non-translational states of the energetic molecules may be significant—a feature which, as we have seen in Chapter 1, is well recognized in unimolecular reactions. Again, the product species emerging from exothermic reactive collisions contain the energy released by the reaction but, as we shall see, this is not initially distributed among the available translational, rotational, and vibrational levels according to the principle of equipartition. Clearly, the manner in which it is, in fact, distributed must be intimately connected with the dynamics of the reactive collision. Hence the possibilities of observing the reactive behaviour of radicals and molecules in specific quantum states and of determining the energy levels occupied in the products as it were at birth are, in both cases, of the greatest theoretical interest.

The kinetic-spectroscopic methods by which, to a greater or lesser degree, these possibilities can be realized are more or less sophisticated versions of the static and flow procedures already described. A completely different type of approach to the same objectives, namely the method of crossed molecular beams, will be outlined in Section 4–5.

4–4–1 Observation of reaction products in specific energy states

It is a commonplace of spectroscopy that photo-dissociation of a molecule can produce one or both fragments in *electronically excited states.* For example, a convergence limit of bands in the spectrum of molecular iodine indicates that absorption of light in the region of 449.5 nm involves the reaction

$$I_2 + h\nu \rightarrow I + I^* \qquad \text{4–XXVII}$$

The symbol $*$ indicates electronic excitation, in this case to the $5^2P_{1/2}$ state, which is 21.7 kcal mole^{-1} above the ground state ($5^2P_{3/2}$). The concentration of the excited atoms following flash photolysis has been determined spectroscopically using both emission in the infrared and absorption in the vacuum ultraviolet. In a similar way, flash photolysis has been used to produce many other atomic species in various known electronic states for kinetic study.[75]

Much less predictable and perhaps more interesting is the production of electronically excited species by thermal reactions. The various 'after-glows' we have seen put to use in determining atomic concentrations are emitted from electronically excited molecules formed by such combination reactions as

$$O + NO + M \rightarrow NO_2^* + M \qquad \text{4–XXVIII}$$

Using flow-tube technique, Clyne and Thrush[76] established that between 50 and 90 per cent of the total number of nitrogen peroxide molecules formed in the overall reaction between nitric oxide and atomic oxygen pass through the electronically excited state.

Since electronically excited species are produced relatively rarely in thermal reactions, rather more general interest attaches to observing the *vibrational and rotational states* of the reaction products. Here flash photolysis experiments have shown that in exothermic transfer reactions of the type:

$$A + BCD \rightarrow AB + CD \qquad \text{4–XXIX}$$

a substantial part of the heat of reaction first appears as vibrational energy in the newly formed bond A—B.[77] Spectra taken at the shortest possible delay times (about 30 μs) after flashing oxygen-containing compounds isothermally at 25 °C have shown that the oxygen molecules produced by the reactions

$$O + NO_2 \rightarrow O_2^{\dagger} + NO \qquad \text{4–XXX}$$

$$O + ClO_2 \rightarrow O_2^{\dagger} + ClO \qquad \text{4–XXXI}$$

$$O + O_3 \rightarrow O_2^{\dagger} + O_2 \qquad \text{4–XXXII}$$

contain up to 8, 8, and 17 vibrational quanta respectively († signifies vibrational excitation). These figures correspond to 74, 56, and 48 per cent of the total energy available from the reaction. Such amounts of vibrational energy are, of course, enormously greater than what would correspond to Boltzmann equilibrium at 25 °C. That they were not produced by a rise in temperature was confirmed by the fact that the distribution of intensities in the rotational structure of the spectra showed the vibrating molecules to be rotationally 'cold'; that is, the population of their rotational states corresponded to Boltzmann equilibrium at 25 °C. Of equal interest to the appearance of excess vibrational energy in the 'new' bond A—B is the fact, established at the same time, that little or none appeared in the 'old' bond C—D. When chlorine or bromine was photolysed in the presence of ozone—but without photolysis of the ozone—the vibrational excitation appeared only in the oxyhalogen radicals formed by the reactions

$$Cl + O_3 \rightarrow ClO^{\dagger} + O_2 \qquad \text{4–XXXIII}$$

$$Br + O_3 \rightarrow BrO^{\dagger} + O_2 \qquad \text{4–XXXIV}$$

In these circumstances the oxygen molecules were not excited, showing again that the new bond received the greater part of the energy of reaction.

Vibrational excitation of unsymmetrical species like ClO can also be detected by the emission of their vibration–rotation spectra in the infrared. In other words, reactions of the type just mentioned give rise to *infrared chemiluminescence.* This provides another and, as we shall see, a more precise way of studying the distribution of vibrational and rotational energy in the products. A striking example of the chemiluminescence is given by the reaction

$$H + O_3 \rightarrow OH^{\dagger} + O_2 \qquad \text{4–XXXV}$$

When hydrogen atoms are mixed with ozone in a flow system, the chemiluminescence from the OH radicals extends into the visible and can be seen as a peach-coloured 'flame'. Analysis of the spectrum shows the presence of radicals containing up to nine vibrational quanta amounting to 90 per cent of the total energy released by the reaction.[78]

Kinetic–spectroscopic experiments of the kind just mentioned have revealed what appear to be two general characteristics of reactions similar to reaction 4–XXXV; first, a substantial part of the energy of reaction is released initially as vibrational energy, and second, there is a preference for this to occur in the newly formed bond. (The theoretical background to these matters will be found in Chapter 5. Here we are chiefly concerned with the means by which basic facts of this kind can be discovered.)

Turning to more specifically kinetic aspects, it is clearly of great interest to determine quantitatively the *detailed rate constants* $k_{v'}$ which characterize the relative rates of formation of molecules excited to different vibrational levels (v'). The chief difficulty in this is caused by the fact that the initial vibrational distribution is soon 'relaxed' towards Boltzmann equilibrium. This occurs by collisions of the vibrating species with each other, with other molecules present and with the walls. Here the flash-photolysis method is at a disadvantage because of the high total pressures required to keep the temperature constant and, probably more seriously, because of the relatively high concentration of the reaction products needed to obtain sufficient optical absorption. With the time-resolution available these factors make it impossible to observe the initial distribution. However, information of a less detailed character can sometimes be obtained. Vibrational quanta are exchanged between similar species much more rapidly than they are converted to rotational or translational energy by the same or other collisions. They therefore tend to become rapidly randomized among the product molecules. In this way, the product molecules may temporarily assume their own Boltzmann-like distribution of vibrational energy before this is degraded to translational and rotational energy in the final state.§ The temporary 'hang-up' and randomization of the vibrational quanta, when they occur, allow one to speak of a *vibrational temperature* of the molecules and, in addition, they allow an estimate to be made of the total amounts of vibrational energy partitioned between the reaction products. Thus, in a flash-photolytic study of the reaction,

$$O + CS_2 \rightarrow SO^{\dagger} + CS^{\dagger} \qquad \text{4–XXXVI}$$

at 25 °C, Smith[80] found the vibrational temperatures of the SO and CS molecules to be 2870 K and 1775 K respectively at the shortest accessible delay times (about 16 μs). These temperatures correspond respectively to the initial release of 18 and 8 per cent of the total heat of reaction into the vibrational levels of the SO and CS.

When steps are taken to maximize the light-gathering power of the apparatus and the sensitivity of the detector, detailed rate constants can be obtained from *studies of chemiluminescent infrared emission* in flow

§ A general account of energy exchange in molecular collisions will be found in the article by Callear.[79]

systems at low pressures. J. C. Polanyi and collaborators[81–83] have used this method to study a series of exothermic reactions of the type

$$H + Cl_2 \rightarrow HCl^\dagger + Cl \qquad \text{4–XXXVII}$$

and

$$F + H_2 \rightarrow HF^\dagger + H \qquad \text{4–XXXVIII}$$

Here again a notable part of the heat of reaction first appears as vibrational energy of the new-born molecules.

The reaction vessel used by Polanyi is a tube with cold walls 60 cm long by 20 cm diameter in which the atomic and molecular reactants are mixed continuously and from which the uncondensed products are pumped out rapidly through an exit located centrally at right angles to the axis. Concave mirrors are installed at both ends of the tube to collect the radiation, which is passed through a slit in one of the mirrors to the spectrometer. The tube is lined internally with a jacket containing liquid nitrogen. Total pressures are maintained at about 10^{-4} torr and the partial pressures of the molecular species at 10^{-6}–10^{-7} torr. The object is to ensure that the product molecules suffer as few collisions as possible before being removed by condensation on the cold wall. With the reactions just mentioned, it could be safely assumed that no relaxation of the vibrational distribution occurred, since relaxation of the rotational distribution—a much faster process—was found to have progressed only a little. Thus the relative values of $k_{v'}$ were obtained directly from the observed relative concentrations of the product molecules in their different vibrational states. For example $k_{v'=3}$, $k_{v'=2}$, and $k_{v'=1}$ for reaction 4–XXXVIII were found to be in the ratios 0.48:1.0:0.31, $v' = 3$ being the highest vibrational level of HF the heat of reaction can attain. Similar maxima in $k_{v'}$ located some way up the vibrational ladder were found with reaction 4–XXXVII and other exothermic reactions of the same general type. By allowing for the partial relaxation of the rotational distribution, it was possible to refine the analysis still further to obtain values of $k_{v',J'}$, the relative rate constants for production of the molecules in specific vibrational–rotational states. Hence, by difference, the relative rates of formation of molecules with different amounts of translational energy were also determined.

It is appropriate to note here that the production of an 'inverted' or partly inverted distribution in the population of the vibrational levels of the products of reactions such as 4–XXXVII and 4–XXXVIII implies the possibility of observing *laser emission* of the chemiluminescence. This, in fact, has been realized both in a pulse fashion and continuously: the former by conducting flash photolysis in a laser cavity,[84] and the latter by causing the reactants to flow through such a cavity.[85] Several reactions, including reactions 4–XXXVII and 4–XXXVIII, have been used as the basis for such *chemical lasers*.[86]

4–4–2 Spectroscopic studies of the reactivity of molecules in specific energy states

Although, as noted in Chapter 1, thermal reactions as a rule do not significantly disturb the Boltzmann equilibrium of the reactants, the dynamics of any particular reaction may cause it to be channelled through particular energy states—for example vibrational states—of the reactants. Population of high vibrational levels is all-important for unimolecular reactions and there is evidence from shock-tube studies that for some bimolecular reactions to occur one of the reacting partners needs to be vibrationally excited.(87)

At present the most satisfactory approach to the study of reactants in specified states is an indirect one, though based on spectroscopic measurements. By application of the principle of microscopic reversibility (see Section 1–3–2), the detailed rate constant $k_{v,J}$ for reaction of a molecule in a specified vibrational–rotational state v, J can be calculated from the measured detailed rate constant $k_{v',J'}$ for its production in the same state by the reverse reaction. In this way Polanyi *et al.*(83) used their measured values of $k_{v',J'}$ relating to the excited AB molecules formed in exothermic reactions of the type $A+BC \rightarrow AB^{\dagger}+C$ to arrive at detailed rate constants for the endothermic reactions $AB^{\dagger}+C \rightarrow BC+A$, $AB^{\dagger}$ being in a specific vibrational–rotational state (v, J) with energy up to the limit of the highest level populated by the exothermic reaction. It was found, for example, that in the reaction

$$HCl^{\dagger}+I \rightarrow HI+Cl \qquad \text{4–XXXIX}$$

increasing the vibrational energy of the HCl molecule from one quantum to three, while retaining constant the rotational energy of the molecule and the relative translational energy of the colliding partners, causes the value of $k_{v,J}$ to increase by two orders of magnitude. And it seems, on the basis of other such evidence, that vibrational energy is much more effective than translational or rotational energy in bringing about endothermic reactions of this type. The effect is explained theoretically in Section 5–1–3.

Direct methods of studying reactions of atoms in *specific electronic states* in many cases present no particular difficulty since, as already mentioned in Section 4–4–1, such species can often be produced by photolysis. To quote a further example of some interest, oxygen atoms can be produced in the 2^1D_2 state by ultraviolet photolysis of ozone.

$$O_3+h\nu \rightarrow O_2+O^* \qquad \text{4–XL}$$

They have been much studied because of their presence in the upper atmosphere (where indeed they are partly produced by reaction 4–XL). Electronically excited polyatomic radicals or molecules are less easily produced for kinetic studies. However, a number of flow-tube investigations have been carried out on oxygen molecules in the $^1\Delta_g$ state. As noted previously these can be generated by an electric discharge; and their concentration can be determined in various ways, including paramagnetic resonance.[88] The Arrhenius parameters obtained by Findlay and Snelling[88] for the reaction

$$O_2(^1\Delta_g)+O_3 \rightarrow 2O_2+O \qquad \text{4–XLI}$$

are of some interest. The reaction is endothermic and its activation energy was found to be little greater than its endothermicity. Together with the fact that the frequency factor is close to the collision frequency this shows that there is no special barrier impeding the electronic excitation energy of the oxygen molecules from passing into the products.

The production of vibrationally excited radicals and molecules by reactions such as 4–XXXV and 4–XXXVIII creates the possibility of directly observing reactions of these species in particular vibrational states. This has been attempted with some success,[89] but the field has yet to be developed.

4–5 The method of crossed molecular beams

The rate of an elementary reaction as observed in a conventional static or flow system is the weighted statistical average of the reaction rates of the reacting species in their many and various energy states and mutual configurations on collision. The ground plan of the 'weighted statistical average' is the Boltzmann distribution and the full realization of this fact, as expressed by activated-complex theory, has led to a deep understanding of chemical reaction as a statistical process. On the other hand, the great number of energy states available makes it practically impossible, on the basis of conventional kinetic measurements, to arrive at an equal understanding of the specific dynamical factors which come into play in individual molecular encounters. The spectroscopic methods for observing the behaviour of molecules in specific energy states described in the previous section represent successful or partly successful efforts to overcome this limitation. But it would be enormously advantageous if, rather than accepting the molecular assemblies presented by Nature, one could experiment with energetically identical molecules isolated from all other influences except the process under study. Conducting a chemical reaction would then consist of engineering molecular collisions of a predetermined kind between species in known states and observing the results before further collisions could occur. Failing direct observation of individual collisions, the ultimate experimental analysis of a bimolecular reaction between the species X and Y might be conceived as follows. A narrow beam of X molecules or radicals, all in the same detailed quantum state as regards internal energy, would be projected with known kinetic energy through an evacuated space so as to come into collision at a fixed angle, say 90°, with a similar beam of Y molecules which likewise would be 'state-selected' as regards internal and kinetic energy. In addition, all the molecules in each beam would in some way be held to a fixed orientation in space in order to fix their mutual orientation on impact. The relative number of reactive and unreactive collisions would be measured as a function of the internal states, relative kinetic energy, mutual orientation of X and Y and so forth; and these data would be supplemented by determining the internal states and kinetic energies of the products.

Although practical realization of the complete gamut of these conditions has not yet been accomplished, experiments of this general type can be successfully carried out as the result of developments in the technology of molecular beams. Several variables of the kind just mentioned have been isolated or partly isolated in studies of fast exothermic atom-transfer reactions of the type

$$A + BC \rightarrow AB + C \qquad \text{4–XLII}$$

Such experiments, though simple in principle, require an elaborate technology in practice, and only an account of their bare essentials can be given here. For further details the student should consult the reviews cited in the Bibliography. The articles by Toennies, which include a number of instructive diagrams, are particularly recommended. Additional references will be found in the course of the following discussion.

Figure 4–11 is a schematic diagram of typical molecular beam apparatus. Collimated beams of the reactants are projected through a vacuum so as to intersect at right angles. Collisions other than between molecules belonging to the different beams are eliminated by reducing the pressure in the vacuum chamber to the point where the mean free path of the residual gas molecules is many times greater than the dimensions of the chamber. The great majority of the beam molecules pass through the small intersection volume undisturbed, but a small proportion from each beam 'collide', that is, come close enough to interact. Collisions that do not lead to reaction scatter the reactant molecules out of the beams and the products formed in the still fewer reactive collisions also emerge from the intersection volume in various directions. Thus both *unreactive* and *reactive scattering* occur. Our concern is with reactive scattering, though it will be appreciated that the study of unreactive scattering, which gives much information on the nature of molecular collisions, is also relevant to gas kinetics.[91] The relative or, with much greater difficulty, the absolute number of the product molecules taking different directions is measured by a detector which can be moved around the crossing-point as shown in the diagram. Alternatively, the apparatus is arranged so that the beam sources can be moved together in a circular fashion and the detector remains fixed. Provision may also be made for vertical motion in order to detect molecules scattered out of the plane of the beams. The essential feature is that each scattered molecule arrives at the detector without having suffered another collision.

The beams are produced initially either by molecular effusion from a narrow orifice or slit or by 'skimmed' supersonic flow from a nozzle as outlined in Section 4–2–4 in connection with molecular-beam mass

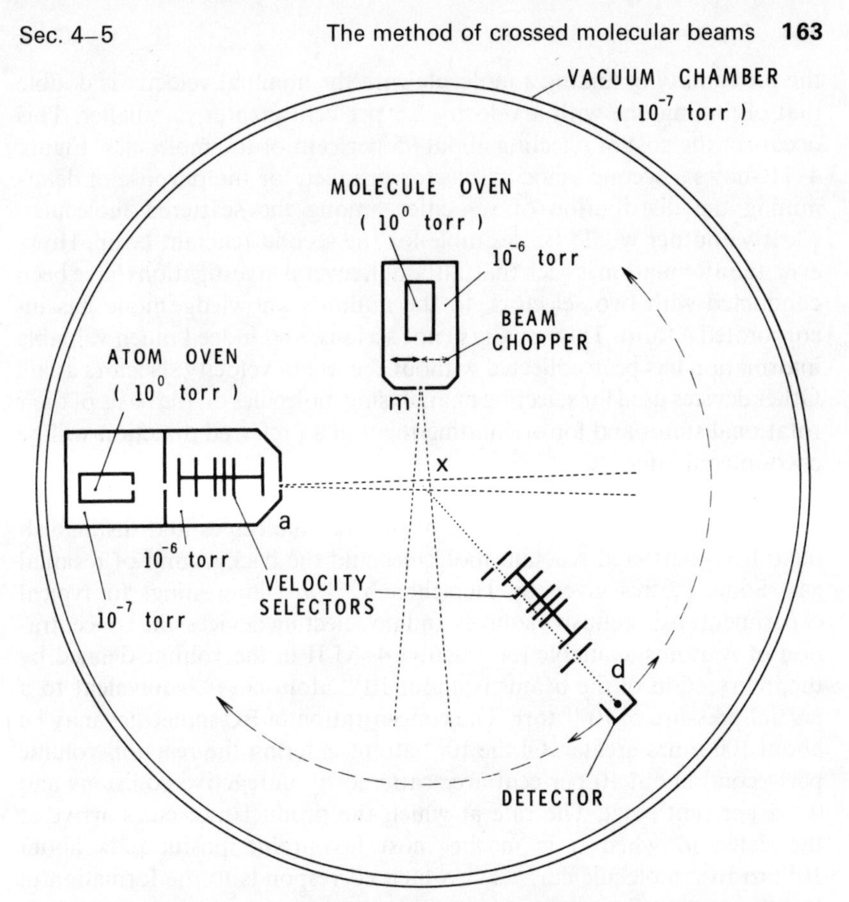

Fig. 4–11 Schematic diagram of apparatus for crossed molecular beams. The apparatus is designed to investigate the angular scattering and velocity distributions of the products from reactions of the type $A + BC \rightarrow AB + C$. Typical dimensions are *a*–*x* and *m*–*x*, 1–10 cm; *x*–*d*, 30–40 cm (5–15 cm without velocity selector); beam angles *a*, 0.5–5°; *m*, 3–30°. For practical details see, for example, reference 90.

spectrometry. The latter method is the more recent development and most investigators hitherto have used effusively generated beams. With this method the source is usually an 'oven' analogous to a Knudsen cell. The appropriate gas or vapour is maintained in thermal equilibrium within and the atoms or molecules emerge with a range of velocities corresponding to the Boltzmann distribution at the temperature of the oven. Thus their most probable velocity can be varied by adjusting this temperature. When sharper resolution of the velocity is required a 'velocity selector' is introduced.[92] This usually consists of a set of slotted wheels rotating on the same shaft, a device similar to that used by Fizeau to measure the velocity of light. The speed of rotation is set to allow only the molecules with the required velocity to pass through all the slots. In this way the Boltzmann distribution is replaced by a narrower triangular distribution of velocities with, of course, a corresponding reduction in the density of molecules in the emergent beam. Typically, the resolution may be such that the probability of finding a molecule with the nominal velocity is double that of finding one with a velocity 2.5 per cent greater or smaller. This occurs at the cost of rejecting about 95 per cent of the molecules. Figure 4–11 shows a second velocity selector installed for the purpose of determining the distribution of velocities among the scattered molecules. Clearly another would be desirable for the second reactant beam. However, the attenuation is such that, although several investigations have been conducted with two selectors, to the author's knowledge none has incorporated a third. This usually is not serious, and indeed much valuable information has been collected without the use of velocity selectors at all. Other devices used for selecting or analysing molecules on the basis of their rotational states and for orientating them in a preferred direction will be encountered later.

The technique is limited basically by the detector. This must be able to detect an extremely small flux of product molecules and distinguish them from scattered reactant molecules and the background of residual gas. Some figures given by Herschbach[(93)] are interesting. In typical experiments using effusive sources and no selecting devices, the concentration of A atoms available for reaction 4–XLII in the volume defined by the intersection of the beams is about 10^{10} atom cm^{-3}, equivalent to a partial pressure of 10^{-6} torr. The concentration of BC molecules may be about 100 times greater. Of the 10^{14} atoms entering the reaction volume per second about 10 per cent are scattered by unreactive collisions and 0.1–1 per cent react. The rate at which the product molecules arrive at the detector, when it is in the most favourable position, is about 10^{10} to 10^{11} molecule cm^{-2} s^{-1}, which corresponds to the formation of less than a monolayer per month.

The only device known to achieve this level of sensitivity in combination with the necessary selectivity is based on surface ionization. When atoms with a low ionization potential, or molecules containing them, impinge on a hot tungsten or platinum filament, the atoms re-evaporate as ions leaving behind the equivalent number of electrons which can be measured. *De facto* the effect is largely confined to alkali metal atoms and their compounds. The crucial development due to Taylor and Datz[(90)] in 1955, which first made the study of chemical reactions in molecular beams possible, was the means of distinguishing between the atoms and compounds by using two filaments one of which is sensitive to both species and the other to the atoms alone. In this way, a number of alkali metal-halogen reactions of the type

$$M + X_2 \rightarrow MX + X \qquad \text{4–XLIII}$$

and

$$M + RX \rightarrow MX + R \qquad \text{4–XLIV}$$

have been studied. Until about 1968 very little success was achieved with any other kind of reaction. But more recently various technical developments, such as the use of nozzle beams to produce higher molecular densities and beam modulation combined with phase-sensitive detection have enabled reactions such as

$$Cl + Br_2 \rightarrow BrCl + Br \qquad \text{4–XLV}$$

and

$$D + X_2 \rightarrow DX + X \qquad \text{4–XLVI}$$

to be studied by means of mass-spectrometer detectors about as precisely as the alkali metal reactions.[94] Nevertheless before referring qualitatively to some examples of the kind of information obtained by crossed-beam technique, it will be as well to note that, of necessity, the reactions to which the facts refer are of rather special types. At present the technique is applicable only to very fast reactions with low activation energies and, of these, the alkali metal-halogen reactions have the unusual feature of converting a covalent bond in the molecular reactant to an ionic bond in the product.

Scattering angles

In determining the angular distribution of the scattering of the reaction products, the crossed-beam technique provides first-hand information about the dynamics of reactive collisions which can be obtained by no other method. Figure 4–12 shows typical experimental results obtained with two reactions of potassium atoms. The relative number of product molecules received by the detector is graphed as a function of scattering angle referred to the direction of the atomic beam as zero. In the reaction with ICl it is immediately evident that most of the potassium atoms are deflected very little from their original trajectory by reaction; in the most probable type of collision the potassium atom detaches a halogen atom from the ICl molecule and proceeds on its way substantially undeflected (with the halogen atom attached). In other less probable collisions greater deflection occurs, but the angular distribution of the product molecules is distinctly anisotropic. For more than qualitative discussion, the collisions and their aftermath must be considered independently of the 'laboratory' geometry imposed by the apparatus. This is done by expressing the motion of the particles in terms of coordinates having their origin at the centre of mass of the molecular system. The velocity through 'laboratory' space of the centre of mass of the colliding particles A + BC, considered as a vector, is entirely determined by the initial velocities and masses of the particles. The laws of dynamics ensure that it is the same before and after collision (even a reactive collision). Thus the new coordinate system rides along with the centre of mass and the motion of the latter in relation to the apparatus is not relevant to the interaction of the particles. In the new coordinates the particles approach the centre of mass, that is, the point of collision in the head-on direction with a relative velocity determined by the initial conditions. This, conceptually, is the meaning of the 'centre-of-

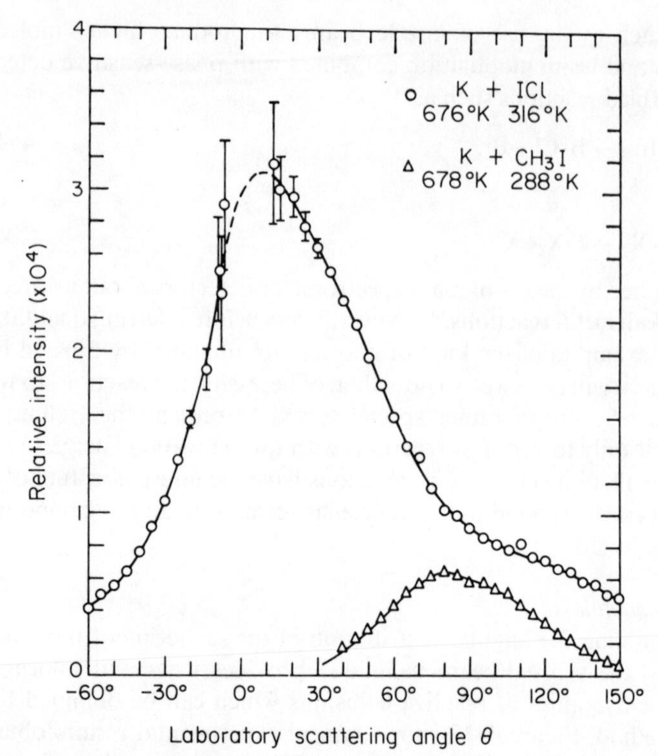

Fig. 4–12 Angular distributions of scattered potassium halide products from molecular beams of K+ICl and $K+CH_3I$ crossed at 90°. The angle θ is referred to the direction of the primary beam of atoms (Herschbach[93]).

mass system' to which the student will find frequent reference in the literature of molecular beams.

Returning to Fig. 4–12, the velocity distribution of the halide product of the K+ICl reaction, when transformed into centre of mass coordinates, shows, as expected, that most of the halide molecules are scattered 'forward' from the collisions; that is, in the original direction of the incident potassium atoms. Such reactions have come to be known as *stripping reactions*: in the limiting case, known as *spectator stripping*, the atomic reactant snatches an atom from the molecule so rapidly (and rushes on with it) that the other atom of the molecule acts as a mere 'spectator' of the event, continuing serenely on its original trajectory 'oblivious' of the loss of its partner. Figure 4–12 also shows the laboratory angular distribution of the halide product of the reaction

$$K+CH_3I \rightarrow KI+CH_3 \qquad \text{4–XLVII}$$

This again is anisotropic, but here the most probable scattering angle is well away from the direction of the incident atomic beam. When the distribution is transformed into centre-of-mass coordinates, it becomes apparent that almost all the product molecules are thrown 'backward' from the collisions. Thus we have a *rebound reaction*. Such are less common than stripping reactions, at least among alkali-halogen reactions. Examples of *intermediate* behaviour with predominantly 'sideways' scattering are known;[95] as also are cases of symmetrical backward and forward scattering.[96] The latter distribution, unlike those mentioned previously, is symptomatic of reactions in which a collision complex remains in existence for the period of a few molecular rotations. In the language of molecular beams, the former reactions are known as *direct* and the latter as *complex*.

Product energy distributions

Other important information to be obtained from the angular distribution relates to the way in which the exothermic reaction energy is distributed as kinetic and internal energy in the products. The amount of kinetic energy naturally is manifested by the relative velocity of the products after collision. In certain circumstances this can be estimated semi-quantitatively on the basis of conservation laws from the most probable laboratory angle at which the products are scattered.[97] (To take a fictitious example, if *all* the energy brought to the collision and generated by it were to be released as rotational or vibrational energy, the products would not move apart; therefore they would be detected only at the laboratory angle corresponding to the direction of motion of the centre of mass.) In this way it was found early in the history of molecular-beam work that very often relatively little of the reaction energy appears as kinetic energy, and therefore most of it must be present as internal energy in the products. This has been confirmed and made more precise by experiments using velocity selectors to measure the velocities of the product molecules directly. Figure 4–11, in fact, refers to apparatus set up for this type of experiment. By such means Entemann and Herschbach[98] found, for example, that less than 5 per cent of the energy of the reaction

$$K + Br_2 \rightarrow KBr + Br \qquad \text{4–XLVIII}$$

appears as kinetic energy, but a somewhat larger fraction (40 per cent) is characteristic of reaction 4–XLVII. In the reaction

$$Cs + CH_3I \rightarrow CsI + CH_3 \qquad \text{4–XLIX}$$

on the other hand, translation takes up the greater part of the energy (namely 65 per cent).

In general the energy distributions found by the beam method reflect the same type of behaviour as is revealed by the infrared chemiluminescence technique described in Section 4–4–1. The two methods are somewhat complementary. The spectroscopic method resolves the vibrational and rotational contributions to the internal energy whereas this, though possible (see below), is more difficult to effect with the beam method. On the other hand, as we have seen, the beam method alone provides detailed information about the angular distribution of the products; and this together with the velocity distribution can be related to a predetermined collisional energy.

Rotational excitation in the products can be estimated by using a 'state selector', based on the Stern–Gerlach effect, in place of the product velocity selector shown in Fig. 4–11.[97] The (polar) molecules are passed through an inhomogeneous electric field which deflects each of them by an amount which depends on its rotational state. The effects are complicated, but it has been possible to show, for example, that the rotational energy content of the KBr molecules from reaction 4–XLVIII is not far from its thermal level and hence that the high degree of internal excitation of these molecules is present as vibrational energy.[99]

Two remarkable experiments

We conclude this short and highly selective sketch of the rapidly developing topic of molecular-beam technique by referring to the attempts made by Brooks and Jones and Beuhler, Bernstein, and Kramer[100] to measure the geometric steric factor of a bimolecular reaction. They represent perhaps the closest approach so far to the 'ideal' kinetic experiments described at the beginning of this section. The reactions investigated were reaction 4–XLVII and the analogous reaction between rubidium and methyl iodide. The experimental objective was to orient the polar methyl iodide molecules in a beam by means of an electric field so that either the iodine atoms or the methyl groups were presented to the impact of another beam of alkali atoms. This was carried out by passing the methyl iodide beam first through an inhomogeneous (six-pole) electric field to select molecules in rotational states which respond most to the orienting effect of a homogeneous field. The selected molecules then passed into a homogeneous field which was maintained across the intersection volume of the beams and could be reversed in polarity so as to orient the dipoles one way or the other to the alkali beam. In both investigations the number of alkali iodide molecules received at the detector was found to be greater when the iodine atoms were presented to the alkali atoms than when they were turned away; and Beuhler *et al.* found the reactivities towards rubidium atoms to be in the ratio of about 1.5:1. Since the methyl iodide molecules were not completely oriented, this represents a lower limit to the true steric effect.

No doubt the result of these experiments will not strike the reader as particularly remarkable; but what surely is remarkable is that such experiments can be done at all. These investigations show in a striking way that the degree of discrimination attainable by beam technique is the key—or potential key—to a level of experimental fact relating to chemical reactivity which cannot be reached by the 'bulk' methods considered earlier in this chapter, however much they may be refined. This is not to suggest that these methods are obsolete for, as noted previously, the applicability of the beam method is restricted at present to very simple and very fast reactions. On the other hand, the phase in the history of gas kinetics wherein the main object of experimental study of elementary reactions is simply to determine their kinetic parameters is well advanced, at least in regard to the chemistry of readily volatile species. It will be sufficiently evident from this and the previous section that the future in this area lies with experimental techniques, such as beam methods, where the effects of 'Boltzmannization' of the molecules before and after reaction can be eliminated.

Some theoretical considerations relating to results of beam experiments will be found in Section 5–1–3.

Bibliography

(A) R. G. W. Norrish, The kinetics and analysis of very fast chemical reactions (by kinetic spectroscopy), in *Chemistry in Britain*, p. 289, 1965.

(B) F. K. McTaggart, *Plasma Chemistry in Electrical Discharges*, Elsevier, 1967.

(C) H. Gg. Wagner and J. Wolfrum, Reactions of atoms, *Angew. Chem. Internat.*, **10,** 604, 1971 (English edn).

(D) A. A. Westenberg, Applications of electron spin resonance to gas-phase kinetics, *Science*, **164,** 381, 1969.

(E) F. P. Lossing, Mass spectrometry of free radicals, Chap. 11 of *Mass Spectrometry*, C. A. McDowell (ed.), McGraw-Hill, 1963.

(F) D. R. Herschbach, Reactive scattering in molecular beams, Chap. 9 of *Adv. in Chemical Physics*, Vol. 10, *Molecular Beams*, J. Ross (ed.), Interscience, 1966.

(G) J. P. Toennies, Molecular beam studies of chemical reactions, in *Chemische Elementarprozesse*, H. Hartmann (ed.), Springer–Verlag, 1968; *idem*, Molecular beam investigations of bimolecular reactions, *Ber. Bunsenges. physik. Chem.*, **72,** 927, 1968. Both articles are in English.

References

1 S. R. Novak and M. W. Windsor, *Proc. Roy. Soc.*, **A308,** 95, 1968; G. Porter and M. R. Topp, *ibid.*, **A315,** 149, 1970.

2 G. Porter in *Photochemistry and Reaction Kinetics*, P. G. Ashmore, F. S. Dainton, and T. M. Sugden (eds), Chap. 5, C.U.P., 1967.

3 G. Porter and F. J. Wright, *Trans. Faraday Soc.*, **51,** 1469, 1955; G. Porter and B. Ward, *Proc. Roy. Soc.*, **A287,** 457, 1965.

4 R. G. W. Norrish and A. P. Zeelenberg, *Proc. Roy. Soc.*, **A240,** 293, 1957; S. J. McGarvey and W. D. McGrath, *ibid.*, **A278,** 490, 1964.

5 G. Porter and F. J. Wright, *Disc. Faraday Soc.*, **14,** 23, 1953; F. J. Lipscomb, R. G. W. Norrish, and B. A. Thrush, *Nature*, **174,** 785, 1954.

6 M. A. A. Clyne and J. F. White, *Trans. Faraday Soc.*, **67,** 2068, 1971.

7 R. M. Fristrom and A. A. Westenberg, *Flame Structure*, pp. 96ff., McGraw-Hill, 1965.

8 F. C. Fehsenfeld, K. M. Evenson, and H. P. Broida, *Rev. Sci. Instr.*, **36,** 294, 1965.

9 T. M. Shaw, *Formation and Trapping of Free Radicals*, Chap. 3, A. M. Bass and H. P. Broida (eds), Academic Press, 1960.

10 F. Kaufman, *Adv. Chem. Series*, **80,** 29, 1969.

11 G. G. Mannella, *Chem. Rev.*, **63,** 1, 1963.

12 E. L. Tollefson and D. J. Le Roy, *J. Chem. Phys.*, **16,** 1057, 1948.

13 F. Kaufman and J. R. Kelso, *Disc. Faraday Soc.*, **37,** 26, 1964.

14 O. R. Lundell, R. D. Ketcheson, and H. I. Schiff, *12th Symposium (Internat.) on Combustion*, p. 307, The Combustion Institute, 1969.
15 J. E. Bennett and D. R. Blackmore, *Chem. Commun.*, **1968,** 1521.
16 J. C. Boden and B. A. Thrush, *Proc. Roy. Soc.*, **A305,** 107, 1968.
17 F. P. Lossing and A. W. Tickner, *J. Chem. Phys.*, **20,** 907, 1952.
18 F. S. Larkin and B. A. Thrush, *10th Symposium (Internat.) on Combustion*, p. 397, The Combustion Institute, 1965.
19 J. R. Dingle and D. J. Le Roy, *J. Chem. Phys.*, **18,** 1632, 1950; W. R. Schulz and D. J. Le Roy, *Canad. J. Chem.*, **42,** 2480, 1964.
20 L. Elias, E. A. Ogryzlo, and H. I. Schiff, *Canad. J. Chem.*, **37,** 1680, 1959.
21 E. A. Ogryzlo, *Canad. J. Chem.*, **39,** 2556, 1961.
22 D. O. Ham, D. W. Trainor, and F. Kaufman, *J. Chem. Phys.*, **53,** 4395, 1970.
23 A. A. Westenberg and N. de Haas, *J. Chem. Phys.*, **50,** 707, 1969.
24 M. A. A. Clyne and B. A. Thrush, *Proc. Roy. Soc.*, **A261,** 259, 1961.
25 M. R. Dunn *et al.*, *J. Phys. Chem.*, **75,** 722, 1971.
26 W. G. Burns and F. S. Dainton, *Trans. Faraday Soc.*, **48,** 52, 1952.
27 L. F. Phillips and H. I. Schiff, *J. Chem. Phys.*, **37,** 1233, 1962.
28 M. A. A. Clyne and H. W. Cruse, *Trans. Faraday Soc.*, **67,** 2869, 1971.
29 F. Kaufman, *Proc. Roy. Soc.*, **A247,** 123, 1958.
30 J. O. Sullivan and P. Warneck, *Ber. Bunsenges. physik. Chem.*, **69,** 7, 1965.
31 M. A. A. Clyne and D. H. Stedman, *Trans. Faraday Soc.*, **62,** 2164, 1966.
32 A. McKenzie, M. F. R. Mulcahy, and J. R. Steven, in press.
33 S. N. Foner, *Science*, **143,** 44, 1964.
34 D. J. E. Ingram, *Free Radicals as Studied by Electron Spin Resonance*, Butterworths, 1958.
35 S. Krongelb and M. W. P. Strandberg, *J. Chem. Phys.*, **31,** 1196, 1959.
36 A. A. Westenberg and N. de Haas, *J. Chem. Phys.*, **40,** 3087, 1964.
37 R. S. Alger, *Electron Paramagnetic Resonance: Techniques and Applications*, Interscience, 1968.
38 A. Carrington, P. N. Dyer, and D. H. Levy, *J. Chem. Phys.*, **47,** 1756, 1967.
39 A. A. Westenberg, *J. Chem. Phys.*, **43,** 1544, 1965.
40 A. A. Westenberg and N. de Haas, *J. Chem. Phys.*, **40,** 3087, 1964.
41 A. McKenzie and B. A. Thrush, *Chem. Physics Letters*, **1,** 681, 1968.
42 M. A. A. Clyne and B. A. Thrush, *Proc. Roy. Soc.*, **A275,** 544, 1963.
43 M. A. A. Clyne and B. A. Thrush, *Trans. Faraday Soc.*, **57,** 1305, 1961.
44 E. Hutton and M. Wright, *Trans. Faraday Soc.*, **61,** 78, 1965.
45 M. A. A. Clyne and D. H. Stedman, *Trans. Faraday Soc.*, **64,** 1816, 1968.
46 J. V. Michael and R. E. Weston, *J. Chem. Phys.*, **45,** 3632, 1966.
47 C.-L. Lin, D. A. Parkes, and F. Kaufman, *J. Chem. Phys.*, **53,** 3896, 1970.
48 M. A. A. Clyne and H. W. Cruse, *Trans. Faraday Soc.*, **67,** 2869, 1971.
49. F. P. Del Greco and F. Kaufman, *Disc. Faraday Soc.*, **33,** 128, 1962.
50 T. G. Slanger and G. Black, *J. Chem. Phys.*, **53,** 3717, 1970.
51 F. P. Lossing, Mass spectrometry of free radicals, Chap. 11 of *Mass Spectrometry*, C. A. McDowell (ed.), McGraw-Hill, 1963.
52 R. W. Kiser, *Introduction to Mass Spectrometry and its Applications*, Prentice-Hall, 1965.
53 S. N. Foner and R. L. Hudson, *Adv. Chem. Ser.*, **36,** 34, 1962.
54 K. Hoyermann, H. Gg. Wagner, and J. Wolfrum, *Ber. Bunsenges. physik. Chem.*, **71,** 603, 1967.
55 S. N. Foner and R. L. Hudson, *J. Chem. Phys.*, **21,** 1374, 1953.

56 F. Kaufman, *Progress in Reaction Kinetics*, G. Porter (ed.), Vol. 1, p. 1, Pergamon Press, 1961.
57 R. V. Poirier and R. W. Carr, *J. Phys. Chem.*, **75,** 1593, 1971.
58 M. Polanyi, *Atomic Reactions*, Williams and Norgate, 1932.
59 J. F. Reed and B. S. Rabinovitch, *J. Phys. Chem.*, **61,** 598, 1957; G. A. Kapralova, A. M. Chaikin, and A. E. Shilov, *Kinetics and Catalysis*, **8,** 421, 1967.
60 V. L. Tal'rose *et al.*, *Adv. Mass Spectrometry*, **3,** 993, 1965; A. F. Dodonov *et al.*, *Kinetics and Catalysis*, **7,** 341, 1966; A. F. Dodonov, G. K. Lavrovskaya, and V. L. Tal'rose, *ibid.*, **10,** 14, 1969.
61 M. F. R. Mulcahy, J. R. Steven, and J. C. Ward, *J. Phys. Chem.*, **71,** 2124, 1967.
62 G. B. Kistiakowsky and C. G. Volpi, *J. Chem. Phys.*, **27,** 1141, 1957; E. L. Wong and A. E. Potter, *ibid.*, **39,** 2211, 1963.
63 S. W. Benson and G. N. Spokes, *J. Am. Chem. Soc.*, **89,** 2525, 1967.
64 D. J. Williams and M. F. R. Mulcahy, *Aust. J. Chem.*, **19,** 2163, 1966.
65 J. C. Greaves and J. W. Linnett, *Trans. Faraday Soc.*, **55,** 1346, 1355, 1959.
66 M. F. R. Mulcahy and D. J. Williams, *Trans. Faraday Soc.*, **64,** 59, 1968.
67 D. R. Warren, *Trans. Faraday Soc.*, **53,** 199, 1957.
68 A. B. Ayling and D. J. Williams, *Chem. Instrument.*, **2,** 149, 1970.
69 A. A. Westenberg and N. de Haas, *J. Chem. Phys.*, **46,** 490, 1967.
70 M. F. R. Mulcahy, J. R. Steven, J. C. Ward, and D. J. Williams, *12th Symposium (Internat.) on Combustion*, p. 323, The Combustion Institute, 1969.
71 K. H. Homann, G. Krome, and H. Gg. Wagner, *Ber. Bunsenges. physik. Chem.*, **72,** 998, 1968.
72 I. M. Campbell and B. A. Thrush, *Proc. Roy. Soc.*, **296,** 201, 1967.
73 (a) N. Basco, D. G. L. James, and R. D. Suart, *Int. J. Chem. Kin.*, **2,** 215, 1970. (b) H. E. van den Berg, A. B. Callear, and R. J. Norstrom, *Chem. Phys. Letters*, **4,** 101, 1969.
74 R. Gomer and G. B. Kistiakowsky, *J. Chem. Phys.*, **19,** 85, 1951; A. Shepp, *ibid.*, **24,** 939, 1956.
75 R. J. Donovan and D. Husain, *Chem. Rev.*, **70,** 489, 1970.
76 M. A. A. Clyne and B. A. Thrush, *Proc. Roy. Soc.*, **A269,** 404, 1962.
77 W. D. McGrath and R. G. W. Norrish, *Proc. Roy. Soc.*, **A254,** 317, 1960.
78 R. E. Murphy, *J. Chem. Phys.*, **54,** 4852, 1971.
79 A. B. Callear, *Photochemistry and Reaction Kinetics*, Chap. 7, P. G. Ashmore, F. S. Dainton, and T. M. Sugden (eds), C.U.P., 1967.
80 I. W. M. Smith, *Disc. Faraday Soc.*, **44,** 194, 1967.
81 J. C. Polanyi *et al.*, *Disc. Faraday Soc.*, **44,** 183, 1967.
82 J. C. Polanyi *et al.*, *J. Chem. Phys.*, **53,** 4091, 1970.
83 J. C. Polanyi *et al.*, *J. Chem. Phys.*, **51,** 5716, 5717, 1969.
84 J. V. V. Kasper and G. C. Pimentel, *Phys. Rev. Letters*, **14,** 352, 1965.
85 D. J. Spencer, T. A. Jacobs, H. Mirels, and R. W. F. Gross, *Int. J. Chem., Kinetics*, **1,** 493, 1969; T. A. Cool, R. R. Stephens, and T. J. Falk, *ibid.*, **1,** 495, 1969.
86 G. C. Pimentel, *Pure Appl. Chem.*, **18,** 275, 1969; C. Wittig, J. C. Hassler, and P. D. Coleman, *Nature*, **226,** 854, 1970.
87 D. Lewis and S. H. Bauer, *J. Am. Chem. Soc.*, **90,** 5390, 1968; T. Baer and S. H. Bauer, *ibid.*, **92,** 4773, 1970.
88 G. A. Hollinden and R. B. Timmons, *J. Am. Chem. Soc.*, **92,** 4181, 1970; I. D. Clark, I. T. N. Jones, and R. P. Wayne, *Proc. Roy. Soc.*, **A317,** 407, 1970; F. D. Findlay and D. R. Snelling, *J. Chem. Phys.*, **54,** 2750, 1971.

89 A. E. Potter, R. N. Coltharp, and S. D. Worley, *J. Chem. Phys.*, **54,** 992, 1971.
90 E. H. Taylor and S. Datz, *J. Chem. Phys.*, **23,** 1711, 1955; Y. T. Lee *et al.*, *Rev. Sci. Instr.*, **40,** 1402, 1969.
91 E. F. Greene, A. L. Moursund, and J. Ross, *Adv. Chem. Phys.*, **10,** 135, 1966.
92 H. U. Hostettler and R. M. Bernstein, *Rev. Sci. Instr.*, **31,** 872, 1960.
93 D. R. Herschbach, Reactive scattering in molecular beams, Chap. 9 of *Adv. in Chemical Physics*, Vol. 10, *Molecular Beams*, J. Ross (ed.), Interscience, 1966.
94 Y. T. Lee *et al.*, *J. Chem. Phys.*, **51,** 455, 1969; H. J. Loesch and D. Beck, *Ber. Bunsenges. physik. Chem.*, **75,** 736, 1971.
95 K. R. Wilson and D. R. Herschbach, *J. Chem. Phys.*, **49,** 2676, 1968.
96 W. B. Miller, S. A. Safron, and D. R. Herschbach, *Disc. Faraday Soc.*, **44,** 108, 1967.
97 E. F. Green and J. Ross, *Science*, **159,** 587, 1968.
98 E. A. Entemann and D. R. Herschbach, *Disc. Faraday Soc.*, **44,** 289, 1967.
99 R. R. Herm and D. R. Herschbach, *J. Chem. Phys.*, **43,** 2139, 1965.
100 P. R. Brooks and E. M. Jones, *J. Chem. Phys.*, **45,** 3449, 1966; R. J. Beuhler, R. B. Bernstein, and K. H. Kramer, *J. Am. Chem. Soc.*, **88,** 5331, 1966.

Problems

4–1 Write an account of the development of flash photolysis technique based on the following and any other papers you may find appropriate: Porter, *Proc. Roy. Soc.*, **A200,** 284, 1950; Christie *et al.*, *ibid.*, **A216,** 152, 1953; Burns and Hornig, *Can. J. Chem.*, **38,** 1702, 1963; Ip and Burns, *Disc. Faraday Soc.*, **44,** 241, 1967; Braun and Lenzi, *ibid.*, **44,** 252, 1967; ref. 1.

4–2 (a) The reaction

$$S(^3P)+C_2H_4 \quad (+M) \rightarrow C_2H_4S \quad (+M)$$

was studied in excess argon by flash photolysis. The S atoms were generated by photolysis of COS and their decay, monitored by resonance fluorescence, was found to follow first-order kinetics (Davis *et al.*, *Int. J. Chem. Kinet.*, **4,** 383, 1972). The following values of the pseudo first-order rate constant k_ψ were obtained with various initial pressures at 298 K.

p_0(Ar) torr	50	100	100	100	200	400
$10^3 p_0(C_2H_4)$ torr	70	20	150	70	50	110
$10^{-2}k_\psi$, s^{-1}	11.2	3.2	25.7	10.9	7.6	18.4

Initial pressures of S were about 10^{-5} torr. Deduce the order and rate constant of the reaction.

(b) The recombination of iodine atoms has been studied by flash-photolysing I_2 vapour and monitoring the subsequent increase in $[I_2]$ by optical absorption. Several investigations have found that the kinetics obey the relation $d[I_2]/dt = \kappa[I]^2[M]$, where M refers to an inert gas present in large excess. Careful measurements, however, show that κ is not quite constant (at constant temperature) but increases with increasing values of $[I_2]_a/[M]$, $[I_2]_a$ being the mean value of $[I_2]$ present during the reaction. The following data were obtained by Christie *et al.*, (*Proc. Roy. Soc.*, **A231,** 446, 1955) from experiments with M = He and Kr respectively.

$10^3[I_2]_a/[He]$	$10^{-15}\kappa$ cm^6 mole^{-2} s^{-1}	$10^3[I_2]_a/[Kr]$	$10^{-15}\kappa$ cm^6 mole^{-2} s^{-1}
0.26	1.45	0.58	4.70
1.14	2.24	1.17	5.11
1.97	2.90	2.22	5.80
2.83	3.32	3.07	6.91

Interpret these results in the light of the discussion on p. 207 and deduce the values of the termolecular rate constants.

4–3 (a) Oxygen atoms were generated in a Pyrex flow-tube of 1.35 cm internal diameter by passing a 2:98 mixture of O_2 in Ar at 0.76 torr pressure through an electric discharge. The total flow rate, measured at 1 atm pressure, was 10.0 cm^3 min^{-1}. NO_2 was added downstream in increasing amount until the 'air afterglow' chemiluminescence was extinguished. This occurred at an NO_2 flow of 0.056 cm^3 min^{-1}. What percentage of the O_2 molecules was dissociated and at what flow rate of NO_2 was the maximum luminescent intensity observed?
(b) When only a minute flow of NO_2 was added in the above circumstances, the luminescence decreased exponentially along the flow-tube with a decay constant of 2.3 s^{-1}. Given that the termolecular rate constants for the O+O and $O+O_2$ gas-phase combinations are both about 3×10^{14} cm^6 mole^{-2} s^{-1}, what was the collision efficiency (γ) for recombination of O atoms at the surface? If the same gas mixture was flowing through a 50 litre Pyrex sphere and the flow was suddenly stopped, how long would it take for the luminescence to fade to one tenth its original intensity?

4–4 (a) The decay of N atoms was studied by measurements of the decrease in intensity of the 'nitrogen afterglow' along a flow-tube combined with NO titrations.[72] The first measurements showed high rates but, with repeated experiments, these gradually decreased to constant values. The decay was found to obey the relation $-2d[\mathrm{N}]/dt = \kappa[\mathrm{N}]^2$ where $\kappa = a+b[\mathrm{M}]$ and b, but not a, depended on the nature of the carrier gas M. The following values were obtained with $M = N_2$.

Temp. K	196	254	327
$10^{-8}a$ cm^3 mole^{-1} s^{-1}	2.50	3.50	4.70
$10^{-15}b$ cm^6 mole^{-2} s^{-1}	3.29	1.97	1.21

Interpret these results and derive Arrhenius parameters for the reactions involved.
(b) The intensity of the nitrogen afterglow obeys the relation $I = I_0[\mathrm{N}]^2$ where I_0 decreases with increasing temperature, is independent of pressure over a wide range, but depends on the nature of the inert gases present. Devise a reaction mechanism to explain these facts.

4–5 (a) When a catalytic probe is used to measure atomic concentrations ([A]) in flow-tube experiments, the effect of the probe on the axial diffusion gradient must be carefully considered since, with total destruction of the atoms, [A] is effectively zero at the catalytic surface. Show that for a pseudo first-order reaction of A beginning abruptly at the position $z = 0$ the concentration measured by the probe at the position $z = L$ is given to a good approximation by

$$\frac{[\mathrm{A}]_L}{[\mathrm{A}]_0} = \frac{2\alpha}{(1+\alpha)} \exp[(1-\alpha)\bar{u}L/2D] \qquad (1)$$

where the symbols are those used in equations 4–4 and 4–5. [Note (i) that the number of atoms arriving (and destroyed) per second per cm^2 at the probe surface (considered as a radial plane) is given by $N_A = -D(d[\mathrm{A}]/dz)_L$ and (ii) that $[\mathrm{A}]_L = N_A/\bar{u}$. Values of $(\bar{u}\alpha L/D)$ may be considered to be always less than about 5].
(b) Supposing $[\mathrm{A}]_L$ was measured at a series of positions $z = L_1, L_2$ etc. ($L \neq 0$) and k_ψ was calculated by the simple plug-flow formula, what extra diffusional error, if any, is introduced by the probe over and above what obtains with a non-destructive *in situ* method of detection?

4–6 Rate constants for the very fast reaction

$$O + OH \rightarrow O_2 + H \qquad (1)$$

were measured by Westenberg *et al.* (*J. Phys. Chem.*, **74**, 3431, 1970) using the (quartz-lined) cavity of an ESR spectrometer as a stirred flow reactor (volume 67 cm^3). Helium containing small concentrations of O and H atoms was passed through the cavity, where OH radicals were generated from added NO_2 by the reaction $H+NO_2 \rightarrow OH+NO$ (which is about 10 times faster than the reaction of O with NO_2). Assuming that reaction (1) is much faster than any other reaction occurring by which O atoms are formed or destroyed, derive approximate Arrhenius constants for reaction (1) from the following experimental measurements. [O] and [OH] are the concentrations in the reactor in the presence and $[O]_0$ in the absence of added NO_2.

Temp. K	Pressure torr	Total flow $cm^3\ s^{-1}$ (at 1 atm and 298 K)	$[O]_0/[O]$	$10^{12}[OH]$ mole cm^{-3}
228	0.70	2.29	12.5	14.2
298	0.55	1.27	3.30	2.85
340	0.40	1.63	1.88	1.90

4–7 The ozone in the Earth's upper atmosphere is produced by the reactions

$$O_2+h\nu \rightarrow 2O \tag{1}$$

$$O+O_2+M \rightarrow O_3+M \tag{2}$$

and is destroyed in various ways. To understand quantitatively the factors contributing to the steady-state concentration of O_3 present, accurate rate constants are required for reaction (2) in the temperature range −60 to 0 °C. Make a literature survey to discover what information is available. Supposing you were to undertake new measurements, describe the method you would adopt and give the reasons for your choice. List the equipment required and make appropriate enquiries to estimate what the investigation would cost (exclusive of your own salary).

5 Theories of the kinetics of elementary reactions

This chapter continues the discussion of theoretical concepts begun in Chapter 1. The function of a theory is to generalize the observed facts and relate them to what may reasonably be regarded as more fundamental facts or concepts. The primary observed facts with respect to the kinetics of elementary reactions are the absolute values of the rate constants determined over a range of temperature; these are found to be well expressed, in the great majority of cases, by the Arrhenius parameters A and E_a. Other facts relate to the dependence of rate constants on the type of reaction and on the molecular structures of the reactants. Others again are concerned with the influence of 'inert' gases. More subtle but no less significant facts, ascertained by molecular beam experiments and in other ways, relate to the influence of the energy states of the reactants on the rate, and to the manner in which energy is distributed in the products. Such are the 'experimental data' which theories of elementary reactions are called upon to explain, and from which they aim to construct (or discern?) a unified pattern of chemical reactivity.

There are, of course, different levels of theoretical explanation. The most fundamental conceivable at present would be an explanation purely in terms of the properties of electrons and nuclei. For this it would be necessary to show that the reaction rates observed experimentally could be calculated *ab initio* on the basis of accepted principles of quantum mechanics without reference to 'empirical', that is, experimental, information about the molecules involved. This, though conceivable, is scarcely approachable at present except for the simplest of all reactions between neutral species, the exchange of a hydrogen atom between atomic and molecular hydrogen. Even here, as we shall see shortly, theoreticians have not yet quite succeeded in reproducing the experimental facts. The level of explanation appropriate to the great mass of kinetic data is in terms of molecular properties. An elementary example is provided by simple collision theory which attempts to calculate the value of A for bimolecular reactions from the viscosity diameters of the reacting molecules by means of molecular statistics. Similarly, theories of unimolecular reactions draw upon knowledge of molecular vibrations, and almost all theoretical work makes much use of spectroscopic and thermodynamic data relating to the species undergoing reaction. On the other hand, we shall shortly encounter some attempts at an intermediate, 'semi-empirical', level of theoretical treatment where certain results from quantum mechanics relating to systems of electrons and nuclei are supplemented by information derived directly from molecular spectroscopy.

It was observed in Chapter 1 that the concept of a critical energy for reaction derived from the Arrhenius relation can be incorporated into two different theories of reaction rate. 'Collision theory', in the elementary form presented there, is based on the frequency of 'hard-sphere' molecular collisions, and 'activated-complex theory' is based on the quasi-equilibrium concentration of activated complexes: the collisions bring together or the activated complexes contain the critical energy. In either case, the critical energy can be associated with a maximum in the path of least potential energy between reactants and products, and the magnitude of this potential energy barrier is found by calculating potential energy surfaces for the reaction complex (though methods for carrying out the calculations were not specified in Chapter 1). When the geometry of the activated complex requiring the least energy is known, the corresponding potential energy surface contains sufficient information, at least in principle, for the rate constant to be calculated by activated-complex theory.§ Results of such calculations presented briefly in Chapter 1 showed that activated-complex theory gives a better account of the facts than elementary collision theory. The superiority lies in the fact that activated-complex theory makes greater use of the information expressed by the potential energy surface. We shall see later, however, that more sophisticated collision theory draws equally upon this information. Indeed it is not too much to say that a great deal of contemporary theoretical discussion of reaction rates is basically discussion of potential energy surfaces.

Our first exercise, therefore, will be to become acquainted with the methods that have been developed for calculating potential energy surfaces. We shall be concerned with bimolecular atom transfer reactions of the type $A+BC \rightarrow AB+C$. Before proceeding, however, we must return briefly to a difficulty which was touched upon earlier in Section 1–4–2. It is clear at the outset that the height of the lowest potential energy barrier E_c 'corresponds' to the experimental E_a defined by the Arrhenius equation; but it soon becomes evident that, in general, the correspondence cannot be exact. Theoretical attempts to relate rate constants to E_c lead almost invariably to expressions of the form

$$k(T) = B(T)\exp[-E_c/RT] \qquad 5\text{–}1$$

This is so even with 'hard-sphere' collision theory, for which $B(T)$ is proportional to $T^{1/2}$. Since the Arrhenius A is regarded as independent of temperature, E_c cannot be the same as E_a. Similar considerations to those given in Section 1–4–2 lead to the general relation

$$E_a = E_c + RT\, d\ln B(T)/d\ln T \qquad 5\text{–}2$$

The difference between E_a and E_c depends on the form of the theoretical

§ Strictly, an average should be taken over the potential energy hypersurface corresponding to all geometrical configurations of the activated complex.

expression for $B(T)$ and in general this varies somewhat with the particular theoretical approach adopted. A significant point, of which $B(T)$ must take account, arises from the fact that potential energy surfaces calculated by standard methods make no reference to the zero-point vibrational levels of either the reactants or the collision complex. Reference to equations 1–57 and 1–64a will show that, when these are considered, equation 5–2 becomes

$$E_a = E_c + [\Delta E_z + RT\, d \ln C(T)/d \ln T] \qquad 5\text{–}3$$

ΔE_z is the difference between the zero-point energies of the activated complex and the reactants and $C(T)$ refers to the quotient of partition functions, including $\bar{k}T/h$, in equation 1–61. Since the terms in the brackets together may amount to ± 1 to ± 3 kcal mole^{-1}, better agreement than this cannot be expected between the calculated value of E_c and E_a. This conclusion, at least in a qualitative sense, can also be derived on general grounds without reference to activated-complex theory.[1]

5–1 Bimolecular reactions of the type A + BC → AB + C

5–1–1 Potential energy surfaces

Methods currently used to calculate potential energy surfaces are of three kinds: 'theoretical' or '*ab initio*', 'semi-empirical', and 'empirical'. A *theoretical method* uses purely quantum-mechanical procedures to calculate the potential energy resulting from the electrostatic forces between the electrons and nuclei of the reacting species. The potential energy (hyper) surface is, of course, the result of calculations carried out for a range of distances and angles between the various nuclei. The procedures used for a reaction system, for example H ... H ... H, do not differ in principle from those applicable to a system of atoms combined in a stable molecule, for example H ... H. A *semi-empirical* method is one in which quantum mechanics is used to establish a relation between energy and configuration which contains one or more unknown parameters. These are evaluated from independently known properties of the reacting species or are simply adjusted to fit some experimental property of the reaction itself. The latter is usually the activation energy and the resulting surface is then used to calculate the reaction rate. Finally, an *empirical* method attempts to construct a surface, or part of one, by more or less intuitive means, drawing upon potential energy relations that are known to apply to stable molecules. The distinction between the semi-empirical and empirical methods relates to the degree to which quantum-mechanical theory has been attenuated in their derivation. As we shall see, it is not very sharply defined in practice.

Because of the great difficulties involved in dealing with more complex systems, *theoretical methods* so far have been applied only to the H_3 system:

$$H_a + H_b - H_c \rightarrow H_a - H_b + H_c \qquad 5\text{–I}$$

The original treatment is that carried out by London in the 1920s.[2] It involves drastic approximations and leads to the following relation for the potential energy relative to the situation when all three atoms are at infinite distance from each other:

$$V_{abc} = Q_{ab}+Q_{bc}+Q_{ac} \pm\{\tfrac{1}{2}[(\alpha_{ab}-\alpha_{bc})^2+(\alpha_{bc}-\alpha_{ac})^2+(\alpha_{ac}-\alpha_{ab})^2]\}^{1/2} \quad 5\text{–}4$$

Q_{ij} and α_{ij} are respectively the 'Coulombic' and 'exchange' integrals deriving from the corresponding interactions between the atoms H_i and H_j. They are complicated functions of the internuclear distance r_{ij} and the electronic charge (only).[2] When one of the atoms is infinitely removed from the other two, the potential energy becomes

$$V = Q \mp \alpha \quad 5\text{–}5$$

The $\pm$ sign is reversed to indicate that the absolute magnitude of α is negative. Thus the positive sign corresponds to the state of lower energy, that is, to molecular hydrogen in its ground state ($^1\Sigma$). It follows that, on the London theory, the energy of the diatomic molecule is simply the algebraic sum of Coulombic and exchange contributions. The negative sign in equation 5–5 gives the repulsive energy of the atoms in their antibonding state ($^3\Sigma$).

The lower root of equation 5–4 represents the state of lowest energy for the H_3 system and is therefore the one in which we are interested. We should note that the less approximate Heitler–London treatment of the isolated hydrogen molecule leads to the relation

$$V = \frac{Q \mp \alpha}{1 \mp S^2} \quad 5\text{–}6$$

where S^2 is the 'overlap' integral.[2] This expression accounts for only two-thirds of the observed binding energy. Equation 5–5, which in effect equates S^2 to zero, must therefore be expected to be even less exact. On the other hand, our concern is with the potential energy of the H_3 system relative to that of the hydrogen molecule in its state of minimum energy $(V_{bc})_{min}$ at infinite distance from the atom. The relevant energy is the difference:

$$E_{abc} = V_{abc} - (V_{bc})_{min} \quad 5\text{–}7$$

Thus some errors may be cancelled by calculating the values of both V_{abc} and $(V_{bc})_{min}$ by the same method.

London concluded that V_{abc} is least, that is most negative, when H_a approaches H_b-H_c end-on, and subsequent work has consistently confirmed this. For example Hirschfelder[3] found the energy of the linear symmetrical configuration to be about 80 kcal mole^{-1} less than that of the equilateral triangle. For this reason, it is generally sufficient to consider the end-on trajectory. However, because of its approximations, equation

5–4 cannot be expected to yield an accurate potential energy surface. Its present importance is chiefly in relation to semi-empirical methods.

Since the advent of high-speed computers, numerous theoretical calculations have been made on the basis of the variational principle. Of these the most elaborate so far is that carried out by Shavitt *et al.*[4] in which two 1*s* and three 2*p* orbitals for each atom were considered. The height of the potential energy maximum E_c above the energy of H_2 calculated by the same treatment was found to be 11.0 kcal mole^{-1}. Another (non-variational) calculation by Conroy and Bruner[5] gave 7.7 kcal mole^{-1}. The values are to be compared with that of about 9 ± 1 kcal mole^{-1} derived from experiment. The results of these and other theoretical investigations[6] have been obtained at the cost of an immense amount of labour and computer time. That they are still not entirely satisfactory for the H_3 system casts a sombre light on the prospect for *ab initio* calculations of potential energy surfaces for more complex systems. Nevertheless the theoretical H_3 surfaces which, as we have seen, come close to yielding the correct barrier height, provide invaluable standards by which to judge the accuracy of the semi-empirical methods; and these are capable of wider application.

The first *semi-empirical* procedure, known nowadays as the LEP method, was developed from the London equation (5–4) by Eyring and Polanyi.[7] In its original form, it involves two main assumptions. First, the Coulombic part of the total interaction energy of each pair of atoms of the triatomic complex is assumed to be constant, irrespective of the internuclear separation and of the nature of the atoms. Thus,

$$\rho = Q/(Q+\alpha) = \text{constant} \approx 0.14 \qquad 5\text{–}8$$

and second, the energy of the two atoms is related to their internuclear distance r by the well-known Morse function:

$$(Q+\alpha) = V = D\{\exp[-2\beta(\Delta r)] - 2\exp[-\beta(\Delta r)]\} \qquad 5\text{–}9$$

where $\Delta r = (r - r_0)$ and r_0 is the equilibrium internuclear distance in the stable diatomic molecule (cf. Fig. 1–2). The classical dissociation energy D and the constants β and r_0 can all be obtained from spectroscopic data relating to the stable molecule. Insertion of equation 5–8 in equation 5–4 gives

$$V_{abc} = 0.14(V_{ab} + V_{bc} + V_{ac}) - 0.61[(V_{ab} - V_{bc})^2 + (V_{bc} - V_{ac})^2 + (V_{ac} - V_{ab})^2]^{1/2} \qquad 5\text{–}10$$

(a vast simplification). V_{abc} is easily evaluated at appropriate values of r_{ab}, r_{bc}, and r_{ac} by means of this relation and equation 5–9. Eyring and his collaborators applied the method to a number of atomic exchange reactions and found a reasonable correlation between the calculated heights of the potential barrier and the experimental activation energies. This is shown in Table 5–1. More recently ρ has come to be regarded as

Table 5–1 Reactions of the type $A+BC \rightarrow AB+C$.

Arrhenius activation energies determined experimentally and potential energy barriers calculated by standard methods. (E_c = height of potential barrier; E_0 = height allowing for zero-point energies. Values to nearest kcal mole^{-1}.)

Reaction	E_a *exptl.* (kcal mole^{-1})	*Calculated values of E_c and E_0 (kcal mole^{-1})* LEP		LEPS			BEBO	
		ρ	E_c	k	E_c	E_0	E_c	E_0
$H+p\text{-}H_2$	9	0.14	14	0.18	5	5	10	9
		0.10	7	0.10	13	—	—	—
		0.07	20	0.00	24	—	—	—
$H+Cl_2$	2	0.14	3	—	—	—	—	3
$H+Br_2$	1	0.14	2	—	—	—	—	3
$F+H_2$	2	0.14	6	0.18	1	—	2	2
$Br+H_2$	19	0.14	22	0.18	17	—	21	—
$I+H_2$	33	0.14	40	0.18	33	—	—	34
$Br+CH_4$	19	—	—	0.18	20	20	18	—
CF_3+H_2	11	—	—	0.11§	14	11	13	9
$H+CH_4$	12	—	—	0.13§	11	12	12	13

§ Adjusted to give agreement between E_a and E_0.

an adjustable parameter. Its effect on the calculated barrier height is apparent from Table 5–1.

It appears, however, that the LEP method has a fundamental defect. With many systems, the calculated potential energy surface shows *two* maxima separated by a shallow depression or 'basin' near the origin. In such a case, Fig. 1–3 would show a basin in the region of the point *. The basin on the surface for H_3 is about 3 kcal mole^{-1} deep and in addition there is a much deeper cavity at very short interatomic distances. These features, which can be traced to the London equation itself, are unlikely to be real. They are not found on the most reliable *ab initio* surfaces for H_3. More generally, the presence of a depression in the saddle-point area implies that the activated state should persist for at least several molecular vibrations. But molecular beam studies show that such behaviour, though not unknown, is rare among reactions of this type.

These defects are avoided by a modified LEP procedure introduced by Sato;[(8)] and his method—the LEPS method—has come into fairly general use. The assumption embodied in equation 5–8 is rejected, and equation 5–6 is used instead of equation 5–5. To be able to separate Q from α, it is assumed first, that the energy of the attractive state of the atoms can be found from the Morse function as before; and second, that the energy of the repulsive state is expressed by an 'anti-Morse' function specially invented for the purpose:

$$(Q-\alpha)/(1-S^2) = {}^3V = 0.5\,D\,\{\exp[-2\beta(\Delta r)]+2\exp[-\beta(\Delta r)]\} \quad 5\text{–}11$$

The form of the anti-Morse function is purely intuitive but it leads to reasonable agreement with well-established quantum mechanical calculations for $H_2(^3\Sigma)$. Equation 5–11, together with its equivalent for the attractive state, enables Q and α to be expressed independently as functions of Δr. The functions contain S^2, which is regarded as a constant for all three pairs of atoms. To complete the operation these functions are substituted in the London equation (5–4) and, in addition, the right-hand side of the latter is divided by $(1+S^2)$. At this stage, S^2 bears little relation to its origin and it is simply treated as an adjustable constant (k). In effect, it takes the place of ρ in the LEP method. The LEPS method is little if at all superior to the LEP method in predicting relative activation energies. Its merits are that it leads to potential energy surfaces without basins or cavities and, as regards H_3, it agrees better with the *ab initio* calculations. Its status as a 'semi'-empirical method, however, is dubious. More authentic mixtures of theory and empiricism have also been developed, mainly with the idea of eliminating some of the approximations involved in the London equation. They are necessarily rather sophisticated.[(9)] The surface for H_3 shown in Fig. 1–4 (which has no basin) was obtained by a method of this type.

When we come to consider *empirical* procedures we find a certain ambiguity in what is meant in the present context by 'empirical'. This arises from the different purposes for which potential energy calculations are made. One purpose with which we are already familiar from the semi-empirical methods is to relate the potential energy of a real reaction system to equilibrium properties of the reactants. In this sense, empirical means that the relation achieved has less to do with quantum-mechanical theory than the 'semi'-empirical methods. Another purpose is to investigate generally how the dynamics of reaction depend upon specific features of the potential energy surface; and for this it has been found convenient to construct surfaces from 'empirical' potential energy functions bearing little or no relation to theory. An example of this type of operation will be given at the end of this section.

A well-known expression of the first type is that proposed by Evans and Polanyi[(10)] to relate the activation energies of exothermic A+BC reactions to their equilibrium enthalpy changes. From qualitative considerations based on the potential energy profiles of similar reactions, A+BC, A+BD, etc., Evans and Polanyi were led to the relation

$$E = c+\alpha(\Delta H) \qquad 5\text{–}12$$

where ΔH is negative and c and α are empirical constants which vary from one series of reactions to another. This is sometimes known as the Polanyi–Semenov relation.[(11)] It may be expressed alternatively in terms of the dissociation energy of the bond attacked by the atom or radical A:

$$E = \alpha D(\text{B—R})-\text{c} \qquad 5\text{–}13$$

The relation holds well when the molecular reactants belong to the same homologous series. This is shown in Fig. 5–1. But it fails, as might be

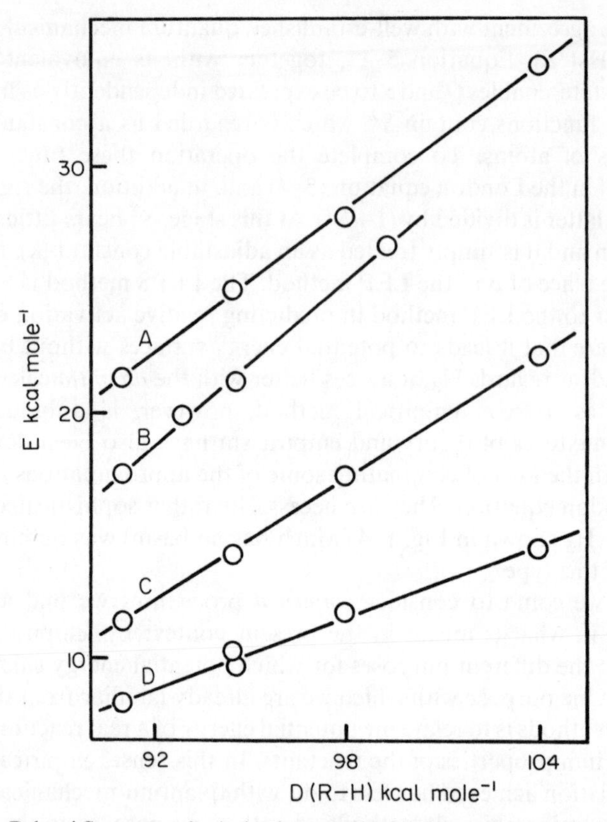

Fig. 5–1 Polanyi Semenov rule. Activation energies for the reaction $Y+RH \rightarrow YH+R$ graphed against the dissociation energy of R–H, where R is an alkyl radical, A, Y = I, B, $Y = NF_2$; C, Y = Br; D, $Y = CH_3$. The values of E in curve C are displaced by $+4\,\text{kcal mole}^{-1}$ (Kerr[13]). (Copyright 1966 by the American Chemical Society, reprinted by permission.)

expected, when there are notable differences in R. Within this limited range, it is useful for evaluating unknown activation energies from thermochemical data and vice versa.

A more ambitious procedure for calculating the heights of potential barriers is the bond–energy–bond–order (BEBO) method developed by Johnston and Parr.[12] By the criterion of content of quantum-mechanical theory, it is to be classified as empirical; but, unlike the Polanyi–Semenov relation, it makes use of information entirely from outside the field of kinetics, and there are no adjustable parameters. In the original form given below, it is applicable only to hydrogen transfer reactions (i.e. B = H), but a modified form is available for the transfer of other atoms.[14] By the standard of comparison with experiment, it has been uniquely successful. A complete potential energy surface is not attempted. The object rather is to calculate the change in potential energy along the reaction coordinate as the atom or radical A approaches end-on to the molecule H—R and R recedes. The essential point is the definition of the reaction coordinate. This is arrived at by noting that in the complex A ... H ... R, A and R are bound to H by 'partial bonds', otherwise bonds of fractional order in the sense of Pauling's concept of bond order.[15] Since the energy required to sever the bond H—R is mostly supplied by the energy released in forming A—H, the changes in order occurring in both bonds during reaction must be strongly correlated. The basic assumption is that the path of least energy from reactants to products is always where the sum of the bond orders is equal to unity.

The energy of each partial bond is related to its order n by an empirical relation of the form

$$E = Dn^p \qquad 5\text{–}14$$

where D is the energy of the single bond, for example A—H. The value of p is obtained from empirical relations between bond energy, order, and length and is characteristic of the particular bond: 1.04 for H ... H, 1.09 for CH_3 ... H and so on. The total potential energy is the algebraic sum of the energies of the bonds A ... H and H ... R and the repulsive energy between A and R. The last arises from the opposed spins of the electrons on A and R. It is estimated from an anti-Morse function similar to equation 5–11, the distance r_{AR} being related to n via the relation between bond order and length mentioned above. Taking into account the assumption that $n_{HR} = (1-n_{AH})$, the potential energy is expressed finally as a function of n_{HR} $(\equiv n)$:

$$E_{AHR} = D_1(1-n^{p_1}) - D_2(1-n)^{p_2} + D_3B(n-n^2)^{0.26\beta_3} \qquad 5\text{–}15$$

where the 1, 2, and 3 refer to the single bonds H—R, A—H, and A—R respectively, and $B \equiv 0.5 \exp\left[-\beta_3(r_{0,1}+r_{0,2}-r_{0,3})\right]$. When the function E_{AHR} is evaluated for incremental values of n between 1 and 0, it is found to pass through a maximum value and this is regarded as the barrier height for the reaction (E_c). In general, the maximum values of E_{AHR} agree with the experimental values of E_a within the reasonable limits of 2 or 3 kcal mole^{-1}. Some comparisons are given in Table 1–1.

As noted earlier, other procedures conventionally labelled 'empirical' have a different purpose from the above. This is to learn how the topography of a potential energy surface—the position and dimensions of the barrier, the contours of the valleys and so on—can influence reaction trajectories, a type of exercise to which we shall return in Section 5–1–3. For such it has been found useful to calculate an artificial surface, or rather a set of surfaces, using a purely empirical function with parameters that can be adjusted to produce the required features in a straightforward and systematic way. Wall and Porter,[16] for example, used a function which, in effect, rotates the Morse function (5–9) through 90° around a point situated at large values of r_{AB} and r_{BC} (that is, at a point beyond the top right-hand corner of Fig. 1–4). By making the Morse parameters depend on the angle of rotation, the channel so generated could be distorted so as to produce a saddle point at any desired value of r_{AB}, r_{BC}

or to change the shape of the surface in other ways. Such a surface, of course, bears at most a tenuous relation to any actual reaction.

5–1–2 Tests of activated-complex theory

The ultimate test of activated-complex theory, as of any theory of reaction rate, must be to establish how far it is capable of predicting or reproducing the observed rates. It has to be recognized, however, that a test of this kind is handicapped from the beginning: the theory relies heavily on knowledge of the potential energy surface and, as we have seen, in practice there is no definitive way of calculating potential energy surfaces. A discrepancy between theory and experiment, therefore, is not easily recognized as due to a deficiency in the theory or in the potential energy surface or both. There is, however, a second kind of test in which the influence of the potential energy surface, though by no means removed, is at a minimum. This is an examination of kinetic isotope effects; that is, the effects on the reaction rate of substituting an atom, in one or other or both of the reacting species by an isotope. In the following, we shall consider the results of some theoretical calculations of rates and isotope effects.

The transition-state expression for the rate constant of the reaction $A+BC \rightarrow AB+C$ is

$$k = \frac{\Gamma^* l^\ddagger \bar{k} T}{h} \frac{Q^\ddagger_{ABC}}{Q_A Q_{BC}} \exp[-(\Delta E_z + E_c)/RT] \qquad 5\text{–}16$$

$$= C(T) \exp[-E_0/RT] \qquad 5\text{–}17$$

where the terms are as defined in Section 1–4. This expression includes the quantum ‘correction’ for tunnelling through the potential energy barrier (Γ^*), but otherwise assumes the transmission coefficient to be unity.

The kinetic isotope effect is defined as the ration k/k', where k' is the rate constant observed when an atom in A or BC is replaced by an isotope. We are concerned with replacement of an atom which is directly involved in the reaction; that is, with primary isotope effects. For these considerations it is convenient to regroup the terms in equation 5–16 as follows:

$$k = \frac{\Gamma^* l^\ddagger Q^\ddagger_{ABC}}{Q_A Q_{BC}} \exp[-\Delta E_z/RT] \left\{ \frac{\bar{k}T}{h} \exp[-E_c/RT] \right\} \qquad 5\text{–}18$$

Since neither the geometry nor the potential energy of the reacting species is changed by isotopic substitution, the potential energy surface remains the same. The terms in the large brackets in equation 5–18 are cancelled in the ratio k/k' and the effect of the substitution is exerted entirely through the change in mass.§ This affects all the remaining terms, but the main effect is likely to be exerted through changes in the zero-point energies; that is, through ΔE_z. This is best explained by referring to the

§ The details of the application of activated-complex theory to isotope effects are given by Bigeleisen and Wolfsberg.[17]

dissociations of H_2 and D_2. The potential energy 'surface' (Fig. 1–2) is the same for both processes and therefore the vibrational force constant is the same. Because of the heavier atoms, however, the vibration frequency, and consequently the zero-point energy, is less in D_2 than in H_2. Thus, to scale the same potential energy barrier, the D_2 molecule must begin from a lower level; the barrier height is effectively higher and the rate of dissociation correspondingly smaller. Returning to A+BC, if, as is usual, the bond B—C is merely weakened rather than broken in the activated complex, the substitution of a heavier isotope for B will still produce a lower rate, though to a smaller extent. Conversely, if B—C were to be strengthened in becoming incorporated into the activated complex, the heavier molecule would react the faster: there would be an 'inverse' isotope effect. Most isotopic work has been carried out with hydrogen–deuterium substitution. Unfortunately, the advantage of the large mass ratio is partly counteracted by the necessity to consider tunnelling, which becomes increasingly important with decreasing atomic mass.

It was noted in Section 1–4–2 that, in principle, the potential energy surface contains all the information required to evaluate equation 5–16. Considering a linear-activated complex, the moment of inertia required to evaluate the partition function $Q^{\ddagger}_{ABC}$ is fixed by the coordinates $r^{\ddagger}_{AB}$, $r^{\ddagger}_{AC}$ of the saddle point, and the three vibration frequencies can be calculated from force constants derived from the curvature of the surface in the immediate vicinity of the saddle point. The latter include the negative force constant and imaginary asymmetric stretching frequency required to evaluate Γ^* by equation 1–60 or other less approximate means.[18] Nevertheless with the exception of calculations by the BEBO method, to which we shall return later, so far it has not been possible to calculate a reaction rate in agreement with experiment without incorporating in the calculation some information derived from experimental rate measurements. The information is generally used to adjust the potential energy surface so as to make the calculated Arrhenius activation energy E_a equal to the experimental value. The theoretical expression 5–16 is then tested by the manner in which it reproduces the experimental pre-exponential factor.

This procedure was followed by Westenberg and de Haas[19] in considering the three-atom reactions listed in Table 5–2. They calculated the potential energy surfaces by the LEPS procedure and used the Sato parameter k as the (only) adjustable parameter for each reaction. The activated complex was assumed to be linear and tunnelling was not considered. Any value of k yields a value of E_c as well as values of $C(T)$ and ΔE_z via the partition functions and vibration frequencies. A value of E_a was found from these by applying equation 1–64a at the mid-point of the experimental temperature range; and the value of k was varied by a computer program until the calculated value of E_a matched the experimental. At this point, the absolute value of A was calculated from $C(T)$ by means of equation 1–64b. Reference to Table 5–2 will show that, on the average, the results are within a factor of 2 of the experimental values.

The closest approach to a completely *ab initio* rate calculation so far is that made by Shavitt[22] for the series of isotopic reactions:

$$H + H_2 \rightarrow H_2 + H \qquad \text{5–IIIa}$$

$$D + D_2 \rightarrow D_2 + D \qquad \text{5–IIIb}$$

$$H + D_2 \rightarrow HD + D \qquad \text{5–IIIc}$$

$$D + H_2 \rightarrow HD + D \qquad \text{5–IIId}$$

The *ab initio* potential energy surface calculated by Shavitt *et al.*[4] for linear H_3 was used. This, however, has a barrier height of 11.0 kcal mole^{-1}, which is too high to be compatible with the observed rates. To overcome the difficulty, the height of the barrier and of the surface along the reaction path was reduced uniformly to give agreement with experiment. The remainder of the surface, however, was left unchanged. Thus the moment of inertia and the two real vibration frequencies of the complex for each reaction were derived directly from the original surface. The reduction of the potential energy along the reaction path (by 11 per cent) was the only empirical element in the calculation. The result for reaction 5–IIId, illustrated in Fig. 5–2, agrees well with the experimental behaviour over

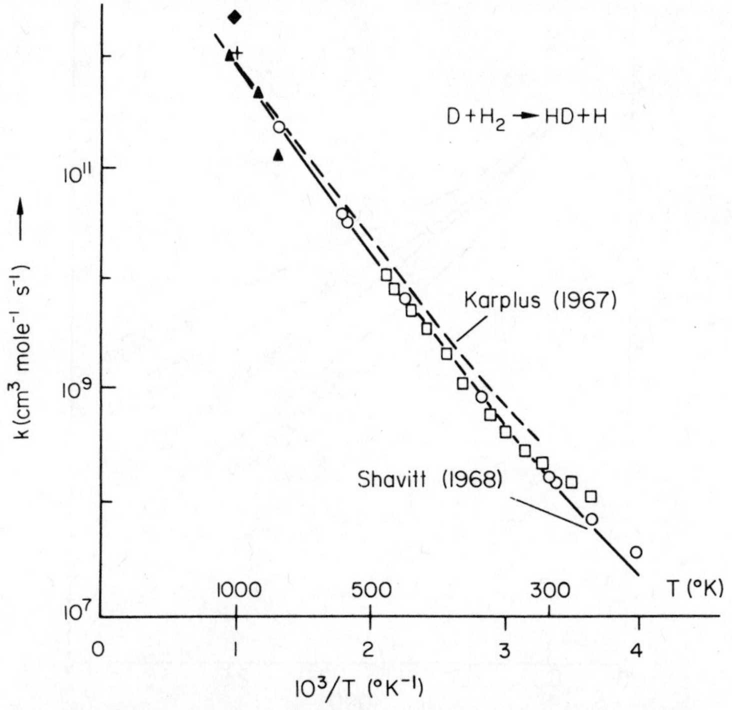

Fig. 5–2 Comparison of theoretical with observed rate constants for the reaction $D + H_2 \rightarrow HD + H$. The *points* are experimental values obtained in several investigations. The *continuous curve* represents values calculated by Shavitt[22] by activated-complex theory. The *broken curve* represents values obtained by Karplus and Porter[23] by Monte Carlo averaging of calculated $D + H_2$ trajectories (Wagner and Wolfrum[24]).

most of the temperature range. Considerably more impressive, however, is the fact that the rates for reactions 5–IIIa, b, and c, calculated without further adjustment to the potential energy surface, also agree as well or almost as well with the experimental. In other words, the calculations reproduce the observed isotope effects.

At the other extreme from Shavitt's calculations with regard to empiricism are rate calculations by the BEBO method. The BEBO potential energy profile, besides providing a value for E_c, contains enough information about the activated complex for absolute rates to be computed via equation 5–16.[12, 25] The method is unique in that the rate constants are calculated—often with remarkable accuracy, as Fig. 5–3 shows—from information drawn entirely from non-kinetic properties of the reactants.

A few words should be said about the rather difficult question of tunnelling. To a first approximation (equation 1–60), $\Gamma^* - 1$ is inversely proportional to T^2. The influence of tunnelling should therefore be manifested by an upwards curvature of the Arrhenius plot which becomes more prominent at low temperatures. Such an effect has been found

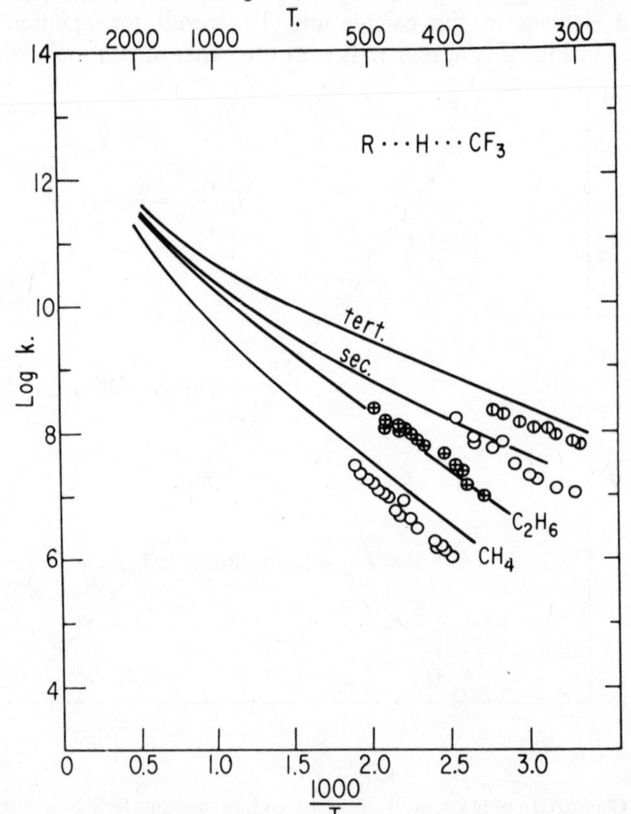

Fig. 5–3 Rate constants calculated by the BEBO method for hydrogen transfer reactions of CF_3 radicals with hydrocarbons (*lines*) compared with observed values (*points*); sec. and tert. refer to the corresponding H atoms of *n*-butane and *iso*-butane (Johnston and Parr[12]). (Copyright 1963 by the American Chemical Society, reprinted by permission.)

experimentally in two independent investigations of the isotopic $H+H_2$ reactions at temperatures below 60 °C.[19,26] At higher temperatures, the Arrhenius curves become indistinguishable from straight lines, but Shavitt's theoretical treatment using the *ab initio* potential energy surface requires appreciable values of Γ^* over the whole temperature range (Fig. 5–2). On the other hand, the inclusion of tunnelling in Westenberg and de Haas's LEPS treatment of the same reactions would destroy the excellent agreement with experiment they obtained. As against this, their treatment was based on the assumption that tunnelling is negligible at the higher temperatures, which is at variance with Shavitt's result. More telling evidence against tunnelling, however, is the experimental finding that the $H+D_2$ reaction shows more pronounced non-Arrhenius behaviour at low temperatures than the $D+H_2$ reaction,[19] whereas any theory of tunnelling requires the opposite effect. On the contrary, Sharp and Johnston[20] concluded from their LEPS study of the isotope effect in reaction 5–II that tunnelling plays a 'vital role'. To summarize a confused situation: on the whole the reality of tunnelling seems to be established for reactions of H and D atoms but its precise significance is obscure.

What then, finally, is the quantitative status of activated-complex theory? Basically the answer is that straightforward application of the theory with LEPS or BEBO potential functions can generally be expected to give *A* factors within an order of magnitude of the observed values and sometimes much better. This can usually be improved by careful attention to tunnelling and various refinements.[20,27] When this is done, discrepancies of the order of a factor of 2 may remain; and for these we may adopt a remark made by Shavitt:[22] '. . . it is difficult to say whether [these are] due to . . . experimental errors, to inappropriate potential-surface parameters, to an inadequate treatment of tunnelling, or to some fundamental deficiency of transition-state theory.'

5–1–3 Reaction dynamics

The elementary dynamical picture of bimolecular reaction presented in Section 1–3–1, in so far as it used the potential energy surface at all, simply identified the height of the energy barrier with the Arrhenius activation energy and then resorted to the frequency of 'hard-sphere' molecular collisions to calculate the reaction rate. But the behaviour of any particular pair of molecules propelled into mutual proximity—that is, their *trajectory*—is determined by the forces impressed upon their constituent atoms in accordance with the laws of dynamics. The magnitudes and directions of the forces at any time during the encounter are expressed by the changes in potential energy with respect to the appropriate spatial coordinates. Hence they can be derived from the potential energy (hyper) surface. In other words, if the potential energy surface is known, the complete trajectory can be calculated by solving the equations of motion, provided the initial relative velocity and other conditions of the two molecules are specified. Furthermore, if the calculations can be averaged correctly over all relevant initial conditions, the proportion of reactive encounters, and therefore the reaction rate, can be estimated.§ Electronic computers make such calculations practicable, and collision theory currently prospers under the new title of *reaction dynamics* or, more generally, *molecular dynamics*. It has been much stimulated by experiments with molecular beams and chemiluminescent emission of the kind described in Chapter 4. Typical calculated trajectories for non-reactive and reactive collisions between H and H_2 are shown in Fig. 5–4 and we shall refer to them again shortly.

A full reaction rate calculation as described in the previous paragraph requires the computation of thousands of trajectories and is a very large undertaking. However, it has been carried through by Karplus and collaborators for the $H+H_2$ reaction and its isotopic variants.(23,28) The semi-empirical surface partly illustrated in Fig. 1–4 was used and the remarkable agreement with experiment shown in Fig. 5–2 was obtained. In view of the uncertainty surrounding the calculation of potential energy surfaces, however, the main objective of trajectory calculations at present is not so much to reproduce reaction rates as to discover more about the details of reactive collisions. One method of achieving this is to study how altering the initial conditions of the molecules in various ways affects their trajectory 'over' a given potential energy surface. Much information of this kind was obtained in the course of the study of the $H+H_2$ system just mentioned, and reference will be made to some of this presently. The converse method is to investigate how the trajectories are affected by changing the shape of the potential energy surface when the initial conditions are fixed. This procedure becomes particularly fruitful when the results are compared with the behaviour observed in molecular-beam and chemiluminescence experiments. In this case, of course, it is really the nature of potential energy surfaces that is being investigated.

A most significant fact with regard to the $H+H_2$ reaction is already evident from Fig. 5–4b. It will be seen that the time taken for the hydrogen atom to transfer from one particle to the other is shorter than the period of a vibration.¶ Since the period of a rotation is longer than that of a vibration, it is clear that the collision complex is too short-lived for the energy to become equilibrated among its internal degrees of freedom. The reaction is of the *direct* type, of which many examples have been observed experimentally by molecular-beam technique (Section 4–5).

The reaction cross-section, impact parameter, and threshold energy

For trajectory and molecular beam studies it has been found convenient to introduce the concept of *reaction cross-section* as a measure of the

§ In activated-complex theory the averaging is done automatically via the partition functions.

¶ It is worth noting that this is what is assumed by activated-complex theory (when the transmission coefficient is near unity).

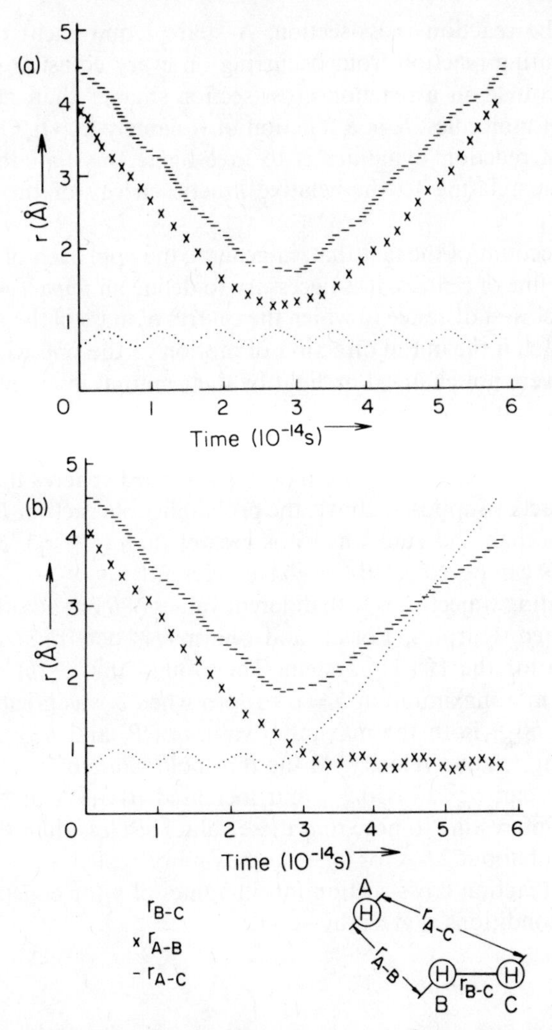

Fig. 5–4 Typical trajectories for collisions between a hydrogen atom and a hydrogen molecule (a) non-reactive, (b) reactive. (Calculations by Karplus, Porter, and Sharma[28]; after reference 24.)

probability of reaction. Its definition can be seen by analogy with elementary kinetic theory. According to this, the number of collisions in unit time between hard-sphere molecules **A** and **B** approaching one another with the relative velocity **v** with respect to their centres of mass is given by

$$z_{\mathbf{v}} = \pi(r_{\mathbf{A}}+r_{\mathbf{B}})^2\mathbf{v}[n_{\mathbf{A}}][n_{\mathbf{B}}] \qquad 5\text{–}19$$

where $r_{\mathbf{A}}$, $r_{\mathbf{B}}$ are the molecular radii and $n_{\mathbf{A}}$, $n_{\mathbf{B}}$ the concentrations of **A** and **B**. If every collision led to reaction, $z_{\mathbf{v}}$ would be the reaction rate. Hence, in this case,

$$\frac{\text{Reaction rate}}{\mathbf{v}[n_{\mathbf{A}}][n_{\mathbf{B}}]} = \pi(r_{\mathbf{A}}+r_{\mathbf{B}})^2 = S_r \qquad 5\text{–}20$$

where S_r is the reaction cross-section. A steric requirement or energy barrier preventing reaction from occurring on every collision would be reflected, of course, in a reaction cross-section smaller than $\pi(r_A+r_B)^2$; and, with real molecules, S_r is a function of **v**, namely $(S_r(\mathbf{v}))$. One of the main tasks of reaction dynamics is to investigate this function or the equivalent one relating to the relative kinetic energy of the colliding particles, $S_r(E)$.

To take account of the fact that, in general, the approach of **A** to **B** is not along the line of centres, it is necessary to define an *impact parameter*, b. This is the closest distance to which the centres of mass of the molecules would approach if the initial direction of motion of the one with respect to the other were not changed in flight by their mutual interaction. Thus b is one of the initial conditions of the trajectory; it is not a property of the molecules. Head-on collision corresponds to $b = 0$ and grazing incidence between hard spheres to $b = (r_A+r_B)$. For hard spheres that always react on contact as supposed above, the probability of reaction P_r is unity when b is less than and zero when it is greater than (r_A+r_B). Naturally, real molecules can be expected to behave differently.

By computing trajectories with different values of b but all other initial conditions fixed, Karplus, Porter, and Sharma[28] determined P_r as a function of b for the $H+H_2$ system. They found that $P_r(b)$ decreased uniformly from a maximum at $b = 0$ to zero when b was greater than a certain value b_{max}. Both the maximum values of P_r and b_{max} increased with increasing relative velocity. At the threshold value of v required for reaction to occur, b_{max} was 0.5 Å and increased to 1.3 Å at twice this velocity. It is interesting to note that these values are less than the kinetic theory value of about 2.5 Å for (r_A+r_B) shown in Fig. 1–1.

The total reaction cross-section for all values of b for constant v and other initial conditions is given by

$$S_r = 2\pi \int_0^{b_{max}} P_r(b) b\, db \qquad 5\text{–}21$$

as can easily be verified by considering the case of hard spheres ($P_r = 1$). Karplus and collaborators studied the dependence of S_r on the relative velocity, as well as on the initial vibrational and rotational states (v, J) of the H_2 molecule. Although the trajectories were based on classical mechanics, the molecule was assumed to be restricted to its quantum-mechanical rotational-vibrational states. The other initial parameters—mutual orientation of H and H_2, vibrational phase of H_2, and so on—were averaged by computing about 300 trajectories for each set of constant values of **v**, v, and J. The values of the other parameters just mentioned were chosen at random for each trajectory by 'Monte Carlo' methods[29] which take into account their probability distributions. In this way, S_r was found to increase continuously from zero below a threshold value of **v** to an asymptotic limit at higher values.

It is of great interest first to compare the relative kinetic *threshold energy* E_θ (corresponding to the threshold $\mathbf{v}$) with the barrier height E_c (9.1 kcal mole^{-1}) and then to examine the effects of vibration and rotation on E_θ. With hypothetical vibrationless H_2 molecules, E_θ was found to exceed E_c by 0.3 kcal mole^{-1}.§ When 6.2 kcal of vibrational energy—corresponding to the zero-point vibration—was added to the H_2 molecule, the threshold kinetic energy fell to 5.7 kcal. Thus some, but not all, of the vibrational energy is available for surmounting the barrier. The remainder passes into the vibrational modes of the collision complex. Some further calculations indicated that raising the vibrational state of the H_2 to $v = 1$ reduced E_θ markedly. Increasing the rotational excitation, on the other hand, showed that rotational energy is not available for crossing the barrier. Later on we shall see that, in the general case, the relative effects of vibrational and translational energy depend on the shape of the potential energy surface.

The rate constants graphed in Arrhenius fashion in Fig. 5–2 were obtained by integrating S_r over the Boltzmann distribution functions for $\mathbf{v}$, J, and v at each temperature, the lowest value of v being the zero-point value. Enthusiasm about the agreement with experiment has to be tempered by the considerations that the potential energy surface used may not be correct and that the quasi-classical calculations took no account of zero-point energy in the collision complex or of quantum effects leading to tunnelling through the potential barrier. The corresponding errors may be negligibly small, or, on the other hand, they may be self-cancelling.

Effects of the potential energy surface

It was observed in Sections 4–4 and 4–5 that chemiluminescence and molecular-beam methods can provide much detailed experimental information about the dynamics of A+BC reactions. Among other accessible effects are the relative distribution of exothermic reaction energy between the translational and internal degrees of freedom of the products, the preferred directions taken by the products, and the relative effects on the probability of reaction of translational and vibrational energy in the reactants. The proper interpretation of such information can lead to unprecedented insight into the reaction dynamics. A productive line of investigation pursued in several laboratories is to attempt to relate effects of this kind to potential energy surfaces by means of trajectory calculations. The (long-term) objective is to define the features of the potential energy surface which reproduce the experimental observations; in other words to 'fit the surface to the data'. This is a quasi-experimental operation, for, apart from gross effects, there is no way of determining the effects of changing the potential energy function except by calculating statistically significant numbers of trajectories before and after the change and comparing the results.

§ The difference is genuine and is to be attributed to the fact that the reaction path tends to mount the outside wall as it turns the corner of the potential energy surface. In colder climates, this is known as the 'bobsled effect'.

One of the first positive results to be obtained relates to the manner of energy release. For highly exothermic A + BC reactions with little or no activation energy—$H+Cl_2$, $Na+ICH_3$, etc.—there is a significant correlation between the proportion of the energy released as vibrational energy of AB and the position on the potential energy surface where the energy is released; that is, where the reaction path leads 'downhill'. When this part of the surface is located as shown in Fig. 5–5a, the greater proportion of the energy is released as vibration of AB. But when the surface first inclines downhill in the exit direction, as in Fig. 5–5b, the energy is mainly released as relative kinetic energy of AB and C. This applies particularly when the mass of atom A is much less than the masses of B and C. The former type of surface is designated *attractive* and the latter *repulsive*, and the terms are also applied to the nature of the energy release.

J. C. Polanyi and his collaborators established the correlation just mentioned for a hypothetical A + BC reaction by creating a 'spectrum' of potential energy surfaces ranging from highly attractive ('early downhill') to highly repulsive ('late downhill').[30] A modified LEPS function was used and three parameters akin to S^2 in equation 5–6 were adjusted to produce the desired features. The Morse parameters assigned to AB and BC, however, were kept constant. Thus the surfaces differed, but the 'reactants' remained the same. In an application to experimental data obtained from chemiluminescence studies, the relatively high proportion of kinetic energy released by halogen reactions of the type

$$H+X_2 \rightarrow HX+X \qquad \text{5–IV}$$

was able to be attributed to the influence of a repulsive energy surface. Similarly Blais and Bunker[31] discovered that the opposite effect—energy largely released as internal energy of AB—found with the alkali–metal reaction,

$$K+ICH_3 \rightarrow KI+CH_3 \qquad \text{5–V}$$

in molecular beams could only be reproduced by the trajectories on the basis of an attractive potential energy surface.

It seems natural that energy released as the atom A is attracted to B should appear as relative motion of AB (i.e. vibration) and that energy released by the repulsion of C from B should appear as translation of C away from AB. However, when A is not of smaller mass than B and C as supposed previously, its approach is slower and C tends to move away from B before A and B have reached their equilibrium separation. Thus the trajectory tends to 'cut the corner' of the potential energy surface and the effect is to generate vibration in AB even on a repulsive surface. This is known as *mixed* energy release.[30]

It appears that the attractive-repulsive character of the potential energy surface also has a strong effect on the directions taken by the

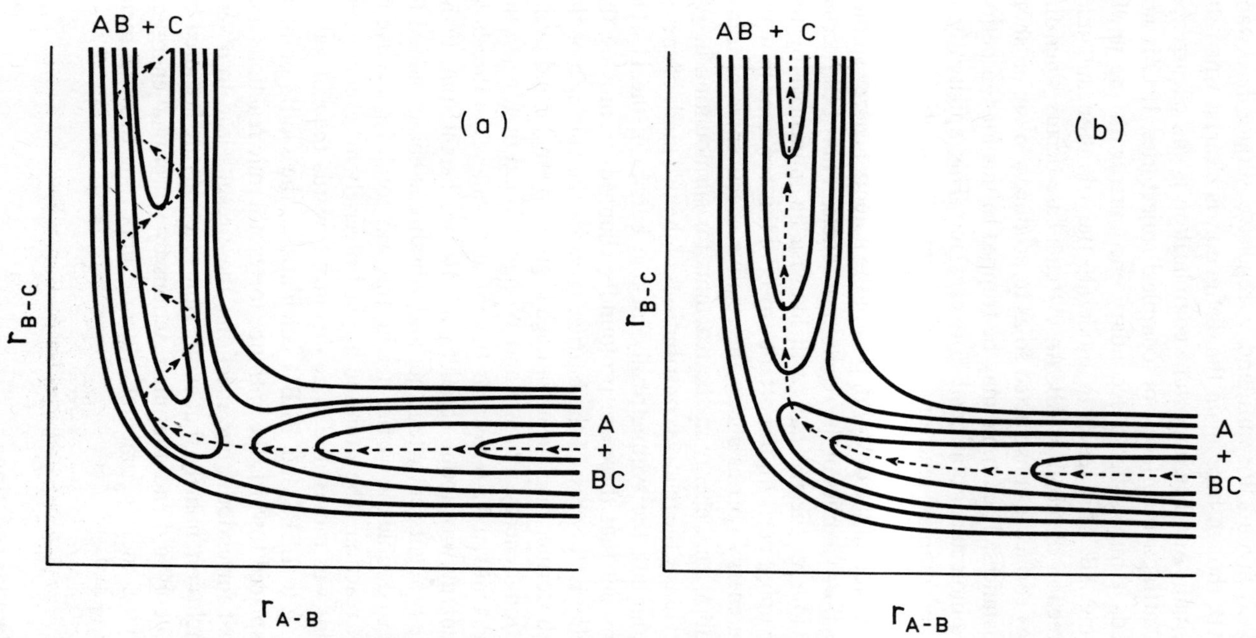

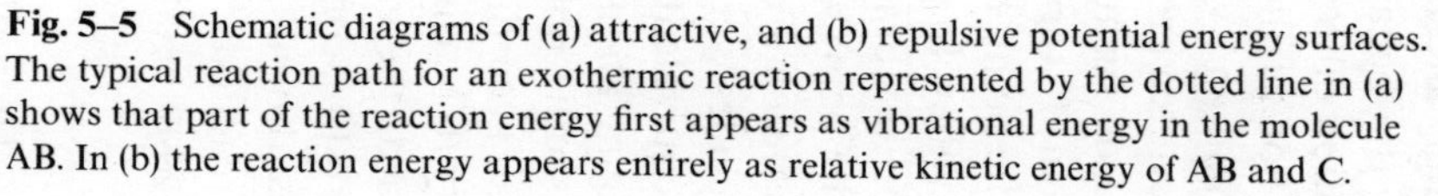

Fig. 5–5 Schematic diagrams of (a) attractive, and (b) repulsive potential energy surfaces. The typical reaction path for an exothermic reaction represented by the dotted line in (a) shows that part of the reaction energy first appears as vibrational energy in the molecule AB. In (b) the reaction energy appears entirely as relative kinetic energy of AB and C.

products with respect to the original line of collision—the centre-of-mass scattering angles to which reference was made in Section 4–5.[32] An attractive surface favours forward and a repulsive surface backward scattering of AB; though here again the distinction is clearest when the mass of A is small. Another important complication is the occurrence, principally on attractive surfaces, of 'complex' trajectories. If C is not rapidly repelled, it may suffer a secondary encounter with the newly formed vibrating AB. This tends to moderate the vibration and may reverse the direction of recoil of AB. An extreme case occurs when the potential energy surface is constructed so as to produce a 'basin' at short internuclear distances. The atoms may be trapped in this for the period of one or more molecular rotations before escaping. This, apparently, is the origin of the 'complex' reactions giving symmetrical scattering mentioned in Section 5–4.

Since the position of the downhill part of the potential energy surface strongly affects the dynamics of the reaction products, it may be expected that the behaviour of the *reactants* will likewise be influenced by the position of the uphill part. An important aspect of this concerns how the location of the energy barrier affects the relative effectiveness of translational and vibrational energy in the reactants for surmounting it. We have noted previously the relative contributions of the two kinds of energy towards surmounting the symmetrically located barrier of the $H+H_2$ system. It turns out that these are profoundly changed by moving the saddle point, that is, the *crest* of the barrier, towards the entrance or the exit valley. In the former case, the advantage is given to kinetic and in the latter to vibrational energy. Polanyi and Wong[33] found that, when the saddle point was moved 0.3 Å towards the entrance valley, the threshold translational energy was about equal to the barrier height; but, if the molecules were brought gently together and vibrational energy added to BC equivalent to double the barrier height, this was totally ineffective in bringing about reaction. The situation applied precisely in reverse when the saddle point was moved the same distance towards the exit valley. Since, according to LEPS and BEBO calculations, the saddle point is likely to be displaced towards the entrance for exothermic reactions and towards the exit for endothermic reactions, these results lead to a very interesting conclusion; namely, that exothermic reactions are favoured by translational energy and endothermic reactions by vibrational energy in the reactants.[34] Some experimental evidence supporting this was mentioned in Section 4–4–2.

5–2 Unimolecular reactions

It was observed in Section 1–3–2 that the broad facts relating to unimolecular reactions

$$A \rightarrow \text{Products}$$

are explained by the Lindemann–Hinshelwood mechanism:

$$A + A \rightarrow A^* + A \tag{a}$$

$$A^* + A \rightarrow A + A \tag{b}$$

$$A^* \rightarrow \text{Products} \tag{c}$$

where the molecules A* contain at least the critical energy E^* necessary for reaction.§ According to the theory, at sufficiently high pressures of A, the equilibrium concentration of A* $(= K[A])$ is perturbed to a negligible extent by reaction (c) and the unimolecular rate constant is given by

$$(k_U)_\infty = k_c K \qquad 5\text{–}22$$

(see equation 1–21); at lower pressures, [A*] becomes less than the equilibrium concentration and k_U 'falls off' with decreasing pressure, in the limit becoming proportional to [A] (equation 1–24). Closer examination of the fall-off region, however, shows that the mechanism as it stands above, though correct in outline, is only an approximation to the actual situation. Rearrangement of equation 1–23 for the fall-off region yields the relation

$$\frac{1}{k_U} = \frac{1}{k_c K} + \frac{1}{k_a[A]} \qquad 5\text{–}23$$

Hence a graph of $1/k_U$ against the reciprocal of the pressure should be linear. In fact, the experimental result is invariably a curve, as is shown schematically in Fig. 5–6; k_U falls off less steeply with decreasing pressure from its 'high-pressure' value than is predicted by equation 5–23.

The origin of the discrepancy lies in an assumption tacitly made with regard to the rate of reaction (c). The basic idea of the theory advanced in Section 1–3–2 is that k_c expresses the rate at which the critical energy, after having been received by the molecule by collision, becomes concentrated into producing the particular distortion of the molecular structure which leads to reaction. The relevant energy is vibrational though this need not exclude a contribution from the rotational energy of the molecule. The theory assumes that k_c is independent of the energy content of the molecule so long as it is equal to or greater than the critical value; in other words, the chance of the critical amount of energy appearing in the requisite location or configuration is supposed to be the same whether the molecule as a whole contains exactly this amount of energy or vastly more. This is most improbable. It is more reasonable to suppose that the chance will become greater the more the total internal energy E exceeds the critical value; that is, to assume that k_c is an increasing function of the total energy $(k_c(E))$ when $E \geqslant E^*$.

To derive an expression for k_U on this assumption we must note that the concentration of A* molecules possessing a particular energy E is also

§ The symbol * is used here to indicate vibrational (or rotational) excitation. The change from † used in Chapter 4 is made simply to conform with the different conventions which unfortunately have become current in the different fields of study.

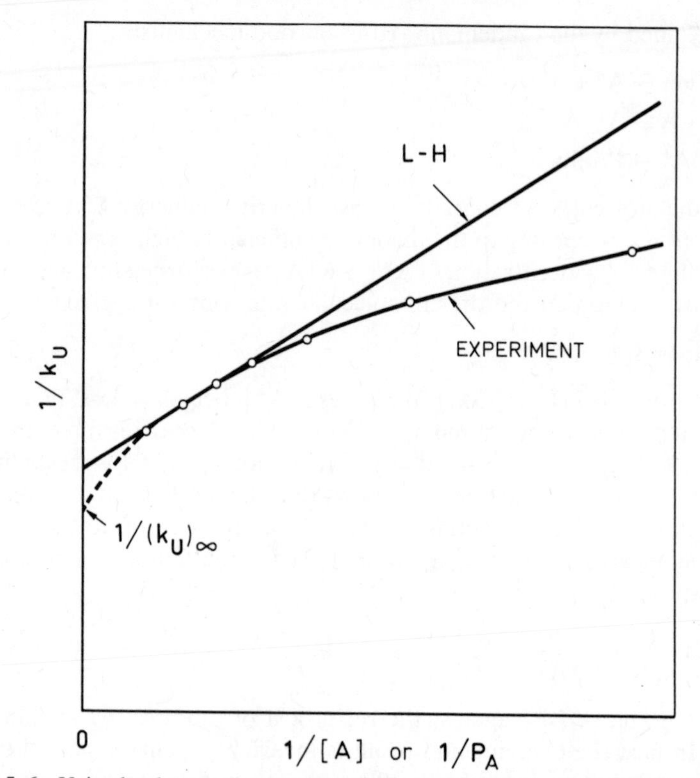

Fig. 5–6 Unimolecular reactions. Graphs of $1/k_U$ against $1/[A]$ or $1/P_A$ as predicted by the simple Lindemann–Hinshelwood mechanism and as typically found experimentally (schematic).

a function of E. Consider first the high-pressure rate constant $(k_U)_\infty$. Following equation 1–16, we have for the equilibrium fraction of molecules with energy between E and dE

$$\frac{d[A^*]_E}{[A]} = K(E)\,dE \tag{5–24}$$

The contribution of this fraction of the energized molecules to the rate constant is obtained by multiplying equation 5–24 by $k_c(E)$ as in equation 1–21. Hence the total rate constant is given by

$$(k_U)_\infty = \int_{E^*}^{\infty} k_c(E)\,K(E)\,dE \tag{5–25}$$

To evaluate the integral, expressions are needed for $k_c(E)$ and $K(E)$. The latter can be derived from statistical mechanics, but to obtain an expression for $k_c(E)$ it is necessary to assume a specific model for the reacting molecule. The original treatment, which is still useful today, is known as *Rice–Ramsperger–Kassel* (*RRK*) *theory*.[35] It has been developed in both classical and quantum forms, the one yielding the limiting rate expression of the other. Though rather less realistic physically, the classical form has been most frequently used and this will be given here. The quantum form is developed in reference (B) of the Bibliography. In either case, the theory is semi-empirical in that an adjustable parameter is required to fit the experimental results quantitatively.

5–2–1 RRK theory

The theory regards the potentially reactive molecule as an assembly of loosely coupled oscillators to which the critical energy is transmitted initially by collision in the ordinary way. 'Loosely coupled' means that the oscillators can be associated, at least in the first instance, with the normal vibrational modes of the molecule, but yet the energy is capable of flowing freely from one oscillator to another.§ All the oscillators have the same frequency. Reaction occurs when, in the course of random fluctuations of the total energy E, the quantity E^* finds itself in one particular distribution or, alternatively, and more easily visualized, in one particular oscillator. $k_c(E)$ is assumed to be proportional to the chance of this happening. It can be shown from probability theory[36] that the chance is equal to $[1-(E^*/E)]^{s-1}$. This shows, as expected, that the chance increases with the total energy available and decreases with the total number of oscillators (or 'distributions' or 'degrees of freedom') present. Hence,

$$k_c(E) = a[1-(E^*/E)]^{s-1} \qquad 5\text{–}26$$

where a is a constant for the molecule under consideration and has the dimensions of frequency.

Substitution of $k_c(E)$ together with the value of $K(E)$ from classical statistical mechanics in equation 5–25 gives

$$(k_U)_\infty = \int_{E^*}^{\infty} a\left[1-\frac{E^*}{E}\right]^{s-1} \cdot \frac{E^{s-1}\exp[-E/RT]\,dE}{(s-1)!(RT)^s} \qquad 5\text{–}27$$

(cf. equation 1–30). Integration[36] gives (remarkably)

$$(k_U)_\infty = a\exp[-E^*/RT] \qquad 5\text{–}28$$

which almost alone among theoretical rate expressions is unequivocally of Arrhenius form. Hence a is to be identified with the Arrhenius A. It should be noted, however, that equation 5–28 does not agree with the expression for $(k_U)_\infty$ derived from elementary activated-complex theory (equation 1–66), which predicts some temperature dependence for the pre-exponential factor.

The simplicity of equation 5–28 disappears when we come to the RRK expression for k_U in the fall-off region. The general form of equation 1–23 for this is

$$k_U = \int_{E^*}^{\infty} \frac{k_c(E)K(E)\,dE}{1+k_c(E)/\lambda Z[\mathrm{A}]} \qquad 5\text{–}29$$

§ This implies that the oscillators are to some degree anharmonic.

After substituting for $k_c(E)$ and $K(E)$ as previously and some manipulation, the result expressed relatively to $(k_U)_\infty$ is

$$\frac{k_U}{(k_U)_\infty} = \frac{1}{(s-1)!}\int_0^\infty \frac{x^{s-1}\exp(-x)\,dx}{1+a\{x/(x+E^*/RT)\}^{s-1}/\lambda Z[A]} \qquad 5\text{–}30$$

It is assumed that only A is present and λ is normally taken to be unity in accordance with the 'strong' collision assumption. To find $k_U/(k_U)_\infty$, the integration has to be carried out numerically for each set of values of s, E^*, [A], etc.[37] The expression is tested against the experimentally observed fall-off of k_U (see Fig. 1–6) by computing $k_U/(k_U)_\infty$ as a function of [A]. The value of $(k_U)_\infty$ is found by extrapolating a graph of the experimental values of $1/k_U$ to zero $1/[A]$ as indicated in Fig. 5–6,§ and the values of E^* and a required for equation 5–30 are obtained from the Arrhenius constants of $(k_U)_\infty$. Z is calculated from kinetic theory. This leaves the value of s to be decided. If, as originally proposed, the oscillators are identified with the normal vibrational modes, $s = 3n-6$ for a non-linear n-atomic molecule. In fact, however, substituting this value in equation 5–30 nearly always predicts too gradual a fall-off in k_U. The practice is therefore to regard s as an adjustable parameter, using trial and error to find the value which gives the best fit to experiment. An example is illustrated in Fig. 5–7. A value of s which, as an empirical parameter need no longer be integral, can usually be found which reproduces the fall-off curves within, or almost within, the experimental error.

In spite of its empirical limitations, equation 5–30 is a notable advance on the unmodified Lindemann–Hinshelwood expression 1–23, and its success supports the inherently plausible idea that the intrinsic lifetime of the energized molecules decreases with their increasing energy content. The best value of s can be considered the number of 'effective' oscillators. It is normally about half the number of normal modes¶ and hence increases with the atomicity of the molecule. As Fig. 5–7 shows, equation 5–30 predicts that, with all other factors equal—or, in fact, not too dissimilar—the fall-off region is moved to lower pressures by increasing the value of s. This agrees with the well-established experimental tendency noted in Chapter 1 for the same effect to occur as the reacting molecule becomes more complex. According to equation 5–30, the value of $k_U/(k_U)_\infty$ at any particular pressure [A] is determined by the second term in the denominator. For similar values of a and s, the fall-off region should move to higher pressures with decreasing values of E^*/T. The fact that this occurs with any single reaction with increasing values of T was noted in Section 1–3–2. It is difficult to find sufficient data to test satisfactorily the significance of E^*/T in regard to the comparative behaviour of different reactions, but the general trend appears to be in the predicted direction. A more definite confirmation of the theory, however, can be found by

§ A more accurate method will be found in reference 38.

¶ Higher values may be obtained when the molecule is small or contains few hydrogen atoms.

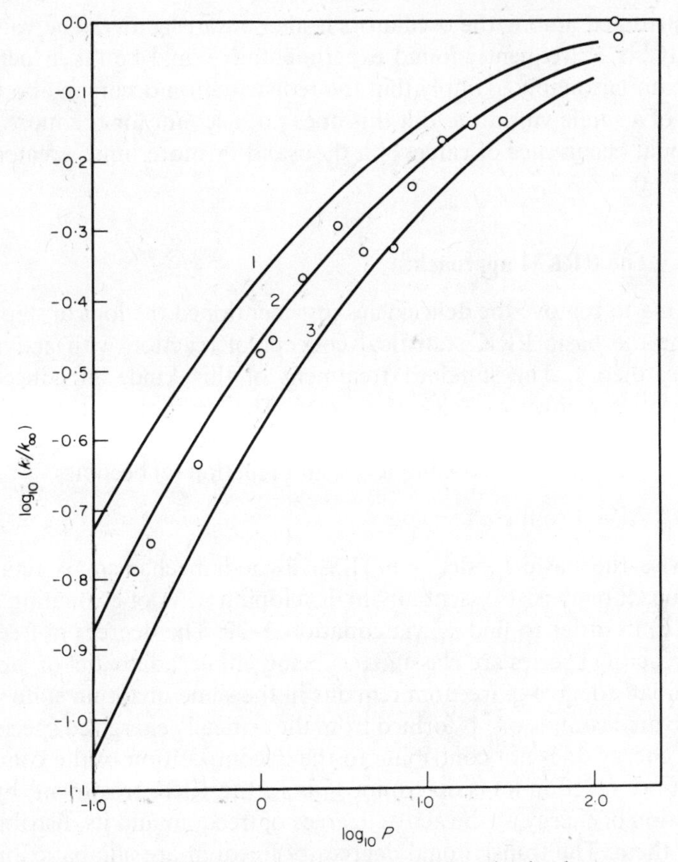

Fig. 5–7 'Fall-off' curves showing the dependence of the unimolecular rate constant on reactant pressure calculated by the RRK expression (5–30) with (1) $s = 13$; (2) $s = 12$; (3) $s = 11$ ($A = 10^{14.03}\ s^{-1}$). The points are experimental values for the reaction $C_2H_5Cl \rightarrow C_2H_4 + HCl$ at 521 °C (P in torr) (Holbrook and Marsh[39]).

considering the unimolecular endothermic decompositions of free radicals:

$$\text{RAB}\cdot \quad (+\text{M}) \rightarrow \text{R}\cdot + \text{A:B} \quad (+\text{M}) \qquad \text{5–VI}$$

In such a reaction, the energy required to break the bond R—A is partly compensated by the formation of a new bond in the stable molecule AB, and the activation energy is generally lower than that of decompositions of stable molecules of similar complexity. In agreement with the theory, the rate constants for several radicals decompositions have been found to be pressure dependent at 'abnormally' high pressures.[40]

Although the RRK theory leads to sufficient agreement with experiment to support the basic ideas behind it, its shortcomings are sufficiently obvious. The featureless coupled oscillator model is unable to be related to any specific molecule beyond what is implied by the value of s; and even here, as we have seen, there is no way of assigning this *a priori*. The significance of the constant a representing the frequency with which energy is redistributed among the oscillators is also unsatisfactory. The value of about $10^{13}\,s^{-1}$ frequently found experimentally could be taken perhaps not too unreasonably to imply that the redistribution occurs in about the period of a single vibration. But this does not account for the more than occasional occurrence of values A a thousand or more times greater (see Table 1–3).

5–2–2 The RRKM approach

In seeking to remove the deficiencies just mentioned the logical step is to combine the basic RRK statistical concept of reaction with activated-complex theory. The standard treatment of this kind, introduced by Marcus[41] is known as 'RRKM theory'. The reaction of the critically energized molecules is considered to occur via the usual quasi-equilibrium concentration of activated complexes, and reaction (c) becomes

$$A^* \rightarrow A^\ddagger \rightarrow \text{Product(s)} \tag{c}$$

Otherwise the basic Lindemann–Hinshelwood mechanism is retained. The procedure consists essentially in developing ways of evaluating $k_c(E)$ and $K(E)$ in order to find k_U via equation 5–29. The degrees of freedom of the reacting species are classified as being either 'adiabatic' or 'active'. An adiabatic degree of freedom remains in the same quantum state when the activated complex $A^\ddagger$ is formed from the critically energized species A^* and its energy does not contribute to the decomposition of the complex. The rate of reaction (c) is determined, as in the RRK treatment, by the acquisition of energy by the active degrees of freedom and its distribution among them. The translational degrees of freedom are adiabatic, but all vibrations and internal rotations are regarded as active and this may also be the case with an overall rotation. For simplicity, however, we shall assume that there are no internal rotations in either A^* or $A^\ddagger$, that all three overall rotations are adiabatic, and that A^* and $A^\ddagger$ have the same moments of inertia. In this case only the $3n-6$ vibrational modes in A^* need to be considered; and, following the usual postulate of activated complex theory, one of these becomes the reaction coordinate in $A^\ddagger$.

To derive the expression for $k_c(E)$,[42] the total energy $E^\ddagger$ of the active modes of the complex over and above their zero-point energy $E_z^\ddagger$ is equated to the sum of the energy contained in the $3n-7$ vibrations $E_v^\ddagger$ and the kinetic energy of translation $E_t^\ddagger$ across the potential energy barrier. $E^\ddagger$, which may have any value from zero to infinity, is distributed between the vibrational and translational quantum states in accordance with Boltzmann statistics. The number of translational states with energy between $E_t^\ddagger$ and $E_t^\ddagger + dE_t^\ddagger$ is regarded as equivalent to that of a particle in a one-dimensional box with the length of the potential energy barrier. The total specific rate of passage over the energy barrier—that is, $k_c(E)$—is obtained by finding the probability that the complex has kinetic energy between $E_t^\ddagger$ and $E_t^\ddagger + dE_t^\ddagger$, dividing the probability by the time taken to cross the barrier at the corresponding velocity and integrating from 0 to $E^{\ddagger}$. This gives the expression

$$k_c(E) = \frac{l^{\ddagger} \Sigma P(E_v^{\ddagger})}{hN^*(E^{\ddagger}+E_0)} \tag{5–31}$$

where $\Sigma P(E_v^{\ddagger})$ is the total number of active states of the complex at the total active energy $E^{\ddagger}$ and $N^*(E^{\ddagger}+E_0)$ is the number of states per unit energy of the active modes of the reactant molecule at the energy $(E^{\ddagger}+E_0)$.

The expression for k_U is obtained by combining the expression for $k_c(E)$ with the quantum statistical expression for $K(E)\,dE$ and substituting in equation 5–29:

$$k_U = l^{\ddagger} \frac{\bar{k}T}{h} \frac{\exp[-E_0/RT]}{Q_v} \int_0^{\infty} \frac{(\Sigma P(E_v^{\ddagger}))\exp[-E^{\ddagger}/RT]\,d[E^{\ddagger}/RT]}{1+k_c(E)/\lambda Z[\mathrm{A}]} \tag{5–32}$$

At the high pressure limit, this becomes

$$(k_U)_{\infty} = l^{\ddagger} \frac{\bar{k}T}{h} \frac{Q_v^{\ddagger}}{Q_v} \exp[-E_0/RT] \tag{5–33}$$

which is the relation given directly by activated-complex theory (cf. equation 1–66). And at the low-pressure limit, equation 5–32 reduces to

$$(k_U)_0 = \frac{\lambda Z[\mathrm{A}]\exp[-E_0/RT]}{Q_v} \int_0^{\infty} N^*(E^{\ddagger}+E_0)\exp[-E^{\ddagger}/RT]\,dE^{\ddagger} \tag{5–34}$$

which is the quantum statistical equivalent of equations 1–24 and 1–31. It will be recalled that equations 5–31 to 5–34 refer to situations where rotational degrees of freedom exert no influence. The complete expressions are given by Wieder and Marcus.[43]

Evaluation of equation 5–32 for comparison with experimental fall-off data is a complex and laborious procedure which will not be entered into here. A useful account will be found in the paper by Elliott and Frey.[44] Complete frequency assignments are required for the reactant molecule and the complex. The first is obtained from spectroscopic data, but the second is more arbitrary and depends in the first place on the choice of the reaction coordinate. The latter is made largely on intuitive grounds, but with informed intuition is less critical than might be expected. Schneider and Rabinovitch,[45] for example, considered the reaction coordinate for the isomerization of CH_3NC to be related to an asymmetric ring deformation:

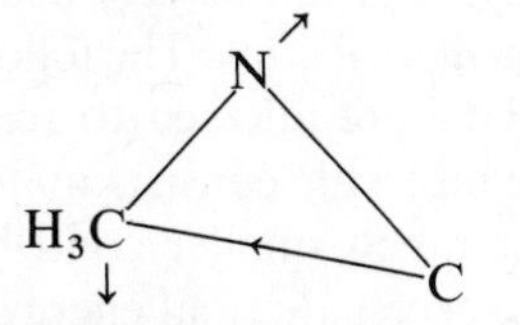

More obviously, the reaction coordinate in the dissociation of ethane takes the place of the C—C stretching frequency. The frequencies of the remaining vibrations of the activated complex are adjusted in a reasonable manner so as to obtain agreement between equation 5–33 and experimental 'high-pressure' Arrhenius parameters (chiefly A) via equations 1–64a, b. Here again the assignment usually is not highly critical. When rotations must be considered, the moments of inertia are calculated from assigned dimensions of the complex; the latter may not be independent of the frequency assignments.

The result of an RRKM calculation for the fall-off of k_U in the isomerization of CH_3NC is shown in Fig. 1–6. Similar agreement was obtained with experiments carried out at temperatures 30 and 60° higher.[45] Calculations made for the isomerization of cyclopropane,[46] the decomposition of the ethyl radical,[47] and several other reactions[48] have been equally successful. With the assumption of 'strong' collisions ($\lambda = 1$), there are no further adjustable parameters after the frequencies have been assigned to fit the experimental value of A. Such results therefore vindicate the theory very satisfactorily. With some other reactions,[49] the agreement with experiment is improved by relaxing somewhat the strong collision assumption—setting $\lambda = 0.2$ for example in the decomposition of ethyl chloride[39]—though the validity of this is more than doubtful. Notwithstanding such discrepancies, the theory is a distinct improvement on RRK theory in that it explains the fall-off behaviour in terms of concrete and realistic molecular models; and, though empirical knowledge of A and E_a is still required, it eliminates the unknown parameter s. It has also been successfully applied to isotope effects[45b, 50] about which RRK theory has nothing to say. The RRKM treatment of fall-off, however, allows remarkably little to be deduced about the nature of the activated complex. Not infrequently it is possible to assign quite different though equally plausible internal motions to the reaction coordinate and yet to find no significant difference in the calculated fall-off behaviour.[44,49] This is not so much a weakness in the theory as an indication that the phenomena are insensitive to the specific details of the intramolecular energy exchange.

5–2–3 Chemical activation

Tests of RRKM theory based on the fall-off in k_U are complicated by the fact that both $k_c(E)$ and the Boltzmann distribution function $K(E)$ in equation 5–32 are complex functions of E. A great advantage would therefore be gained if the reactant molecules could all be endowed with the same energy content above the critical energy. This can be very nearly achieved by the experimental technique of *chemical activation*. The following is an example. If the radicals CH_3 and CH_2Cl are allowed to react, the energy of the initially formed ethyl chloride molecule consists mainly of the exothermic heat of the combination plus much smaller contributions from the activation energy (if any) and the excess thermal energy of the radicals:

$$CH_3 + CH_2Cl \rightarrow C_2H_5Cl^* \qquad \text{5–VII}$$

There are two alternative fates for the energized molecules: stabilization by collisions or decomposition,

$$C_2H_5Cl^* + M \rightarrow C_2H_5Cl + M \qquad (S) \qquad 5\text{–VIII}$$

$$C_2H_5Cl^* \rightarrow C_2H_4 + HCl \qquad (D) \qquad 5\text{–IX}$$

Decompositions to other products can be neglected since they require much higher critical energies. Assuming strong collisions (though this assumption can be tested by the technique), the unimolecular rate constant for the decomposition of the energized molecules is found experimentally from the relation

$$k_{\bar{c}} = (Z[M])D/S \qquad 5\text{–35}$$

where D and S are the rates of formation of the decomposition products and stabilized molecules respectively. If the energy of all the molecules were precisely the same, $k_{\bar{c}}$ would be independent of pressure. In this particular example, this is the case to within 10 per cent; there is a large difference between the heat of reaction 5–VII and the critical energy for the decomposition, consequently the spread in the thermal energy of the molecules has little effect. In other systems where the energies are less different, for example,

$$H + C_4H_8 \rightarrow s\text{-}C_4H_9^* \begin{cases} \xrightarrow{M} C_4H_9 \\ \rightarrow CH_3 + C_3H_6 \end{cases} \qquad 5\text{–X}$$

there is some decrease in $k_{\bar{c}}$ between infinite and zero pressure, but the spread in energy which this signifies is vastly smaller than the spread in thermally activated molecules.

The reaction system 5-X above may be used to illustrate another important feature of the technique. By producing the same activated reactant by different reactions, the molecules can be generated with different but relatively constant energies. Thus the s-$C_4H_9^*$ radical can be generated with different known energies from but-1-ene and the *cis* and *trans* forms of but-2-ene. The procedure furnishes clear information on the variation of $k_{\bar{c}}$ with E, and this can be applied with much effect to the theoretical studies.

RRKM theory is able to interpret the behaviour of chemically activated molecules very successfully. For quantitative details the interested reader is referred to publications by Rabinovich and Setser.[(51)] Two important qualitative conclusions, however, should be mentioned here. The first is that the assumption of fast intramolecular transfer of energy, which is fundamental to the RRK and RRKM theories, is vindicated: that is, the energy received by the reactant molecule has time to become randomized among the active degrees of freedom before the molecule reacts, at least for the values of $Z[M]$ that normally apply.[(52)] And secondly, when the energy of the activated molecule is within a few kcal $mole^{-1}$ of the critical energy—a condition applying to most thermally activated molecules—a single collision with another reactant molecule is normally sufficient to cause deactivation. In other words, the strong collision assumption is a valid approximation for thermally activated reactions (and for some chemically activated reactions as well).[53] Furthermore, chemical activation experiments with different foreign gases present confirm the type of result illustrated in Table 1–2, namely that the full range of relative deactivating efficiencies is usually covered by an order of magnitude.

5–3 Bimolecular-termolecular combination reactions

It was observed in Section 1–3–3 and Fig. 1–6 that the second-order rate constant k_B for a combination reaction

$$A+B \quad (+M) \rightarrow AB \quad (+M)$$

falls off with decreasing pressure and that this can be explained by what amounts to the reverse of the Lindemann–Hinshelwood mechanism for unimolecular reactions. Indeed, assuming the equilibrium assumption is valid, as discussed in Section 1–3–4, the fall-off curve for k_B is identical with that of k_U the first-order rate constant for the reverse reaction. Thus RRK and RRKM theories are equally applicable to reactions of this type as to unimolecular reactions. Accurate experimental data, however, are much more difficult to obtain: consequently not many theoretical calculations have been carried out. Two such examples of RRKM treatments relate to the reactions

$$CH_3+CH_3 \quad (+M) \rightarrow C_2H_6 \quad (+M) \qquad \text{5–XI}^{(54)}$$

and

$$OH+NO_2 \quad (+M) \rightarrow HNO_3 \quad (+M) \qquad \text{5–XII}^{(55)}$$

Reactions of this type are characterized by 'loose' activated complexes with long extension of the incipient bond as discussed in Section 1–5–1.

The inverse dependence of rate on temperature commonly observed in the termolecular region can usually be expressed equally well by the Arrhenius equation, with a negative pseudo-activation energy, or by a relation of the form $k_t \propto T^{-n}$. With reactions of simple species, $-E_a$ is normally about 1 or 2 kcal $mole^{-1}$ and n about 0.5 to 3. Qualitatively, the effect can be accounted for theoretically by relating k_t to the bimolecular rate constant k'_a for the reverse decomposition. Thus

$$k_t = k'_a/K' = k'_a/F'(T)\exp(-E'_0/RT) \qquad \text{5–36}$$

where K', E'_0, and $F'(T)$ are respectively the equilibrium constant, endothermicity, and partition–function quotient for the decomposition. Substituting equation 1–31 (and E'_0 for E_c) in this expression yields

$$k_t = \{\lambda Z/(s-1)!F'(T)\}(E'_0/RT)^{s-1} \qquad \text{5–37}$$

For analogous reasons to these given in Section 1–4–2, $F'(T)$ is likely to be almost independent of T, as also is λZ. Hence k_t depends inversely on temperature. Equation 5–37 also predicts that, for reactions with similar entropy changes, the termolecular rate constant should increase with increasing exothermicity. An example of this has been found in reactions of the type

$$O + XO + M \rightarrow OXO + M \qquad \text{5–XIII}$$

which become progressively faster and more exothermic in the order X = O, N, S.[56]

There is an alternative mechanism to that of the 'energy-transfer' type so far discussed which also gives rise to third-order kinetics. This is as follows:

$$A + M + M \rightleftharpoons AM + M \qquad \text{5–XIV}$$

$$AM + B \rightarrow AB + M \qquad \text{5–XV}$$

If it is supposed that the reverse decomposition of the intermediate 'complex' AM is much faster than reaction 5–XV, the rate of formation of AB from A and B is third order. This mechanism is likely to be an effective rival to the energy-transfer mechanism only when A and B are atoms or small species—the lifetime of AB* in the energy-transfer mechanism is then short—and there is strong van der Waals' or other attraction between A and M. These conditions are realized in the recombination of iodine atoms, particularly when M is a large molecule. They are recognized by exceptionally large variations both of the absolute value of k_t and its temperature coefficient with different third-body molecules. The value of k_t increases by 25 times when argon is replaced by benzene and by 250 times when it is replaced by molecular iodine. The latter, however, can be classed as a specific effect caused by the intermediate formation of the quasi-stable I_3 molecule. Most of the facts with regard to ordinary molecules can be explained at least semi-quantitatively in terms of the formation of complexes by van der Waals attraction and charge transfer.[57] It seems, however, that for the reasons stated earlier, the iodine recombination is exceptional and, in general, the energy-transfer mechanism is the more important except when large or specifically reactive molecules are involved as third bodies.§

Bibliography

(A) K. J. Laidler, *Theories of Chemical Reaction Rates*, McGraw-Hill, 1969.

(B) H. S. Johnston, *Gas Phase Reaction Rate Theory*, Ronald Press Co., New York, 1966.

(C) D. L. Bunker, *Theory of Elementary Gas Reaction Rates*, Pergamon Press, 1966.

(D) M. Menzinger and R. Wolfgang, The meaning and use of the Arrhenius activation energy, *Angew. Chem., Internat. Edn*, **8**, 438, 1969.

§In the case of iodine recombination, it has been possible to separate the relative contributions of the two mechanisms by studying the pressure-dependence of the rate constant over a wide range.[58]

(E) J. C. Polanyi, Some concepts in reaction dynamics, *Accounts Chem. Research*, **5**, 161, 1972.

(F) P. J. Robinson and K. A. Holbrook, *Unimolecular Reactions*, Wiley Interscience, 1971.

References

1 M. Menzinger and R. Wolfgang, *Angew. Chem., Internat. Edn*, **8**, 438, 1969.
2 S. Glasstone, *Theoretical Chemistry*, pp. 66ff., Van Nostrand, 1944.
3 J. O. Hirschfelder, *J. Chem. Phys.*, **6**, 795, 1938.
4 I. Shavitt, R. M. Stevens, F. L. Minn, and M. Karplus, *J. Chem. Phys.*, **48**, 2700, 1968.
5 H. Conroy and B. L. Bruner, *J. Chem. Phys.*, **47**, 921, 1967.
6 Reviewed by K. J. Laidler, *Theories of Chemical Reaction Rates*, pp. 18–24, McGraw-Hill, 1969.
7 S. Glasstone, K. J. Laidler, and H. Eyring, *The Theory of Rate Processes*, McGraw-Hill, 1941.
8 S. Sato, *J. Chem. Phys.*, **23**, 592, 2465, 1955.
9 R. N. Porter and M. Karplus, *J. Chem. Phys.*, **40**, 1105, 1964.
10 M. G. Evans and M. Polanyi, *Trans. Faraday Soc.*, **34**, 11, 1938; A. F. Trotman-Dickenson, *Gas Kinetics*, p. 228, Butterworths, 1955.
11 N. N. Semenov, *Some Problems in Chemical Kinetics and Reactivity*, Vol. 1, p. 30, Princeton U.P., 1958.
12 H. S. Johnston and C. Parr, *J. Am. Chem. Soc.*, **85**, 2544, 1963; H. S. Johnston, *Gas Phase Reaction Rate Theory*, Ronald Press Co., New York, 1966.
13 J. A. Kerr, *Chem. Rev.*, **66**, 465, 1966.
14 S. W. Mayer, L. Schieler, and H. S. Johnston, *11th Symposium (Internat.) on Combustion*, p. 837, The Combustion Institute, Pittsburgh, 1967.
15 L. Pauling, *J. Am. Chem. Soc.*, **69**, 542, 1947.
16 F. T. Wall and R. N. Porter, *J. Chem. Phys.*, **36**, 3256, 1962.
17 J. Bigeleisen and M. Wolfsberg, *Adv. Chem. Phys.*, **1**, 15, 1958.
18 H. S. Johnston, *Gas Phase Reaction Rate Theory*, Ronald Press Co., New York, 1966.
19 A. A. Westenberg and N. de Haas, (a) *J. Chem. Phys.*, **47**, 1393, 1967; (b) *ibid.*, **47**, 4241, 1967; **50**, 2512, 1969; (c) *ibid.*, **48**, 4405, 1968.
20 T. E. Sharp and H. S. Johnston, *J. Chem. Phys.*, **37**, 1541, 1962.
21 T. Yokota and R. B. Timmons, *Int. J. Chem. Kinetics*, **2**, 325, 1970.
22 I. Shavitt, *J. Chem. Phys.*, **49**, 4048, 1968.
23 M. Karplus and R. N. Porter, *Disc. Faraday Soc.*, **44**, 164, 1967.
24 H. Gg. Wagner and J. Wolfrum, *Angew. Chem., Internat. Edn*, **10**, 604, 1971.
25 S. W. Mayer, L. Schieler, and H. S. Johnston, *J. Chem. Phys.*, **45**, 385, 1966.
26 B. A. Ridley, W. R. Schultz, and D. J. LeRoy, *J. Chem. Phys.*, **44**, 3344, 1966.
27 D. J. LeRoy, B. A. Ridley, and K. A. Quickert, *Disc. Faraday Soc.*, **44**, 92, 1967.
28 M. Karplus, R. N. Porter, and R. D. Sharma, *J. Chem. Phys.*, **43**, 3259, 1965.
29 D. L. Bunker, *Theory of Elementary Gas Reaction Rates*, p. 42, Pergamon Press, 1966.
30 P. J. Kuntz *et al.*, *J. Chem. Phys.*, **44**, 1168, 1966.
31 N. C. Blais and D. L. Bunker, *J. Chem. Phys.*, **37**, 2713, 1962.
32 P. J. Kuntz, M. H. Mok, E. M. Nemeth, and J. C. Polanyi, *Disc. Faraday Soc.*, **44**, 229, 1967; *J. Chem. Phys.*, **50**, 4607, 4623, 1969.
33 J. C. Polanyi and W. H. Wong, *J. Chem. Phys.*, **51**, 1439, 1969.

34 M. H. Mok and J. C. Polanyi, *J. Chem. Phys.*, **51,** 1451, 1969.
35 L. S. Kassel, *The Kinetics of Homogeneous Gas Reactions*, Chap. 5, Chemical Catalog Co., New York, 1932.
36 K. J. Laidler, *Theories of Chemical Reaction Rates*, pp. 118ff., McGraw-Hill, 1969.
37 E. W. Schlag, B. S. Rabinovitch, and F. W. Schneider, *J. Chem. Phys.*, **32,** 1599, 1960; G. Emanuel, *Int. J. Chem. Kinetics*, **4,** 591, 1972.
38 I. Oref and B. S. Rabinovitch, *J. Phys. Chem.*, **72,** 4488, 1969.
39 K. A. Holbrook and A. R. W. Marsh, *Trans. Faraday Soc.*, **63,** 642, 1967.
40 M. F. R. Mulcahy and D. J. Williams, *Aust. J. Chem.*, **17,** 1291, 1964; L. F. Loucks and K. J. Laidler, *Canad. J. Chem.*, **45,** 2767, 2795, 1967.
41 R. A. Marcus, *J. Chem. Phys.*, **20,** 359, 1952; R. A. Marcus and O. K. Rice, *J. Phys. and Colloid Chem.*, **55,** 894, 1951.
42 Helpful details are provided by K. J. Laidler, *Theories of Chemical Reaction Rates*, p. 122, McGraw-Hill, 1969; and H. S. Johnston, *Gas Phase Reaction Rate Theory*, p. 282, Ronald Press Co., New York, 1966.
43 G. M. Wieder and R. A. Marcus, *J. Chem. Phys.*, **37,** 1835, 1962.
44 C. S. Elliott and H. M. Frey, *Trans. Faraday Soc.*, **62,** 895, 1966.
45 (a) F. W. Schneider and B. S. Rabinovitch, *J. Am. Chem. Soc.*, **84,** 4215, 1962; (b) *ibid.*, **85,** 2365, 1963.
46 M. C. Lin and K. J. Laidler, *Trans. Faraday Soc.*, **64,** 927, 1968.
47 M. C. Lin and K. J. Laidler, *Trans. Faraday Soc.*, **64,** 79, 1968.
48 K. J. Laidler, *Theories of Chemical Reaction Rates*, p. 145, McGraw-Hill, 1969.
49 K. A. Holbrook *et al.*, *Trans. Faraday Soc.*, **66,** 868, 1970.
50 H. M. Frey and B. M. Pope, *Trans. Faraday Soc.*, **65,** 441, 1969.
51 B. S. Rabinovitch and M. C. Flowers, *Quart. Rev. Chem. Soc.*, **18,** 122, 1964; B. S. Rabinovitch and D. W. Setser, *Adv. Photochem.*, **3,** 1, 1964; J. C. Hassler and D. W. Setser, *J. Chem. Phys.*, **45,** 3246, 1966.
52 R. E. Harrington, B. S. Rabinovitch, and H. M. Frey, *J. Chem. Phys.*, **33,** 1271, 1960; J. D. Rynbrandt and B. S. Rabinovitch, *J. Phys. Chem.*, **74,** 4175, 1970.
53 R. E. Harrington, B. S. Rabinovitch, and M. R. Hoare, *J. Chem. Phys.*, **33,** 744, 1960; G. H. Kohlmaier and B. S. Rabinovitch, *ibid.*, **38,** 1709, 1963; see also S. C. Chan *et al.*, *J. Phys. Chem.*, **74,** 3160, 1970; R. Atkinson and B. A. Thrush, *Chem. Phys. Letters*, **3,** 684, 1969.
54 E. V. Waage and B. S. Rabinovitch, *Int. J. Chem. Kinetics*, **3,** 105, 1971.
55 C. Morley and I. W. M. Smith, *J. Chem. Soc. Faraday Trans.*, **II, 68,** 1016, 1972.
56 C. J. Halstead and B. A. Thrush, *Proc. Roy. Soc.*, **A295,** 363, 1966.
57 G. Porter and J. A. Smith, *Proc. Roy. Soc.*, **A261,** 28, 1961; M. Eusuf and K. J. Laidler, *Trans. Faraday Soc.*, **59,** 2750, 1963.
58 H. Hippler, K. Luther, and J. Troe, *Chem. Physics Letters*, **16,** 174, 1972.

Problems

5–1 Potential energy surfaces have been calculated for the reaction $Cl + H_2 \rightarrow HCl + H$ by the LEP and LEPS methods. The following properties were derived for the activated complex, which was assumed to be linear:

	$r^{\ddagger}_{H-H} \times 10^8$ cm	$r^{\ddagger}_{H-Cl} \times 10^8$ cm	$I^{\ddagger} \times 10^{40}$ g cm^2	$\omega^{\ddagger}_s$ cm^{-1}	$\omega^{\ddagger}_b$ cm^{-1}	$\omega^{\ddagger}_r$ cm^{-1}
LEP*	1.40	1.30	14.2	2496	551	720*i*
LEPS**	0.98	1.40	12.1	1359	721	1397*i*

$\omega_s^\ddagger$, $\omega_b^\ddagger$, and $\omega_r^\ddagger$ are respectively the sym. stretching, bending, and asym. stretching frequencies. [* Wheeler *et al.*, *J. Chem. Phys.*, **4**, 178, 1936; ** Westenberg and de Haas, *ibid.*, **48**, 4405, 1968.]

(a) If the rate constant is expressed as $k = C\exp(-E_0/RT)$, write down the activated complex theory expression for C_{LEP}/C_{LEPS} (without tunnelling) and then explore quantitatively the relative effects of the various properties on C at 400 K. (Previous solution of Problem 1–7 will be helpful.)

(b) The LEPS method, in general, gives a 'thinner' potential energy barrier than the LEP method. Show that this is so in the present case by calculating the relative tunnelling effects by the (approximate) equation 1–60.

5–2 Determine graphically the height of the 'classical' potential energy barrier E_c for the above reaction by the BEBO method with the aid of the following information

	D_e(kcal mole^{-1})	r_0(Å)	β(Å)$^{-1}$	p
H—H	109.4	0.74		1.041
H—Cl	106.4	1.27	1.87	0.914

(Use a desk calculator to compute E_{ClHH} as a function of n at intervals of 0.05 from $n = 0.1$ to 0.3.) Calculate the dimensions of the activated complex by means of Paulings' rule ($r = r_0 - 0.60\log_{10} n$). Compare your results with those obtained by the LEPS method above ($E_{c,LEPS} = 6.7$ kcal mole^{-1}).

5–3 (a) Competitive experiments have shown that the activation energies for the reactions between chlorine atoms and isotopic hydrogen molecules differ by the following amounts (Persky and Klein, *J. Chem. Phys.*, **44**, 3617, 1966)

$$E_{DD} - E_{HH} = 1.13\text{ kcal }(4.7\text{ kJ})\text{ mole}^{-1} \quad\text{and}\quad E_{TT} - E_{HH} = 1.69\text{ kcal }(7.1\text{ kJ})\text{ mole}^{-1}$$

Assuming that the differences are due substantially to zero-point energies, that the height of the potential energy barrier E_c is 6.7 kcal (28 kJ) mole^{-1} and that the zero-point energy of $ClH_2^\ddagger$ is 3.9 kcal (16 kJ) mole^{-1}, draw a two-dimensional potential energy diagram with the vertical axis to scale indicating the energies of the initial states and activated complexes of the three reactions. ($\omega_{H_2}, \omega_{D_2}, \omega_{T_2} = 4395, 3109, 2540$ cm^{-1}).

(b) The kinetic isotope effect k_H/k_D in the reaction $Br + PhCH_3 \rightarrow PhCH_2 + HBr$ is abnormally large. Can you suggest an explanation based on the following facts: $M_{Br} \approx M_{CH_2Ph}$ and $D(H{-}Br) \approx D(PhCH_2{-}H)$?

5–4 Derive an expression to show that the relative rate constants of intramolecular isotopic reactions of the type

$$A + BB'C \rightarrow AB + B'C$$

$$A + BB'C \rightarrow AB' + BC$$

are determined entirely by the moments of inertia and vibration frequencies of the activated complexes.

Calculate the relative rates at which HF and DF are formed by the reaction of atomic fluorine with HD at 300 K using the following properties of the activated complexes derived* from the LEPS potential energy surface.

	$I^\ddagger \times 10^{40}$ g cm^2	$\omega_s^\ddagger$ cm^{-1}	$\omega_b^\ddagger$ cm^{-1}	$\omega_r^\ddagger$ cm^{-1}
FHD‡	19.9	3422	365	263i
FDH‡	15.7	3518	323	256i

Is the assumption made in Problem 3(a) above valid in this case?

(* Muckerman, *J. Chem. Phys.*, **54**, 1155, 1971.)

5–5 (a) In a study of the reaction $K + CH_3I \rightarrow KI + CH_3$ in crossed molecular beams Gersh and Bernstein (*J. Chem. Phys.*, **56**, 6131, 1972) measured the total flux (F) of KI scattered out of the crossing zone (volume 0.018 cm^3) by traversing the detector over the surface of an imaginary surrounding sphere. Estimate the total reaction cross-section S_r at 9 kJ mole^{-1} relative translational energy from the following data: densities of K and CH_3I beams: 2.5×10^{10} and 1.2×10^{10} mole cm^{-3}; relative centre-of-mass velocity: 7.7×10^4 cm s^{-1}; $F = 2.1 \times 10^9$ molecule s^{-1}.
(b) The fraction of molecular collisions with relative translational energy in the range ε to $\varepsilon + d\varepsilon$ in a gas in thermal equilibrium at temperature T is $[\varepsilon/(kT)^2]\exp(-\varepsilon/kT)\,d\varepsilon$. (See ref. C of Chap. 1 p. 493). If $S_r(\varepsilon)$ is the reaction cross-section for the energy ε, write down the expression for the average reaction cross-section at T and hence, with the aid of equation 1–12, show that the following relation holds between the bimolecular rate constant and $S_r(\varepsilon)$:

$$k_r = [8/\pi m^*(kT)^3]^{1/2} \int_0^\infty S_r(\varepsilon)\varepsilon \exp(-\varepsilon/kT)\,d\varepsilon \qquad \text{(A)}$$

What assumptions have been made in this derivation?

5–6 The form of the energy dependence of the reaction cross-section $S_r(\varepsilon)$ is not known theoretically. A plausible assumption is: $S_r(\varepsilon) = 0$ when $\varepsilon < \varepsilon_0$ combined with $S_r(\varepsilon) = C(\varepsilon - \varepsilon_0)/\varepsilon$ when $\varepsilon \geqslant \varepsilon_0$, where C is a constant. [cf. equation 5–26]. Use equation (A) above to obtain an expression for the rate constant on this basis. Compare your result with equations 1–12 and 1–13. What significance can be attached to C? Supposing $S_r(\varepsilon)$ were simply constant above the threshold energy ε_0, how would the expression for k_r differ from the previous one?

5–7 It has been suggested that elementary four-centre reactions of the type $H_2 + D_2 \rightarrow 2HD$, which are believed to occur at shock-wave temperatures, take place at an appreciable rate only when one or both of the reactant molecules is vibrationally excited. To test this Morokuma *et al.* (*J. Am. Chem. Soc.*, **89**, 5064, 1967) derived detailed rate constants k_{v_1,v_2} for the $H_2 + H_2$ reaction from Monte Carlo trajectory calculations using an LEPS-type potential energy surface. k_{v_1,v_2} refers to reaction between molecules with fixed numbers of vibrational quanta v_1 and v_2 respectively but otherwise in thermal equilibrium. The following relative figures relate to 1600 K:

v_1, v_2	0,0	0,1	1,1	0,2	1,2	0,3
$k_{v_1,v_2}/k_{0,0}$	(1)	29	475	442	4465	4450

Assuming that in the actual $H_2 + H_2$ reaction the molecules behave as simple harmonic oscillators in thermal equilibrium, estimate approximately the per cent contribution of vibrationally unexcited molecules to the total reaction rate at 1600 K ($\omega_{H_2} = 4395$ cm^{-1}).

5–8 Beginning from equation 5–29 show that the average rate constant for the reaction of the collisionally excited molecules in a unimolecular reaction is given by the relation, $\bar{k}_c = \lambda Z[A]\{(k_\infty/k) - 1\}$, where k and k_∞ are the measured rate constants for the overall reaction at finite and infinite pressures. To what extent does this result depend on theories (e.g. RRK) used to estimate the reactivity and concentration of the excited molecules? At what values of k/k_∞ is the reaction rate of the excited molecules (a) one tenth, (b) equal to, and (c) 10 times the rate of their deactivation. From Fig. 1–6 estimate $\bar{k}_c$ for the isomerization of CH_3NC at 10 torr and 472 K. On the average about how many vibrations take place in the excited molecule before reaction occurs?

5–9 Derive the RRK expression for k_c (equation 5–26) in the following way. Find (a) the number of ways of distributing n identical balls (i.e. quanta) among s boxes (i.e. oscillators), (b) the number of ways of doing this such that m balls are always distributed in one particular way (e.g. all in one particular box), and hence (c) find the probability P of situation (b) arising at random. Use the approximation $n! = n^n/\exp n$ to derive an expression for P when n is very large and $n - m \gg s - 1$.

5–10 Use a desk calculator and mathematical tables to investigate the effect of s on the classical distribution function $K(E)$ given in equation 5–27. (Transform $K(E)$ into $K(y)$ where $y = E/RT$ and plot $K(y)$ against y for $s = 3$ and $s = 15$ on the same axes. Note that $K(y)$ has its maximum value at $y = s-1$). In a similar way, investigate the effect of s on $k_c(E/E^*)$ according to RRK theory. (Note that $k_c(E/E^*)$ has a point of inflexion at $E/E^* = s/2$.)

5–11 The following are values of the 'Kassel integral' $(k/k_\infty)_s$ defined by equation 5–30. They are expressed in logarithmic form as a function of the parameter $\theta \equiv \lambda Z[A]/a$ for the values $s = 7, 11, 15$ and the particular value $E^*/RT = 41$.

$-\log\theta$	5	6	7	8	9
$-\log(k/k_\infty)_7$	0.26	0.62	1.16	1.84	2.60
$-\log(k/k_\infty)_{11}$	0.03	0.12	0.34	0.69	1.18
$-\log(k/k_\infty)_{15}$	0.00	0.02	0.10	0.27	0.56

Use these figures to make dimensionless 'fall-off' graphs of $\log(k/k_\infty)_s$ *vs* $\log\theta$.

An experimental investigation of the unimolecular isomerization of *cyclo* C_3H_5F gave the Arrhenius parameters $\log A_\infty = 14.58$, $E_\infty = 61.0$ kcal (255 kJ) mole^{-1} and the following 'fall-off data' for pressures between 460 and 0.2 torr at 748 K:

$\log(p$ torr$)$	2.67	2.18	1.98	0.73	0.00	−0.15	−0.25	−0.57	−0.77
$-\log(k/k_\infty)$	0.00	0.01	0.03	0.15	0.36	0.44	0.43	0.56	0.69

By means of the above graphs, find the best fit of the Kassel integral and hence determine approximately the value of s for these conditions ($\lambda Z/a = 0.35$ cm^3 mole^{-1}; $p(\text{torr}) = 10^{7.67}[C_3H_5F]$).

5–12 (a) S. W. Benson has suggested that the number of 'effective' RRK oscillators s can be obtained from the heat capacity C_p of the reactant molecule at the reaction temperature. Bearing in mind that the heat capacity of a single classical oscillator is equal to R, derive a simple relation between s and C_p.

(b) When the rates of formation of ethane, *iso*butane and 2,3-dimethylbutane ($\mathscr{R}_e$, $\mathscr{R}_b$, and $\mathscr{R}_d$) during photolysis of mixtures of azomethane $(CH_3N)_2$ and azo*iso*propane were measured at 373 K, the ratio $\mathscr{R}_b^2/(\mathscr{R}_e\mathscr{R}_d)$ was found to increase steadily as the total pressure was reduced from 3 to 0.03 torr. Suggest an explanation for this effect. How would you expect the corresponding ratio for the photolysis of mixed azomethane and perdeutero-azomethane to behave in similar circumstances?

(c) Isothermal fluorination of ethane at 298 K gives rise to notable amounts of ethene as well as ethyl and hydrogen fluorides. How do you explain the fact that the ratio $\mathscr{R}_{C_2H_4}/\mathscr{R}_{C_2H_5F}$ increases in inverse proportion to the total pressure?

Explosions and branching chains

Every chemist knows, and not a few have had it impressed upon them by unwelcome experience, that a mixture of chemicals that reacts quietly or even imperceptibly in some circumstances may explode in others. And it is equally well known that certain types of pure compound that are capable of exothermic decomposition—nitro compounds, azides, peroxides, and the like—are also prone to explode. It is easier to recall the characteristics of a chemical explosion than to give a precise definition of it and at this stage it will suffice to state the obvious. An 'extremely rapid' or 'virtually instantaneous' reaction takes place which equally rapidly produces heat; this gives rise to a violent pressure pulse with attendant mechanical effects and emission of sound and light. The act of initiation of explosion is known as *ignition.* It is common knowledge that, in appropriate circumstances, ignition can be brought about by introducing a localized source of energy such as a spark, a hot wire or a pulse of radiation. In other circumstances, notably when the whole reaction mixture is heated to an appropriate temperature, it may occur spontaneously throughout the bulk of the reactants. In either case, there is evidently a critical condition or set of conditions which must in some way be exceeded for ignition to occur. Since ignition is a rate phenomenon, it is the concern of chemical kinetics to discover what these conditions are and to express them in terms of a general theory. In the following we shall be concerned only with spontaneous ignition and, of course, with gaseous systems. An account of ignition by localized energy sources will be found in the book by Lewis and von Elbe.[1]

In the course of our discussion it will emerge that spontaneous ignition can come about in three ways, two of which are radically different. In the first way, the critical condition arises from the manner in which the reaction is accelerated by the heat it generates. The result is known as a *thermal explosion.* In the second way the critical condition arises from the reaction mechanism. We shall see later that the requirement is for continuous branching of reaction chains. Such are *chain-isothermal explosions.* The third type of ignition involves both the heat of reaction and chain-branching, and so we have *chain-thermal explosions.* The principles on which each type of ignition depends will be presented in separate sections and illustrated by appropriate reaction systems. Our discussion of reactions giving rise to chain-isothermal explosions will include, in addition to briefer references to other systems, a somewhat more extensive account of the hydrogen–oxygen reaction. This is because the long history of investigation of this reaction has provided much of the present insight into the kinetics of branched-chain reactions in general; and the mechanism of the reaction itself is understood more thoroughly than that of any other reaction in the same category.

Later on in the chapter we shall be concerned with a class of reactions known as *degenerate explosions.* These, in fact, are not explosions at all, though they may lead to them. The outstanding examples are reactions of hydrocarbons and other organic compounds with oxygen. At the same time organic oxidation reactions present a number of other extraordinary phenomena which are not yet completely understood. The chapter will conclude with a summary of the principal established facts relating to the kinetics of reactions of this type, and a presentation of some current ideas on the rather complex reaction mechanisms involved.

6–1 Thermal explosions

As noted in Section 2–1–1, the temperature of a gas undergoing exothermic reaction must necessarily be greater, if only slightly greater, than the temperature of the surroundings. In the steady state, it is determined by the balance between the rate of production of heat by the reaction and the rate of loss of heat to the surroundings. We are concerned here with reactions occurring in closed vessels where the heat loss is to the walls. In the types of experiment described in Chapter 2 the object is to keep the temperature rise to a minimum by working with low reaction rates and, when possible, by increasing the rate of heat loss, for example, by stirring the gas (cf. Section 2–3). However, the fact is that the reaction rate and thereby the rate of production of heat increase in an exponential fashion with increasing temperature, whereas the rate of heat loss by conduction and convection increase much less steeply. Hence there is the possibility that heat may continue to accumulate in the gas faster than it can be conveyed away. In these circumstances, the temperature would also continue to rise, thus provoking further acceleration of the reaction with still faster accumulation of heat, and so on. In other words an explosion would ensue. Thus it seems, in accord with experience, that the same exothermic reaction may take place slowly at a temperature little different from the ambient, but under different conditions it may explode. The object of the theory of thermal explosions is to characterize the alternative conditions.

Consider an exothermic reaction

$$\mathrm{X + Y \rightarrow Products}$$

taking place in a closed vessel of volume V. If there were no loss of heat from the gas the total rate of generation of heat would be given by the expression

$$\dot{q}_{\mathrm{G}} = QVf([\mathrm{X}], [\mathrm{Y}], T) \qquad 6\text{–}1$$

where Q is the heat of reaction expressed per mole of X or Y consumed. We may assume that f is at least approximately of the form

$$f = k[\mathrm{X}]^x[\mathrm{Y}]^y = A[\mathrm{X}]^x[\mathrm{Y}]^y \exp(-E/RT) \qquad 6\text{–}2$$

and, for simplicity, we shall suppose that, although the initial total pressure is variable from one experiment to another, the ratio $[\mathrm{X}]_0/[\mathrm{Y}]_0$ is fixed. Thus if c is the sum of the initial concentrations $([\mathrm{X}]_0+[\mathrm{Y}]_0)$, it follows that

$$\dot{q}_\mathrm{G} = QVc^zA' \exp(-E/RT) \qquad 6\text{–}3$$

where $z = x+y$ and A' is a simple temperature-independent function of A, z, and the ratio $[\mathrm{X}]_0/[\mathrm{Y}]_0$. Graphs of $\dot{q}_\mathrm{G}$ as a function of T for various values of c (or p^* the partial pressure of X+Y corresponding to c) are shown schematically by curves I–IV of Fig. 6–1.

In considering the rate of heat loss $\dot{q}_\mathrm{L}$, two limiting situations can be imagined. In the first, the gas has the same temperature throughout and $\dot{q}_\mathrm{L}$ is determined by the rate of transfer of heat across a thin boundary layer between the bulk of the gas and the wall. This would be the case with sufficiently vigorous stirring of the gas by natural or forced con-

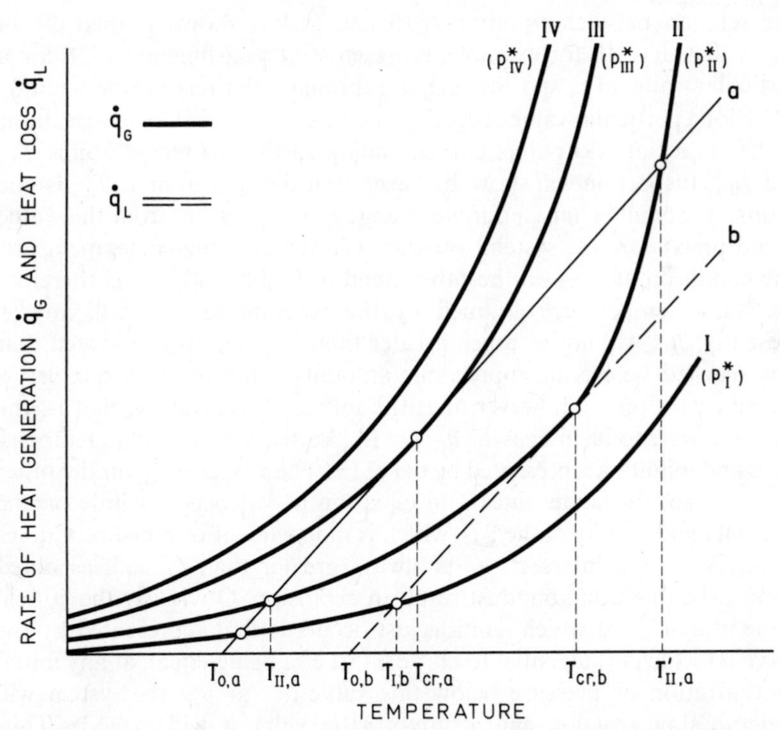

Fig. 6–1 Exothermic reaction in a closed vessel. Rates of generation and loss of heat graphed (schematically) as functions of temperature. The conditions for the graphs of $\dot{q}_\mathrm{G}$ and $\dot{q}_\mathrm{L}$ to be tangential are the critical conditions for explosion. p^* = pressure of reactants ($p^*_\mathrm{I} < p^*_\mathrm{II} < p^*_\mathrm{III} < p^*_\mathrm{IV}$); T_0 = initial = wall temperature.

vection. In the second situation, the gas is perfectly still and the heat is transferred to the wall solely by conduction; there is a continuous temperature gradient from the centre of the vessel, where the gas is hottest to the periphery, where it is at the temperature of the wall. Most experimental studies hitherto have been carried out under conditions approximating better to purely conductive than to purely convective heat loss; but the two situations are not radically different and, since the convective case provides a clearer introduction to essential principles, we shall consider it first. The treatment which follows is usually known as the Semenov theory.[2]

The rate of heat loss is given by

$$\dot{q}_L = S\alpha(T - T_0) \qquad 6\text{–}4$$

where S is the total area of the wall and α is the 'heat-transfer coefficient' per unit area (α is actually defined by this relation but it is substantially independent of temperature). T_0 is the temperature of the wall and is assumed to remain constant throughout the course of events. The condition for the gas to be thermally in a steady state is

$$\dot{q}_L = \dot{q}_G \qquad 6\text{–}5$$

The relation between equations 6–3 and 6–4 is shown graphically in Fig. 6–1. Curve II, for example, represents $\dot{q}_G$ as a function of T for a particular value of c, and the line, a, represents the relation between $\dot{q}_L$ and T for a particular value of T_0. Equation 6–5 is satisfied by intersections of the curves at two points corresponding to the gas temperatures $T_{II,a}$ and $T'_{II,a}$. Inspection will show, however, that the condition at $T'_{II,a}$ is one of unstable equilibrium: a minute change in temperature from this value would precipitate the system to either a lower or a higher temperature. The condition at $T_{II,a}$, on the other hand, is stable and $T_{II,a}$ is therefore the stable temperature attained by the reacting gas. As will appear presently, $T_{II,a}$ will not be much greater than $T_{0,a}$, and it is assumed that it is reached before an appreciable amount of the reaction mixture is consumed. Suppose, however, that the initial concentration, that is, the pressure, were to be increased; $\dot{q}_G$ would likewise increase at all temperatures and might be represented by curve IV. The values of $\dot{q}_L$ on the other hand remain the same, since α in equation 6–4 depends mainly on the thermal conductivity of the gas, which is independent of pressure. Curves a and IV do not intersect; $\dot{q}_G$ is always greater than $\dot{q}_L$ and, as noted previously, this situation must result in explosion. Obviously the critical value of c or p^* at which ignition just occurs is that represented by the curve III which is tangential to curve a. All else being equal, at any initial concentration or pressure below this value (c_{cr} or p^*_{cr}) the system will undergo slow reaction and at any greater value it will explode. Thus there is a sharp *explosion limit* (of pressure); and this is, in fact, observed experimentally.

Figure 6–1 also shows that the limit (i.e. p^*_{cr} or c_{cr}) decreases with increasing initial temperature, a fact also in accord with experience. Thus if T_0 is increased to $T_{0,b}$, the value of p^*_{cr} is that which corresponds to curve II. When $T_0 = T_{0,b}$, the gas at the initial pressure p^*_{I} would react slowly (having attained the temperature $T_{I,b}$) but, at the higher pressure p^*_{III} or p^*_{IV}, it would ignite.

Mathematically the critical condition for ignition is given by the tangential condition

$$d\dot{q}_G/dT = d\dot{q}_L/dT \qquad 6\text{–}6$$

and this relation, together with equation 6–5, enables the critical condition to be found in terms of the initial conditions, T_0, c, and so on. A quantity of some interest is T_{cr}, the temperature which must be attained by the gas for ignition to occur. Equations 6–3, 6–4, 6–5, and 6–6 readily yield a quadratic in T_{cr}, of which the solution is

$$T_{cr} = (E/2R)[1 \pm (1 - 4RT_0/E)^{1/2}] \qquad 6\text{–}7$$

The higher root can be neglected. It exists because the graph of $\exp(-E/RT)$ against T has a point of inflexion which, however, occurs at a value of T that is too high to be physically attained by the system.§ Thus we need only consider the lower root (as illustrated in Fig. 6–1). When typical values are substituted in equation 6–7, the rather surprising result emerges that T_{cr} is generally very little higher than T_0. For example, with $T_0 = 500$ K and $E = 40$ kcal mole^{-1}, we find that $(T_{cr} - T_0) = 13.2$ K. Thus, when the critical initial conditions occur, very little self-heating of the system is required to precipitate explosion. The critical initial conditions are easily found by substituting the value of T_{cr} from equation 6–7 in the solution of equations 6–5 and 6–6. To a good approximation they are expressed by the relation

$$(\alpha S/V) = eQ(E/RT_0^2)c^z A' \exp(-E/RT) \qquad 6\text{–}8$$

All the quantities in this expression can be determined independently. But before making a comparison with experiment, we need to consider the effect of adopting the alternative assumption of purely conductive heat transfer from the gas to the wall.

The theory elaborated to deal with this situation, which will be presented only in outline, is due to Frank-Kamenetskii.[(3)] Because of the temperature gradient from the centre of the gas to the wall, the steady state represented by equation 6–5 must be expressed in terms of an infinitesimal element of volume of the gas. It is given by[(4)]

$$-\lambda\nabla^2 T = Qc^z A' \exp(-E/RT) \qquad 6\text{–}9$$

where λ is the thermal conductivity of the gas, which is considered to be

§When we come to consider ignition in *flowing* gas (in Chapter 7) we shall find that the higher root does have physical significance (corresponding, in fact, to extinction). However, in this case, the inflexion of the heat-generation curve does not originate from the form of $\exp(-E/RT)$.

independent of temperature as well as pressure, and ∇ is the Laplacian operator. The boundary conditions are $T = T_0$ at the wall and, for a symmetrical vessel such as a sphere or infinite cylinder (of radius $\mathbf{R}$), $dT/dr = 0$ at the centre. The solution of equation 6–9 is in the form of T as a function of r from 0 to $\mathbf{R}$. To obtain an analytical solution, a more tractable alternative function of T must be found which gives a good approximation to the heat-generation term on the right. Adopting the conclusion from the Semenov theory that $(T - T_0)$ is much smaller than T_0 in the neighbourhood of the critical condition, Frank-Kamenetskii made the following approximation:

$$\frac{E}{RT} = \frac{E}{RT_0[1+(T-T_0)/T_0]} \approx \frac{E}{RT_0}\left(1 - \frac{(T-T_0)}{T_0}\right)$$

$$= \frac{E}{RT_0} - \frac{E}{RT_0^2}(T-T_0) \equiv \frac{E}{RT_0} - \Theta \qquad 6\text{–}10$$

Expressed in dimensionless quantities, equation 6–9 now becomes

$$-\nabla_\rho^2 \Theta = \delta \exp \Theta \qquad 6\text{–}11$$

where $\rho = r/\mathbf{R}$, and Θ is as defined above; δ is given by

$$\delta = (Q/\lambda)\mathbf{R}^2(E/RT_0^2)A'c^z \exp(-E/RT_0) \qquad 6\text{–}12$$

and contains only quantities which are or can be known *a priori*. Solutions of equation 6–11 exist only for a certain range of values of δ which never surpass a certain maximum value.§ Since this maximum value of δ corresponds to the maximum value of Θ obtainable under stable conditions at the centre of the vessel, it is taken to be the critical condition for ignition (δ_{cr}). Its value differs according to the geometry of the reaction vessel. The values for gas enclosed between infinite parallel plates, in an infinite cylinder and in a sphere are respectively 0.88, 2, and 3.32. Thus, for a sphere, if the quantities on the right of equation 6–12 are such that the value of δ is greater than 3.32, no solution of equation 6–10—and therefore no stable temperature distribution—is possible. In these circumstances the gas ignites. Values of δ less than 3.32, on the other hand, give rise to steady reaction.

The essential continuity between this theory and the Semenov treatment given previously can be seen by comparing the respective critical conditions embodied in equations 6–8 and 6–12. These equations become identical if the heat-transfer coefficient α in equation 6–8 is given the value

$$\alpha_{\text{eff}} = (e\delta_{cr}V/S\mathbf{R})(\lambda/\mathbf{R}) \equiv \beta(\lambda/\mathbf{R}) \qquad 6\text{–}13$$

where $\beta = 3.01$ for a sphere or 2.72 for an infinite cylinder. Hence the simpler Semenov treatment can be legitimately applied to the case of

§ It may assist the student to grasp this point to note (by way of analogy only) that solutions to equation 6–7 exist only for values of the dimensionless quantity (RT_0/E) less than 0.25. A detailed discussion is given by Gray and Lee.[6]

purely conductive heat transfer if the appropriate value of α_{eff} given by equation 6–13 is used in equation 6–4 or 6–8. Naturally, this value does not apply when convection contributes appreciably to the heat transfer. According to Gray and Lee,[6] 'convection plays a part in heat transfer if the reactant pressure is of the order of one atmosphere, but has a negligible effect below about 0.1 atm'. Again, the Frank-Kamenetskii theory predicts a parabola-like temperature distribution in the non-explosively reacting gas but this is disregarded by the Semenov theory.

Experimental studies

These have usually been conducted by admitting predetermined pressures of the mixed reactants or of a single explosive vapour to an evacuated vessel in a thermostat. The occurrence of explosion is observed by the sharp increase in pressure and sometimes by the rise in the temperature of the gas recorded by a sensitive thermocouple. Initial pressures are commonly between 1 and 100 torr, in which circumstances the explosive effects are normally quite feeble. There is merely an audible click accompanied by a weak flash of light. At higher pressures, the effects may be more drastic and, in the author's experience, a glass vessel can be completely shattered by the explosion of a few hundred torr of gas. Since at low pressures the heat loss is expected to be mainly conductive, the first object of most investigations has been a test of the Frank-Kamenetskii theory. Reactions with simple kinetics have naturally been chosen for investigation, chain reactions being avoided because effects of the surface on the rate could obscure the effects of volume and pressure on ignition predicted by the theory. Most, though not all[7] studies have been conducted with first-order exothermic decompositions, for example, those of methyl nitrate,[8] azomethane,[9] and diethyl peroxide.[10, 11] Few if any of these occur completely without some contribution from a straight-chain mechanism, but it is sufficient that their isothermal kinetics are uninfluenced by the surface. The ignition of diethyl peroxide, which has been investigated by Harris[10] and more recently by Fine, Gray, and Mackinven[11] (the latter in the pressure range 0.5–10 torr) will form the basis of our discussion.

It follows from the previous discussion that, for such a reaction in a spherical vessel, the critical condition for ignition expressed in terms of the initial partial pressure of the reactant (p^*) and other measurable quantities is

$$\frac{Q\mathbf{R}^2 p^* A \exp(-E/RT_0)}{\lambda \mathbf{R}^2 T_0^3} = \delta_{cr} = 3.32 \qquad 6\text{–}14$$

and from this it follows that for the same gas composition and reaction vessel (whether spherical or not) the following relation applies:

$$\log(p^*/T_0^3) = E/RT_0 + \text{constant} \qquad 6\text{–}15$$

Figure 6–2 shows that equation 6–15 is obeyed experimentally. The slope

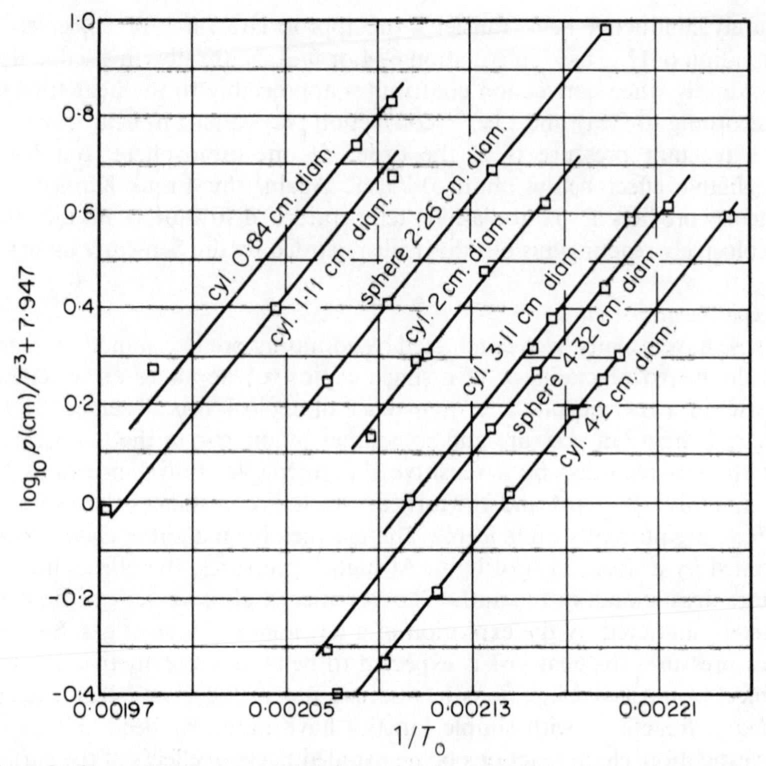

Fig. 6–2 Thermal explosions of diethyl peroxide in vessels of various sizes and shapes; $\log_{10} p^*_{cr}/T_0^3$ graphed against $1/T_0$ (Harris[10]).

of the graphs corresponds to a value of E of 28.3 kcal mole^{-1} which is not far from the activation energy determined under isothermal conditions[12] namely 34.1 kcal mole^{-1}. Fine *et al.*[11] using more advanced technique to study the ignition obtained 35.4 ± 1.0 kcal mole^{-1} from equation 6–15. Figure 6–2 also shows that at the same value of T_0 the value of p^* increases with decreasing size of the vessel; in fact, with the spherical vessels the critical pressures are accurately in the inverse ratio of the square of the radii as required by equation 6–14.

According to the Frank-Kamenetskii theory, the maximum temperature rise the reaction can attain at the centre of a spherical vessel without exploding is $1.61\, RT_0^2/E$ (which may be compared with the Semenov value of RT_0^2/E for the maximum average temperature obtained from equation 6–7). Fine *et al.* were able to verify this maximum central temperature rise (about 20 °C) by means of a thermocouple probe, and furthermore found the shape of the temperature gradient from centre to wall to be as predicted by the theory. Finally, the same workers used their experimental values of $\mathbf{R}^2 p^*/T_0^3$ and other independently determined data to calculate the value of δ_{cr} from the right-hand side of equation 6–14. Their value is 4.3 ± 0.4 which, considering the number of quantities to be inserted in equation 6–14 and the neglect of reactant consumption in the theory, must be considered to agree remarkably well with the theoretical value (3.32).

These experiments, as well as others performed with different reactions,[2] show that the Frank-Kamenetskii theory is essentially correct when its assumptions are realized in practice. However, as the pressure or the size of the reaction vessel is increased, natural convection eventually adds significantly to the purely convective heat transfer assumed by the theory and, for a given pressure, the ignition temperature T_0 is greater than expected.[13] This gives rise to curvature in plots of the type shown in Fig. 6–2.[14] The most fundamental deficiency of the theory, however, is the assumption that the relevant thermal condition is established immediately the gas enters the reaction vessel and before any proportion of it is consumed. In fact, the gas takes a finite time to attain first to the temperature of the vessel (as discussed in Section 2–1–1) and then to the 'self-heated' temperature. During this time some reaction occurs. These effects give rise to induction periods of about 0.1 to a few seconds before ignition which are quite characteristic but necessarily are ignored by the theory. They must be discussed in terms of non-steady-state theory, of which a full account has been given by Gray and Lee.[15]

6–2 Branching-chain explosions: isothermal ignition

The previous discussion of spontaneous ignition centred largely on the existence of a pressure limit: explosion occurred if the pressure of the reactive gas exceeded a critical value. It was shown that for a number of exothermic reactions the thermal imbalance created by the heat of reaction provides a satisfactory explanation of the way in which the critical pressure depends on the initial temperature, the size of the vessel, and so on. But when we come to examine the conditions for ignition to occur with a number of well-known exothermic reactions such as those of oxygen with hydrogen, carbon monoxide, and various other gaseous elements and compounds (P_4, S_2, SiH_4, CS_2, etc.), we find phenomena which cannot be explained by the thermal theory. Under appropriate initial conditions these systems show not one but three ignition limits. Figure 6–3 represents the facts schematically. At the lowest pressures of the reactive gas-mixture the reaction rate is small, often, indeed, undetectable even by sensitive methods. If, however, the pressure is raised to p_1^* or somewhat higher, the gas ignites. So far the effect is similar superficially to that of a thermal explosion. On the other hand, there is a higher pressure range between p_2^* and p_3^* in which stable reaction occurs again. Here the behaviour is remarkable: gradually lowering the pressure causes the reaction rate to fall—as expected from ordinary kinetical considerations—until, however, the pressure p_2^* is reached, whereupon the gas suddenly ignites. Thus p_2^* is a pressure limit *beneath* which ignition occurs. Finally, increase of pressure to p_3^* also produces an explosion. The critical pressures are known as the first, second, and third ignition limits in order of increasing

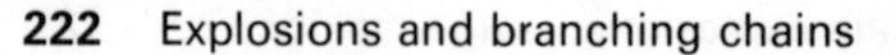

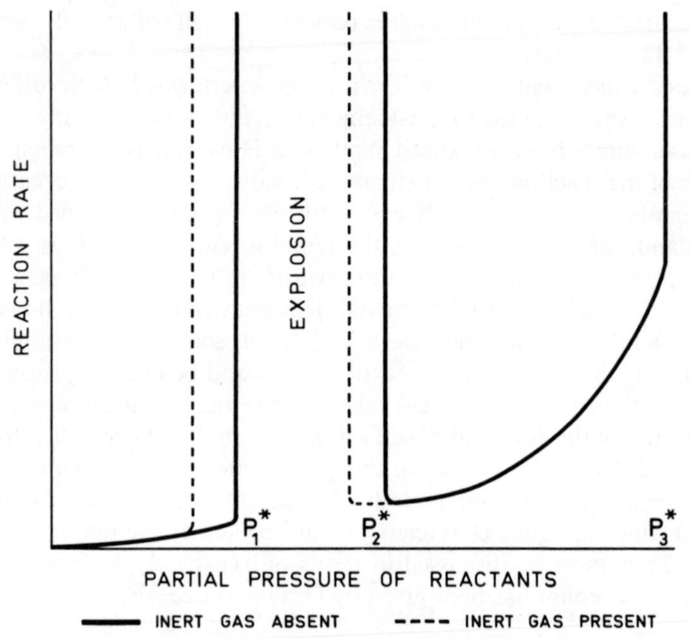

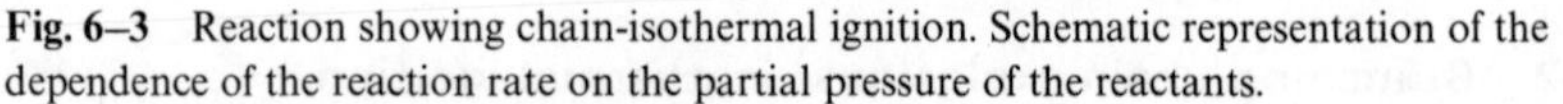

Fig. 6–3 Reaction showing chain-isothermal ignition. Schematic representation of the dependence of the reaction rate on the partial pressure of the reactants.

pressure, though in the older literature p_1^* and p_2^* are commonly referred to as 'lower' and 'upper' limits respectively. The next few paragraphs will show that the first and second limits do not arise from the thermal properties of the system, but from the presence of a reaction mechanism involving branched chains. The third limit p_3^*, on the other hand, usually originates at least partly from thermal imbalance and will receive relatively little attention in the following discussion.

6–2–1 The chain character of the first and second ignition limits

That reaction chains are important at least for the first limit is shown by the fact that p_1^* is sensitive not only to the size and shape of the reaction vessel but also to the condition of the surface, and usually to the presence of inert gas (cf. Section 3–3–3). With a gas of constant composition and vessels of the same material and shape, p_1^* is generally found to be inversely proportional to the radius **R**. Typical behaviour is illustrated in Table 6–1 which indicates the pressure range in which p_1^* is commonly found. Results are frequently erratic and the surface of the reaction vessel generally has to be 'conditioned', for example by repeated experiments, to obtain reproducibility. The effect of adding inert gas is always to lower p_1^* (which now refers to the *partial* pressure of the reactive mixture). This is different from the behaviour of a thermal explosion where the added gas raises or lowers the critical partial pressure, depending on whether it increases or decreases the thermal conductivity of the mixture. With certain types of surface, however, p_1^* is found to be very small and the influence of inert gas disappears though the radius effect remains. This indicates that under these conditions the reaction rate, though influenced by the surface, is no longer limited by diffusion. All these effects lead to the conclusion that the first limit is determined by the termination of reaction chains at the surface.

The response of the second limit to the same factors is quite different. Neither the geometry of the vessel nor the condition of its surface has much influence on p_2^* (see Table 6–1). Hence the second limit is determined by events in the gas phase, the surface playing little part. The effect of inert gas is to lower p_2^*, that is to make the mixture less explosive, in contrast

Table 6–1 Ignition limits of hydrogen–oxygen mixtures. Effects of the radius of the vessel and the nature of its internal surface

Temp. (°C)	*Radius* (**R**) (cm)	*Surface*	*1st limit* (p_1^*) (torr)	*2nd limit* (p_2^*) (torr)	p_1^*R
Active surfaces ((H_2+2O_2) spherical vessels; after Warren[16]).					
500	3.7	KCl	2.8	65	10
500	1.8	KCl	4.5	65	8
500	1.1	KCl	10.0	59	11
460	3.7	KOH	8.8	27	—
460	3.7	KCl	3.8	29	—
460	3.7	KH_2PO_4	0.5	31	—
Inactive surface ((H_2+O_2) cylindrical vessels; after Semenova[17]).					
440	1.50	§	0.33	—	0.50
440	1.00	§	0.49	—	0.49
440	0.78	§	0.66	—	0.51
440	0.51	§	0.92	—	0.47
440	0.29	§	1.38	—	0.40

§ Freshly fused pyrex.

to its effect on p_1^* which is to make the gas more explosive (see Fig. 6–3). Since the inhibiting influence of an inert gas cannot be chemical, the chain carriers must be eliminated in three body collisions. Thus qualitatively, it appears that the first and second limits are determined by heterogeneous and homogeneous chain termination respectively. However, there is nothing in the theory of straight chains developed in Chapter 3 that will explain the existence of more than a single (thermal) explosion limit. For this the concept of branched chains is required.

The existence of branched reaction chains was mentioned in Chapter 3, but the kinetic characteristics which distinguish them from straight chains were not discussed. A complete propagation cycle of a straight-chain reaction can be represented as follows,

$$\text{R} + \text{Reactants} \rightarrow \text{Products} + \text{R} \qquad \text{(straight chain)}$$

Each chain carrier R entering the cycle is replaced by another but no more.

With a branched chain, on the other hand, a complete cycle produces more chain carriers than it consumes:

$$R + \text{Reactants} \rightarrow \text{Products} + jR \qquad \text{(branched chain)}$$

where $j > 1$. The following is an example of a branching cycle relevant to the hydrogen–oxygen reaction (to which we shall return later):

$$H + O_2 \rightarrow OH + O \qquad \text{6–I}$$

$$O + H_2 \rightarrow OH + H \qquad \text{6–II}$$

$$2OH + 2H_2 \rightarrow 2H_2O + 2H \qquad \text{6–III}$$

In this case two elementary branching reactions (6–I and 6–II) are involved, each producing two active particles for one consumed. The complete cycle can be represented stoichiometrically by

$$H + 3H_2 + O_2 \rightarrow 2H_2O + 3H \qquad \text{6–IV}$$

and thus results in a net gain of two hydrogen atoms.§

Naturally the active species participating in branched as in unbranched chains are liable to be neutralized in various ways and, as already indicated, the explosion limits are closely connected with chain-terminating reactions. At this point it is convenient to introduce a simplification. It was found in Chapter 3 that the kinetics of a straight-chain reaction are often completely or at least substantially determined by the concentration of one particular radical species, even though two or more are involved in propagating the chains. We may suppose, therefore, that we are dealing with a branched-chain reaction for which a similar situation holds. If then [R] denotes the concentration of the essential chain carrier, its rate of increase from zero at the beginning of the reaction will be given by

$$\frac{d[R]}{dt} = \rho_i + f[R] - g'[R] - h[R]^2 \qquad \text{6–16}$$

where the terms on the right represent in order the rates of chain initiation, branching, and first- and second-order termination respectively. The coefficient f is proportional to the concentration of one of the reactants and the term $f[R]$ is directly related to the rate of formation of the products. If, for example, hydrogen atoms are the rate-determining species in the branching cycle represented by reactions 6–I to 6–III, f becomes equal to $2k_1[O_2]$ and the rate of formation of water is $2k_1[O_2][H]$. We also assume that the constant h is too small for second-order chain termination to be appreciable with the concentrations of R under consideration. Hence termination is represented by the term $g'[R]$ and we shall see shortly that in general g' is also a function of the reactant concentrations. Thus, equation 6–16 becomes

$$\frac{d[R]}{dt} = \rho_i + (f - g')[R] \qquad \text{6–17}$$

§ It should not pass unnoticed that the cycle as summarized by reaction 6–IV is exothermic to the extent of 11 kcal.

The behaviour of the reaction depends strongly on the value of the *net branching factor* $\phi \equiv (f-g')$. The concentration of the chain carriers at any time t from the beginning of the reaction is obtained by integrating equation 6–17. If ϕ is negative, that is, if $g' > f$, it is given by

$$[\mathrm{R}] = \frac{\rho_i}{g'-f}\{1-\exp[-(g'-f)t]\} \quad (g' > f) \qquad 6\text{–}18$$

It is represented graphically by curve A in Fig. 6–4. As t increases, [R] rapidly approaches the steady-state value $\rho_i/(g'-f)$; and the reaction rate, being proportional to [R], behaves in a similar manner. The effect is basically similar to that of a straight-chain reaction, as described in Section 3–6. The result is quite different, however, if f is greater than g'. Integration of equation 6–17 then yields

$$[\mathrm{R}] = \frac{\rho_i}{(f-g')}\{\exp[(f-g')t]-1\} \quad (g' < f) \qquad 6\text{–}19$$

as represented by curve B in Fig. 6–4. As t increases, the value of the exponential term soon exceeds unity and thereafter the concentration of chain carriers, as also the reaction rate, rises exponentially. Thus we have an ever accelerating reaction; in other words, an explosion. This is

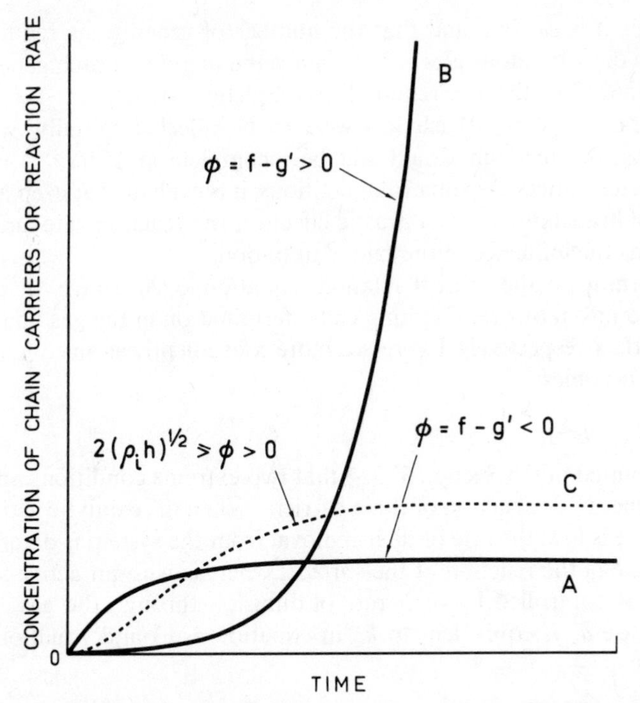

Fig. 6–4 Development of branching chains. Curves A and B show the influence of the net branching factor ϕ when the chains are terminated by reactions first-order in the concentration of the chain carriers. Curve C shows the influence of second-order termination.

brought about solely by the multiplication of active species and not by the evolution of heat.

Obviously the condition for ignition is

$$(f - g') > 0 \qquad 6\text{–}20$$

Before elaborating on this relation it is worth pausing to note the enormous effect a positive value of $(f - g')$ can have on the reaction rate. Following Pease,[18] we shall suppose that a single chain carrier is introduced into 1 cm^3 of the reactive gas (10^{19} molecules) and, initially, that $(f - g')$ is exactly zero. This is equivalent to a straight chain: when the carrier reacts with a reactant molecule it is simply replaced by another carrier and so on. If it is assumed that this occurs on every collision—once in about every 10^{-8} s—the time taken to consume the 10^{19} molecules would be about 3000 years. Suppose, however, that each collision on the average results in the production of 1.01 chain carriers, that is $(f - g')$ is positive and equal to $0.01f$. The number of chain carriers in each 'generation' will now be 1.01 times what it was in the previous one, and the total number of reactant molecules consumed after n generations will be

$$1 + 1.01 + (1.01)^2 + (1.01)^3 + \cdots + (1.01)^n = [(1.01)^{n+1} - 1]/0.01$$

From this it is easily found that the number of generations required to consume the 10^{19} molecules is 3933. Since the lifetime of each generation is 10^{-8} s as before, the time required to complete the reaction is 4×10^{-5} s. By contrast, even if 10^8 carriers were to be injected initially, without branching, the reaction would still be incomplete in 15 min. Although these circumstances are somewhat fictitious, it is evident that even a small degree of branching exerts a drastic effect on the reaction rate and soon outweighs the influence of the rate of initiation.

Returning to the critical relation, equation 6–20 above, g' can be separated into terms representing chain termination in the gas phase and at the surface respectively. Expressed more conveniently as an equality, the relation becomes

$$f = g_g + g_w \qquad 6\text{–}21$$

It was pointed out in Section 3–3–3 that two extreme conditions apply to heterogeneous destruction of chain carriers. When the catalytic activity of the surface is low, the rate of their removal from the system is determined by the rate of the reaction at the surface; whereas with an active surface the rate is controlled by their rate of diffusion through the gas. In the former case g_w is equivalent to k_w in equation 4–13 and equation 6–21 becomes

$$f = g_g + B\bar{c}\gamma/\mathbf{R} \qquad 6\text{–}22$$

where $\bar{c}$ is the mean molecular speed of the chain carrier, γ is the probability of its being destroyed on collision with the surface, and B is equal to 3 for a spherical reaction vessel or 2 for a long cylinder. When we come to the more usual case of rate control by diffusion, a difficulty appears in that equation 6–20 was derived on the assumption that the concentration of the chain carriers is uniform, whereas the presence of a diffusion gradient implies that it decreases from the centre to the wall. However, the problem is more apparent than real; Bursian and Sorokin[19] showed long ago that, if the critical condition is considered as applying to the average concentration of the chain carriers, equation 6–21 becomes

$$f = g_g + (\Gamma\pi^2 D)/\mathbf{R}^2 \qquad 6\text{–}23$$

where Γ is a numerical constant (1 for a sphere or 0.59 for a cylinder) and D is the diffusion coefficient. This equation supplies the proportionality factor ($\Gamma\pi^2$) for the result obtained intuitively from the Einstein–Schmolukowski law in Section 3–3–3. The following considerations show that either equation 6–22 or 6–23 leads naturally to the existence of two pressure limits for ignition, each with the correct properties.

The first limit

It may be supposed that there are conditions in which the chains terminate at the surface much more frequently than in the gas phase and vice versa; and, as we have already seen, the indications are that the former is the case at the first limit. Hence, as an approximation, g_g can be omitted from equations 6–22 and 6–23 for these conditions. *For a gas of constant composition*, f is directly and D is inversely proportional to the pressure. Hence equation 6–23 leads to the relation

$$p_1^*\mathbf{R} = (\Gamma\pi^2 D_0/\kappa_b)^{1/2} = \text{constant} \quad \text{(first limit)} \qquad 6\text{–}24$$

where D_0 is the diffusion constant at unit pressure of the mixed reactants and κ_b is a simple function of the mixture composition and the rate constant for the branching reaction. Thus, in agreement with the observed behaviour of the first limit (e.g. with P_4+O_2[20] and H_2+O_2[22]), we have a maximum pressure for stable reaction which is inversely proportional to **R**. Furthermore, since an addition of inert gas effectively reduces the value of D_0 it must likewise reduce the value of p_1^* (the critical *partial* pressure of the mixed reactants). The effects of different inert gases on the first limits in the phosphorus–oxygen[21] and hydrogen–oxygen[22] reactions are, in fact, inversely related to their diffusivities. Equation 6–22, which applies only to very inactive surfaces, also predicts an inverse relation between p_1^* and **R** but shows no effect of inert gas. Such a surface is difficult to realize in a reproducible condition but Nalbandyan and collaborators[23] obtained results corresponding to equation 6–22 with hydrogen–oxygen mixtures by treating the surface with a solution of sodium tetraborate. As expected, the limit occurred at unusually low pressures (<1 torr).

The second limit

We now consider the applicability of equation 6–21 to the situation where the chains are terminated exclusively in the gas phase. If it is assumed that the chain carriers are destroyed in three-body collisions with molecules of either reactant, the termination rate becomes proportional to the square of the pressure; and equation 6–21 takes the form:

$$\kappa_b p_2^* = k_t'(p_2^*)^2$$

Hence

$$p_2^* = \kappa_b/k_t' = \text{constant} \quad \text{(second limit)} \qquad 6\text{–}25$$

where k_t', a pseudo rate constant, depends on the composition of the gas. Thus p_2^* is a minimum pressure for stable reaction and it is independent of the size and surface of the vessel. The experimental finding that p_2^* is reduced by the addition of inert gas is accounted for by an increase in k_t'. In general the relative effects of different gases are found to be in the order expected for their relative efficiencies as third bodies.[24]

The explosion peninsula

The effect of temperature on the ignition limits is of much interest. If the relatively small temperature dependence of $D_0^{1/2}$ is neglected, equation 6–24 leads to the relation

$$p_1^* \propto \exp[E_b/2RT] \quad \text{(first limit)} \qquad 6\text{–}26$$

where E_b is the activation energy for the branching reaction. Similarly equation 6–25 yields

$$p_2^* \propto \exp[(-E_b+E_t)/RT] \approx \exp[-E_b/RT] \quad \text{(second limit)} \qquad 6\text{–}27$$

where E_t, which refers to the three-body terminating reaction, is small and negative (cf. Section 5–3). According to these relations, the temperature dependence of p_1^* is smaller and in the opposite sense from that of p_2^*. Since p_1^* increases and p_2^* decreases with decreasing temperature, mathematically their values must coincide at some low value of T. This, of course, is incompatible with the mutually exclusive assumptions about the kinds of chain termination operating at p_1^* and p_2^*. Physically, however, one can see that as p_1^* and p_2^* converge both assumptions will become less and less valid; and the blending of surface and gas-phase termination will have the same effect, qualitatively, of bringing p_1^* and p_2^* into coincidence. Thus a graph of ignition pressure p^* against T shows an *explosion peninsula.* This is exemplified by the lower half of the graph in Fig. 6–5 which refers to hydrogen–oxygen explosions.[25] The peninsula is a highly characteristic phenomenon and has been found with a number of oxidation systems: $CO:O_2$; $PH_3:O_2$; $CS_2:O_2$, etc.[26]§ The pressures at

§Figure 6–5, a celebrated and indispensable diagram first published by Lewis and von Elbe,[25] illustrates the temperature dependence of the *three* ignition limits. Its frequent reproduction may lead the student incorrectly to suppose that such behaviour is unique to the $H_2:O_2$ system. A similar diagram showing the three limits for the $H_2S:O_2$ system will be found in reference 27.

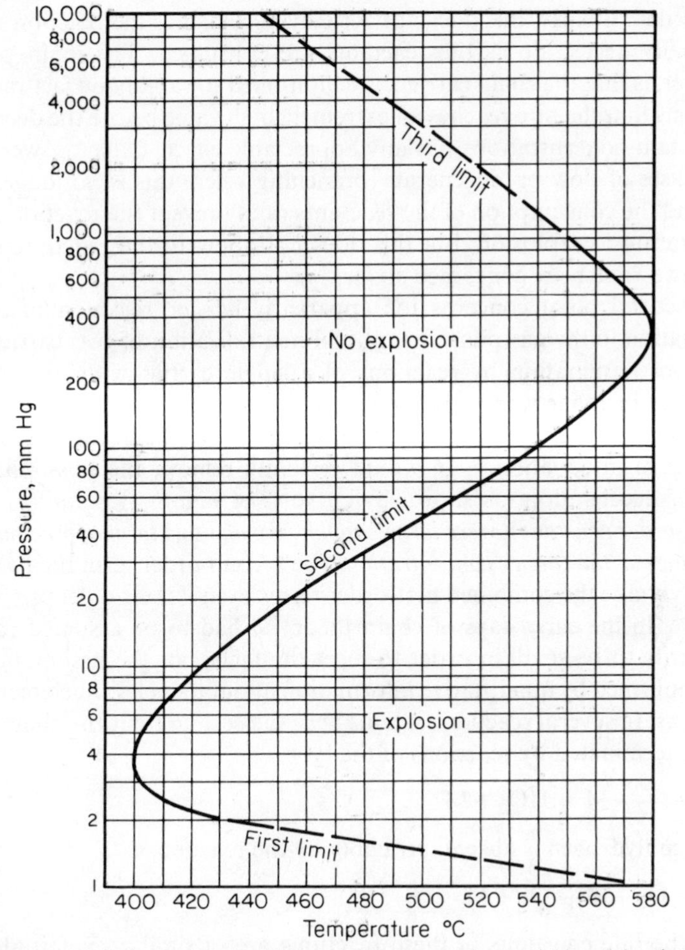

Fig. 6–5 Ignition diagram for the hydrogen–oxygen reaction (stoichiometric mixtures in a KCl-coated spherical vessel of radius 3.7 cm (Lewis and von Elbe[25])).

which the peninsula is cut by a line parallel to the ordinate represent the first and second ignition limits at the particular temperature; and the apex represents the lowest temperature at which spontaneous ignition can occur for the particular gas-composition and reaction vessel.

6–2–2 Further remarks on chain ignition

The previous discussion has shown that the existence and properties of ignition limits are adequately accounted for by chain-branching, the first limit on the assumption that the chains are mostly broken at the wall, and the second that they mostly come to an end in the gas phase. Some difficulties may be felt, however, with regard to the original application of equation 6–17. Curve B of Fig. 6–4, which is based on equation 6–17, represents the rate as accelerating to infinity when the critical condition is exceeded. Clearly, this does not actually occur, since the reaction must eventually decelerate as the reactants are consumed. The main point, however, is that the chain carriers are multiplied so rapidly by fast radical reactions that the rate reaches an extremely high value before the decrease in reactant concentrations has any appreciable effect. Later on we shall meet cases of slow or 'degenerate' branching where this is no longer the case and the consumption of the reactants does prevent the reaction from accelerating to explosion. But this does not apply to the reactions with which we have been concerned so far.

A second point concerns the apparently *ad hoc* rejection of chain termination in the gas phase by mutual neutralization of two carriers in favour of termination by reactions of a single carrier with a reactant molecule; in other words the choice of 'first-order' instead of 'second-order' termination should be justified. Retention of the negative second-order term in equation 6–16 would certainly remove the possibility of 'infinite' acceleration (as shown by curve C of Fig. 6–4) *if the (pseudo) second-order constant* h *were large enough to make this term appreciable at low values of the carrier concentration.* The second-order term then would soon overtake the combined first-order terms even if their difference were positive. In the early days of chain theory, h had to be assumed rather arbitrarily to be small in order to meet the facts, but the assumption is now confirmed by quantitative information about the relevant elementary reactions. In several reaction systems near the second limit the chains are mostly terminated by reactions of the type

$$H+O_2+M \rightarrow HO_2+M \qquad 6\text{–V}$$

or, when hydrogen is absent, probably by the reaction

$$O+O_2+M \rightarrow O_3+M \qquad 6\text{–VI}$$

Since the rate constants of these reactions are of similar magnitudes to those of the pseudo second-order terminating reactions:

$$H+H+M \rightarrow H_2+M \qquad 6\text{–VII}$$

$$H+OH+M \rightarrow H_2O+M \qquad 6\text{–VIII}$$

$$O+O+M \rightarrow O_2+M \qquad 6\text{–IX}$$

etc.

it follows that the rates of the latter reactions do not become comparable with those of reactions 6–V and 6–VI until the concentrations of the chain carriers H, OH, and O become comparable with the concentration of the molecular oxygen present; that is, not until after ignition has actually occurred. The inhibiting effect of reaction 6–V arises from the fact that the HO_2 radicals or their equivalent are much less reactive than the other radicals present and mostly survive until they are destroyed at the wall or by mutual disproportionation in the gas phase. We shall see in Chapter 7, however, that in flames, where a state of continuous explosion prevails and radical concentrations are high and there is no wall, the chains are terminated by radical–radical combination.

The possibility of *branching* by a reaction between two chain carriers was also omitted from equation 6–17. The result of such a reaction could be to produce more than two carriers or, more likely, two new carriers that are more reactive than the old ones towards chain propagation or less reactive towards chain termination. When such *quadratic branching* takes place, ignition can occur even when $(f-g')$, the net factor for *linear* branching, is negative. In these circumstances, the critical condition becomes

$$\frac{d[\mathrm{R}]}{dt} = 0 = \rho_i - (g'-f)[\mathrm{R}] + F[\mathrm{R}]^2 \qquad 6\text{–}28$$

Solution of this quadratic shows that the stationary state is possible only when $(g'-f)^2 \geqslant 4\rho_i F$. Hence the condition for ignition to occur as the result of quadratic branching is

$$[\rho_i F/(f-g')^2] > 0.25 \qquad (f < g') \qquad 6\text{–}29$$

Unlike the case of ignition by linear branching, the critical condition here depends on ρ_i; the greater the rate of initiation the more likely is the system to ignite. In the next section, we shall see that in certain circumstances quadratic branching is responsible for ignition in the hydrogen–oxygen reaction.

Finally, a reference should be made to the term *isothermal* used, perhaps not too aptly, to describe ignition at p_1^* and p_2^*. Since reactions of the type under consideration are generally exothermic, the evolution of heat inevitably contributes to the acceleration of the reaction *after* ignition; 'isothermal' simply refers to the fact that heat plays no part in determining the conditions for an explosion to be *initiated*. Perhaps the most striking example of isothermal ignition was the first to be discovered, that of phosphorus vapour and oxygen.[28] In vessels of normal size, p_1^* occurs at considerably less than 1 torr, at which pressures the heat generated is exceedingly small. It is a paradox difficult to contemplate without mingled emotions that the existence of branched-chain explosions, so familiar to us now from nuclear catastrophes, was first recognized in the feeble glow of phosphorus.

6–2–3 The hydrogen–oxygen reaction

It is safe to say that more experimental investigations have been carried out on this than on any other single reaction. Its key role in the development of chain theory; the vivid contrast between its intricate kinetics and simple stoichiometry; its tendency to irreproducible behaviour; the accessibility of the reactants; and the technological importance of combustion, all together have sustained interest in its kinetics for the last 50 years. The result is that, although some obscurities undoubtedly remain, the mechanism is understood in much quantitative detail. For the moment, however, we are concerned less with the details than with the broad kinetic features. These are themselves sufficiently complex, as a glance at Fig. 6–5 will partly show. Nevertheless, we shall see that several of the most interesting features are the natural, with hindsight one might say the inevitable, consequences of fundamentally simple considerations.[29] Figure 6–5 shows that a complete account of the kinetics would embrace seven phenomena or regimes of reaction: the three ignition limits, the slow reactions occurring below the first and above the second limits, and the explosive reactions within the explosion peninsula and above the third limit. However, we have space only for a short discussion of the first and second limits together with a few words about the third limit and the intervening slow reaction.

Early workers were much troubled by erratic behaviour caused by varying catalytic activity of quartz or glass surfaces towards radical recombination. Subsequently it became clear that this can be overcome either by using a highly active surface, thus throwing most of the control of heterogeneous chain termination upon diffusion through the gas phase, or by using a highly inactive surface, which virtually eliminates heterogeneous termination altogether. A coating of potassium chloride has commonly been used for the first purpose and one of boric oxide for the second. Not unnaturally such drastic changes in surface properties produce some distinct changes in the kinetics; and indeed these cast much light on the reaction mechanism. Our discussion begins with ignition in the presence of an *active* surface. Following the general account of ignition limits given previously, it will be convenient to proceed from the reaction mechanism to the phenomena rather than vice versa. A mechanism for chain branching was given on p. 224. The same and other elementary

Table 6–2 Principal elementary reactions in the hydrogen–oxygen reaction at about 500 °C
(listed in order of appearance in the text but numbered according to Baldwin *et al.*[30]).

Reaction	No.
Primary initiation $\longrightarrow$ OH or H	(i)
$H+O_2 \longrightarrow OH+O$	(2)
$OH+H_2 \longrightarrow H_2O+H$	(1)
$O+H_2 \longrightarrow OH+H$	(3)
$H+O_2+M \longrightarrow HO_2+M$	(4)
$H \xrightarrow{\text{wall}}$ destruction	(5c)
OH or O $\xrightarrow{\text{wall}}$ destruction	(5b)
$HO_2 \xrightarrow{\text{wall}}$ destruction	(5a)
$HO_2+H_2 \longrightarrow H_2O_2+H$	(11)
$HO_2+H \longrightarrow 2OH$	(8)
$H_2O_2+M \longrightarrow 2OH+M$	(7)
$2HO_2 \longrightarrow H_2O_2+O_2$	(10)
$H+H_2O_2 \longrightarrow H_2O+OH$	(14)
$H+H_2O_2 \longrightarrow H_2+HO_2$	(14a)
$OH+H_2O_2 \longrightarrow H_2O+HO_2$	(15)

reactions are shown in Table 6–2, where they are given numbers which have become customary since the work of Baldwin.[30] The control of branching by reaction **2** (followed by reactions **1** and **3**) was mentioned previously as a more or less hypothetical possibility; but, in fact, when combined with gas-phase termination by reaction **4** and chain breaking at the wall by reaction **5c** it provides an almost complete explanation of the behaviour of the first and second limits. This is best seen by reference to the effects of the hydrogen and oxygen concentrations on the critical condition for ignition.

The basic assumptions or approximations are that branching by reaction **2** is much slower than by reaction **3** and that reaction **1** is much faster than reaction **2** (as is very probable *a priori*). Thus H will be in greater concentration than either O or OH and the only terminating reactions which need be considered are those involving H. Assuming a few radicals are generated spontaneously to initiate the branching mechanism, the critical condition for ignition is obtained by equating the rates of branching and termination as in equation 6–23. This becomes

$$2k_2[O_2] = k_4^{H_2}[O_2]\{[H]_2 + a[O_2]\} + GD \qquad 6\text{–}30$$

where a is the ratio of $k_4^{O_2}/k_4^{H_2}$ referring to the relative efficiencies of O_2 and H_2 as third bodies in reaction **4**, and $G = \Gamma\pi^2/\mathbf{R}^2$. It is assumed that no foreign gases are present. D, the diffusion coefficient for H (in low concentration) in the mixed gases, can be obtained from the expression:[24]

$$D = \left\{\frac{[H_2]}{D_0^{H_2}} + \frac{[O_2]}{D_0^{O_2}}\right\}^{-1} \qquad 6\text{–}31$$

where D_0^i is the coefficient for diffusion through the pure gas i at unit concentration. Combining equations 6–30 and 6–31 yields the relation,

$$\{[H_2] + a[O_2]\} + \frac{G'}{[O_2]\{[H_2] + b[O_2]\}} = \frac{2k_2}{k_4^{H_2}} = K \qquad 6\text{–}32$$

where

$$G' = GD_0^{H_2}/k_4^{H_2} = \{\Gamma\pi^2 D_0^{H_2}/k_4^{H_2}\mathbf{R}^2\} \qquad 6\text{–}33$$

and $b = D_0^{H_2}/D_0^{O_2}$. At *high* pressures the first term on the left of equation 6–32 predominates and a graph of $[H_2]$ against $[O_2]$ approximates to a straight line inclined to both axes. This is shown by the line AB in Fig. 6–6. In these circumstances equation 6–32, which it will be seen is well reproduced by the experimental facts, represents the reciprocity which must exist between the concentrations of the two reactants for ignition to occur at the second limit. In the diagram the line divides explosive mixtures (below) from non-explosive (above). Rearrangement of equation 6–32, however, will show that, at sufficiently *low* total pressures, the critical concentrations are represented by a relation of the form

$$[H_2] = (GD_0^{H_2}/2k_2)[O_2]^{-1} - b[O_2] \qquad 6\text{–}34$$

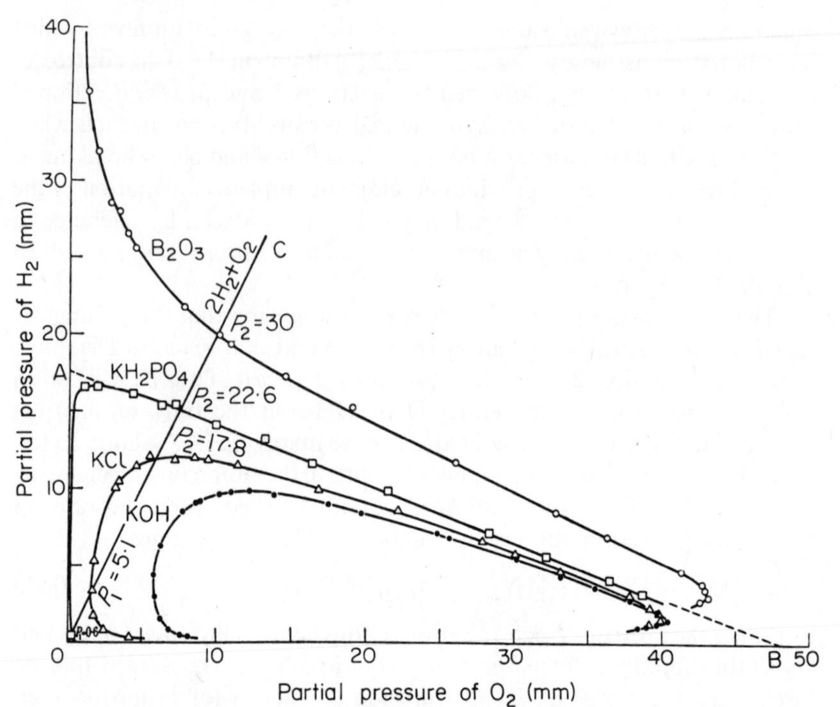

Fig. 6–6 The hydrogen–oxygen reaction. Dependence of the ignition limits on the concentrations of hydrogen and oxygen. KOH, KCl, etc., refer to materials applied to the surface of the vessel (Warren[16]).

Omitting for the present the experiments with the boric oxide surface, this, too, describes the facts, as is shown by the steep fall in $[H_2]$ with increasing $[O_2]$ in the lower left-hand corner of Fig. 6–6. Here it is the first limit that is described, and points on the diagram above and below the line refer to explosive and non-explosive conditions respectively. For quantitative tests of the complete equation 6–32, the interested reader is referred to the very clear papers by Warren.[16] These show that the experimental values of the constants a and b are as expected from the physical properties of the gases and that G' is inversely proportional to the square of the radius of the reaction vessel, as required by equation 6–33. For each surface condition the explosive 'region' is that enclosed by the appropriate line in Fig. 6–6; and the first and second pressure limits (indicated for the mixture $2H_2+O_2$ by the intersections with the line OC) are seen as different manifestations of the same simple mechanism, namely reactions **1 to 5c**.§

Much less is known about the events occurring at the *third limit*. It seems, however, that, as ordinarily observed, the ignition is largely thermal

§ But when the ratio $[O_2]/[H_2]$ is made sufficiently large, the concentrations of O and OH become comparable with that of H as reaction **3** takes over from reaction **2** as the rate-determining branching reaction. Reaction **5b** then can no longer be neglected. This is the origin of the 'turning back' at the right-hand end of the 'limit lines' in Fig. 6–6.[16]

in character; that is, it is precipitated by the heat generated by the slow reaction (cf. Section 6–3). On the other hand when the rate is suppressed by a *highly* active surface (e.g. potassium chloride) the limit becomes principally determined by chain-branching. This comes about as the result of reaction **11**. As already indicated, the suppression of branching at the second limit is due to the fact that the HO_2 radical formed in reaction **4** is too inactive to continue the chain before it is destroyed. As the pressure of the reactants is increased, however, the resistance to diffusion increases, thus increasing the chance of regenerating a hydrogen atom by reaction **11**. The result is that the net branching factor eventually becomes positive again and the mixture ignites.

We must now return to the conditions which apply to ignition when the surface is *inactive* catalytically. In this case new kinetic features appear which indicate a significant modification of the mechanism. Boric oxide provides a surface of this type (cf. Section 4–3–1) which moreover becomes extremely inert (and stable) after it has been 'aged' by repeated exposure to the reaction.[31] Figure 6–6 shows that the behaviour of the second limit is different in these circumstances from that found in the presence of an active surface. In the first place the limit occurs at higher pressures, which shows that in general it is not entirely uninfluenced by chain-breaking at the surface. More significantly, the turning down of the 'limit line' corresponding to the approach to the first limit is replaced by the opposite effect. The cause of the phenomenon can be traced to the higher concentrations of H and HO_2 present as the result of the greater survival of the radicals from collisions with the surface. It transpires that the gas is more explosive than it otherwise would be at low oxygen concentrations because an additional branching reaction now makes its effect:

$$HO_2 + H \rightarrow 2OH \qquad \mathbf{8}$$

Elucidation of the kinetics in terms of a reaction mechanism compatible with the behaviour of the slow reaction and other kinetic data has required a great deal of subtle analysis which cannot be reproduced here. It will suffice to state the conclusion reached by Baldwin, Mayor, and Doran.[31] In the absence of wall termination the critical condition for ignition becomes

$$\{[H_2]+a[O_2]\} - B\big[\{[H_2]+a[O_2]\}\{[H_2]+d[O_2]\}/[O_2]\big]^{1/3} = K \qquad 6\text{–}35$$

(approximately) where

$$B = [27k_{\mathbf{2}}k_{\mathbf{7}}k_{\mathbf{8}}^2/4k_{\mathbf{10}}k_{\mathbf{14}}(k_{\mathbf{4}}^{H_2})^2]^{1/3} \qquad 6\text{–}36$$

and d refers to relative two-body efficiencies in reaction **7**. Equation 6–35 accurately reproduces the behaviour of the second limit with a boric acid surface shown in Fig. 6–6.

It will be seen from equation 6–36 that the increased radical concentrations bring other reactions into prominence in addition to reaction **8**, namely reactions **7**, **10**, and **14** (Table 6–2). Reaction **8** is a branching reaction because it converts an HO_2 radical, which does not lead inevitably to the branching reaction **2**, into an OH radical, which does. It is an example of quadratic branching which, as noted in Section 6–2–2, has the effect of making the ignition condition depend on the rate of initiation of the chains. The evidence is that initiation is brought about by the decomposition of hydrogen peroxide (reaction 7),[31,32] but the situation is complicated by the fact that hydrogen peroxide is produced and destroyed in the course of chain propagation (by reactions **10** and **14**). Indeed equation 6–35 is based on the assumption that its concentration reaches a steady state before ignition occurs. Thus a minimum of eight elementary reactions **2** to **4** and **8** to **14** in the order listed in Table 6–2 is needed to explain the kinetics at the second limit when the surface no longer acts as a radical sink.

Finally a brief reference will be made to the *non-explosive reaction* between the second and third limits. A characteristic feature observed with an inactive surface is an induction period or initial period of acceleration which usually extends for one or two minutes before the reaction reaches its maximum rate. This is not caused by an inhibitor but is due to the 'auto-catalytic' formation of hydrogen peroxide, of which small amounts are found in the products. The function of hydrogen peroxide as chain-initiator was mentioned in our discussion of the second limit, but the evidence for it comes from the slow reaction. The most direct evidence is the fact that the maximum rate of reaction coincides with the maximum concentration of peroxide.[31] From a very low rate of 'primary' initiation by some unspecified means,§ the peroxide builds up auto-catalytically to a maximum concentration at which the rate of its formation becomes balanced by the rate of its destruction. Except near the very beginning of the reaction, the rate of the primary process is negligible compared with the rate of initiation by the peroxide. Hence the reaction in effect provides its own initiation. (This feature puts the reaction in the class of 'degenerate explosions' which we shall examine more generally in Section 6–4.)

The overall mechanism found necessary to explain the behaviour of both the slow reaction and the second limit in the presence of an inactive surface is given in Table 6–2, omitting reactions **5c–a**.[30,31] Various elementary reactions such as

$$HO_2 + H_2 \rightarrow H_2O + OH \qquad \mathbf{11a}$$

and

$$O + H_2O_2 \rightarrow H_2O + O_2 \qquad \mathbf{13}$$

which might otherwise be considered plausible, are excluded either as being incompatible with the observed kinetics or as exerting only minor

§ This is probably the reaction.[33]

$$H_2 + O_2 \rightarrow 2OH \qquad \text{6–X}$$

influence. The mechanism is now sufficiently well established quantitatively (for the inactive surface conditions mentioned above) to allow various rate-constant ratios to be determined from the kinetics (e.g. k_{14}/k_2[30]). For the same reason, hydrogen–oxygen mixtures can also be used as sources of H, OH, and O for studies of elementary reactions of these radicals with other molecules.[34]

6–2–4 Chain-branching in the carbon monoxide–oxygen and hydrogen–fluorine reactions. Sensitized ignition and branching by energy transfer

We saw in Chapter 3 that straight-chain reactions may be greatly accelerated by adding a small amount of a substance which initiates the chains faster than the pure reactants. A still greater effect can be expected if an agent is introduced which causes *chain-branching*, particularly if branching either does not occur otherwise or occurs relatively infrequently. The phenomenon is illustrated in a remarkable way by the effect of hydrogen and its compounds on the oxidation of carbon monoxide. It is the main reason for including this reaction in our discussion of branching chains. A second reason is that the reaction offers an opportunity to introduce the concept of chain-branching brought about by highly energized species rather than free radicals. The $CO:O_2$ reaction is frequently, though as we shall see, perhaps erroneously, quoted as the outstanding example of this phenomenon. This will lead us to a second example provided by the hydrogen–fluorine reaction. But first we need to consider some general features of the carbon monoxide–oxygen system.

6–2–5 Ignition of carbon monoxide–oxygen mixtures

The reaction of oxygen with carbon monoxide has been known to be a branched-chain reaction for almost as long as its reaction with hydrogen. The two reactions are of comparable interest for both science and technology, but in spite of much research the kinetics and mechanism of the carbon monoxide reaction are not so well understood. Experiments are made very difficult by irreproducible surface effects even more acute than in the hydrogen–oxygen system and, above all, by the potent sensitizing effect exerted by traces of water or any substance containing hydrogen. Over almost a century the mutual reactivity of the reactants has been found to decrease as efforts to reduce their water content have been intensified; and, as we shall see later, even today it is not at all certain that a genuine 'dry' or hydrogen-free reaction has ever been observed. When the reactants have not been intensively dried, the explosion peninsula occurs at temperatures about 150° higher than that of the hydrogen–oxygen reaction (cf. Fig. 6–5). In general, the first and second ignition limits have the expected properties but there is no sign of the behaviour attributed to quadratic branching in the hydrogen–oxygen system; nor is there a third limit below atmospheric pressure. According to Hoare and Walsh,[35] the explosion peninsula lies inside another peninsula which extends to a lower temperature in the $p^* - T$ diagram. The boundary of the outer peninsula marks the onset of emission of a visible blue glow which, however, occurs with little consumption of the reactants. The glow does not give way to true ignition unless the explosion peninsula proper is entered. It has been suggested that the 'glow limits' mark conditions at which the net branching factor first becomes positive, but the reaction is rapidly self-inhibited unless sufficient heat is developed to precipitate explosion.[35] On this idea, the true ignition is of a chain-thermal character (cf. Section 6–3), but this is by no means proved.

The 'wet' reaction

The phenomena take on a different aspect in the presence of hydrogen or water. Addition of as little as 0.1 per cent hydrogen has a profound influence, widening the ignition limits and extending the explosion peninsula to lower temperatures. Further additions increase these effects until, when about 1 per cent is added, the explosion peninsula occupies about the same region of the $p^* - T$ plane as that of the hydrogen–oxygen reaction.[36] When 10 per cent of the carbon monoxide is replaced by hydrogen, the behaviour at the second limit is almost the same quantitatively as that of an equivalent hydrogen–oxygen–nitrogen mixture.[37] Thus, kinetically, the carbon monoxide behaves almost as if it were an inert gas in the mixture of hydrogen and oxygen. Nevertheless it is consumed in the explosions, and these become more sharply defined and complete with progressive addition of hydrogen. The dominance of the hydrogen–oxygen kinetics is shown further by the appearance of a third limit.[38] When water vapour is added instead of hydrogen the effects are generally similar.

In the absence of quadratic branching, the strong influence on the explosion limits shows that the function of the hydrogen or water is to introduce a facile means of chain-branching. This must be closely similar to the branching mechanism of the hydrogen–oxygen reaction. In fact, with molecular hydrogen present, most of the kinetic details in the region of the second limit and above are explained by simple modifications of the basic hydrogen–oxygen mechanism:[36,37]

Branching	$H + O_2 \rightarrow OH + O$	**2**
	$O + H_2 \rightarrow OH + H$	**3**
Propagation	$OH + CO \rightarrow CO_2 + H$	6–XI
Termination	$H + O_2 + M \rightarrow HO_2 + M$	**4**
	$O + CO + M \rightarrow CO_2 + M$	6–XII

Reaction 6–XI replaces the analogous reaction

$$OH + H_2 \rightarrow H_2O + H \qquad \mathbf{1}$$

in the pure hydrogen system and there is an extra terminating reaction, namely reaction 6–XII. Since, however, this reaction is 10 to 50 times slower than reaction **4** and neither reaction 6–XI nor **1** affects the concentration of chain-carriers in either system, the condition for ignition is still largely determined by the branching and terminating reactions introduced with the hydrogen. More precisely, by equating the rates of branching and terminating, it can be shown that the condition for the second limit is

$$[M] = (2k_2/k_4)/\{1+k_{12}[CO][M]/k_3[H_2]\} \qquad 6\text{–}37$$

where k_4 and k_{12} are mean values referring to the total concentrations of third-body molecules present ([M]). If this equation is compared with equation 6–32, it will be seen that the limiting condition approaches that applicable to the oxidation of pure hydrogen as the ratio ($[H_2]/[CO][M]$) is increased. How little hydrogen is required to reduce the effect of the carbon monoxide to a low level can be appreciated by substituting the values of k_3 and k_{12} in equation 6–37. At 550 °C, $k_3 = 10^{10.8}$ cm^3 mole^{-1} s^{-1}, $k_{12} \approx 10^{14}$ cm^6 mole^{-2} s^{-2}, and p_2^* is of the order of 100 torr. It follows that replacement of 1 per cent of the carbon monoxide by hydrogen in the mixture $2CO+O_2$ is sufficient to increase p_2^* to within about 10 per cent of its value with pure hydrogen (assuming H_2 and CO to be equally effective as third bodies in reaction **4**). This, as noted earlier qualitatively, is what is observed.

The same mechanism applies to the oxidation of carbon monoxide in explosions at 1200–2000 °C. When ignition is brought about by a shock wave, the exponential rise in the concentration of carbon dioxide due to chain-branching can be followed photometrically.[39,40] Even when the hydrogen content of the gas is as low as 0.01 per cent the results obtained can be accurately reproduced by calculation based on independently determined values of the rate constants of the elementary reactions given above.[41] Thus the sensitizing effect of hydrogen is still very much in evidence at the higher temperatures, and it is clear that even here the carbon monoxide is not capable of providing chain-branching comparable to that introduced by quite minute amounts of hydrogen—or of water.

The effect of water vapour on the kinetics is most probably exerted through the reaction

$$O+H_2O \rightarrow 2OH \qquad 6\text{–XIII}$$

This introduces branching in place of reaction **3** in the hydrogen-sensitized reaction. Over the temperature range 550–1550 °C reaction 6–XIII varies from being about 60 to about 3 times slower than reaction **3**, though it may be that, under non-explosive conditions, water exerts its influence partly by forming hydrogen by a surface-catalysed reaction with carbon monoxide.

The 'dry' reaction

The questions naturally arise whether absolutely hydrogen-free carbon monoxide and oxygen will explode and, if so, what is the mechanism of the ignition. There is no doubt that the gases become less reactive as they are more rigorously purified; nevertheless even with the driest gases—and we shall take 'dry' to include absence of hydrogen—feeble explosions can be observed which have the characteristics of isothermal ignitions. Reproducible ignition limits, however, are difficult to achieve, and, as might be expected, the behaviour becomes extremely sensitive to the presence of water. Gordon and Knipe[42] found that addition of 40 parts per million (ppm) of water to very dry gas lowered the ignition temperature at a given pressure by 100°; and, conversely, Dickens, Dove, and Linnett[43] found that the ignition temperature of intensively dried gas increased by 100° when the gas was dried still further by storage overnight at −183 °C. Under these conditions, which were probably the driest yet achieved, the second limit was displaced about 200° to higher temperatures from that observed with gas dried by standard methods.

The ignitions observed in such circumstances as these have been widely assumed to represent the behaviour of the pure gases and much speculation has taken place about the reaction mechanism. Oxygen atoms are the most likely chain carriers but the origin of chain-branching is less obvious. There is no acceptable way for the branching reaction to take place by a rearrangement of unpaired electrons as in the reactions hitherto considered, and most of the mechanisms proposed have been based on the following reactions:[42,44]

Primary initiation	$CO + O_2 \rightarrow CO_2 + O$	6–XIV
Propagation	$O + CO \rightarrow CO_2^*$	6–XV
Branching	$CO_2^* + O_2 \rightarrow CO_2 + 2O$	6–XVI
Termination	$CO_2^* + M \rightarrow CO_2 + M$	6–XVII
	$CO + O + M \rightarrow CO_2 + M$	6–XII

together with deactivation of O and CO_2^* at the wall. The essential new feature is the branching reaction 6–XVI. Carbon dioxide molecules are supposed to be formed in a metastable electronically excited state by reaction 6–XV. The corresponding reaction to ground state carbon dioxide is extremely exothermic and the excitation energy of CO_2^* is sufficient to bring about the dissociation of oxygen molecules in reaction 6–XVI. (CO_2^* is usually assumed to refer to the $^3B_2$ state, in which case reaction 6–XVI is exothermic by about 8 kcal mole^{-1}.) Thus we have *branching by energy transfer*.

Many references to this mechanism will be found in the literature of carbon monoxide oxidation but, although the concept of energy-branching is plausible, serious objections can be raised to this particular reaction scheme. The chief of these perhaps relates to the postulated second-order character of reaction 6–XV. The basic problem, however, is whether it is necessary to find a mechanism for the dry reaction at all; in other words, whether the reactivity observed under dry conditions is still due to residual impurities. It has been shown that the chain-branching induced by shock waves can be attributed entirely to the wet mechanism if 20–40 ppm hydrogen or water are assumed to be present in ostensibly dry gas.[40,41] Branching by a dry mechanism evidently requires a higher activation energy than by the wet mechanism and it should therefore be even less in evidence at the lower temperatures of ignition-limit experiments. By substituting the best available values for the rate constants in equation 6–37, Brokaw[41] was able to account for the ignition limits observed by Dickens, Dove, and Linnett on the basis of the wet mechanism if it was assumed that the gas contained about 10 ppm of water or 30 ppm hydrogen. The actual concentrations in the experiments were almost certainly lower than this, but Brokaw has also suggested that the presence of 0.2 ppm of methane might account for the results. If this is so, the possibility of ever reaching certainty on the matter by direct experiments seems remote. On the whole, the question of the existence of a branched-chain dry reaction—and hence the energy-branching mechanism for it—though still undecided, seems likely to be answered in the negative.

6–2–6 Energy-branching in the hydrogen–fluorine reaction

A less equivocal example of branching by energy transfer occurs in the reaction of hydrogen with fluorine. This is a highly exothermic reaction which takes place basically by straight chains propagated by the halogenation mechanism discussed in Chapter 3:[45]

$$F + H_2 \rightarrow HF + F \qquad \text{6–XVIII}$$

$$H + F_2 \rightarrow HF + H \qquad \text{6–XIX}$$

The mono-valence of both kinds of atoms involved leaves no room for chain-branching by the ordinary means and the explosions to which the reaction readily gives rise were long considered to be inevitably thermal in origin. It has emerged more recently that this is not necessarily the case. Addition of a few per cent of oxygen retards but does not extinguish the reaction and in these circumstances an ignition limit of the 'second limit' type can be observed.[46] That is, the mixture ignites when its pressure is reduced below a critical value. This, of course, is highly symptomatic of the presence of branched chains. Similarly there is no appreciable rise in the temperature of the gas immediately before the explosion, thus confirming the isothermal character of the ignition.

The kinetics of the steady reaction above the ignition limit show that the chains are terminated by the familiar reaction

$$H + O_2 + M \rightarrow HO_2 + M \qquad \mathbf{4}$$

but, as just indicated, branching must occur in a less conventional way. Reaction 6–XIX is extremely exothermic and spectroscopic studies of the type discussed in Chapter 4 have shown that the hydrogen fluoride molecule produced is likely to contain a substantial part of the heat of reaction as vibrational energy. This may be much greater than the heat of dissociation of fluorine. The vibrational excitation is normally dissipated in a few molecular collisions but, if one of these happens to be with a fluorine molecule, there is a chance that the transfer of energy will bring about its dissociation. Thus, if the reaction

$$H + F_2 \rightarrow HF^{\dagger} + F \quad \text{6–XIXa}$$

is followed by

$$HF^{\dagger} + M \rightarrow HF + M \quad \text{6–XX}$$

the result is merely a straight chain; but, if by

$$HF^{\dagger} + F_2 \rightarrow HF + 2F \quad \text{6–XXI}$$

then branching occurs. The ignition limit is the pressure at which the termination reaction **4** just fails to balance the degree of branching resulting from the competition between reaction 6–XX and 6–XXI. Chaikin and collaborators[46] have shown that the influence of inert gases and other properties of the limit are compatible with this mechanism.§ Thus it appears that the energetic chain carriers are vibrationally rather than electronically excited as supposed in the more problematical ignitions of 'dry' carbon monoxide–oxygen mixtures. Vibrationally excited species are also likely to be involved in branching in other exothermic reactions of fluorine.[48]

An interesting point arises here in connection with chemical lasers. It was observed at the end of Section 4–4–1 that these become possible when the reaction produces more molecules in higher vibrational states than lower. Whether or not the excited species are actually involved in the branching step, their formation in a branched-chain reaction means that they can be produced in high concentration, thus, in principle, increasing the power of the laser emission. It has been suggested that this situation occurs in the reaction between hydrogen and ClF_3.[49]

6–3 Branching-chain explosions: chain-thermal ignition

Hitherto we have been at pains to distinguish between two fundamentally different types of critical condition for ignition, the one related to the failure of thermal equilibrium and the other to the continued multiplication of chain carriers by chain-branching. Nevertheless the reactions in which chain-branching occurs are invariably exothermic and one may expect to find circumstances in which the conditions for ignition are influenced simultaneously by factors of both kinds. Such circumstances give rise to *chain-thermal* ignition and they can occur in two principal ways. As to the first, we have seen in Section 6–2–2 that isothermal ignition cannot take place when the chains are mainly terminated by mutual

§ Certain quantitative details suggest that the vibrational energy of $HF^{\dagger}$ may be transmitted to hydrogen molecules before bringing about the branching,[46,47] but this does not alter the basic fact that branching is caused by energy transfer.

destruction of the carriers. This is so even when the net branching factor ϕ resulting from competition between first-order branching and terminating processes is positive. The variation of the reaction rate with time under these conditions is shown by curve I in Fig. 6–7, which is similar to curve C in Fig. 6–4. The concentration of the chain carriers—and therefore the reaction rate—rises exponentially at the beginning of the reaction but the reaction never attains to explosion. On the other hand the rate of generation of heat is proportional to the reaction rate; consequently a change in the initial conditions may bring about the situation whereby, before reaching its steady value, the rate becomes too great for thermal equilibrium to be sustained. That is, an explosion ensues. This is shown by curve II of Fig. 6–7. Thus, for ignition to occur in these circumstances

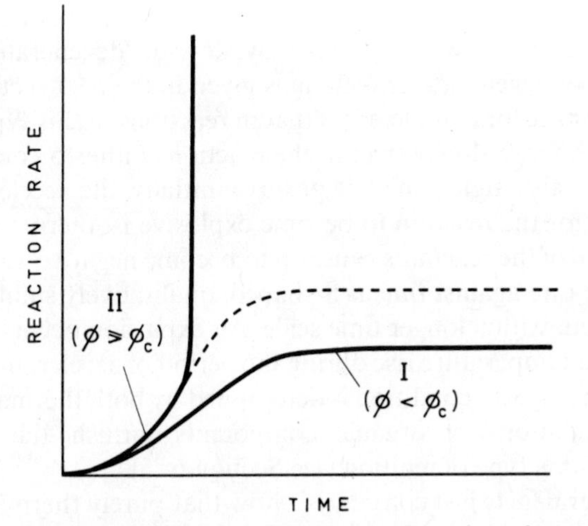

Fig. 6–7 Schematic diagram showing the condition for *chain-thermal ignition*. The net branching factor ϕ must exceed a critical value ϕ_c.

it is not sufficient that ϕ should merely be infinitesimally positive. It must have a finite value (ϕ_c); and the magnitude of ϕ_c is influenced by the thermal properties of the system. It follows that the ignition condition is related to both mechanistic and thermal factors.

In practice it is not always easy to distinguish chain-thermal ignition of this kind from purely isothermal ignition. When the chains are terminated mainly by three-body collisions, however, the presence of the thermal factor may be recognized by the effects of different inert gases on the explosion limit. An isothermal ignition would be less inhibited by addition of, say, helium than by an equal amount of carbon dioxide, but the relative thermal conductivities of these gases would place their effects on a thermal ignition in the opposite order. Thus 'anomalies' found in the order of the quenching efficiencies of inert gases may help the diagnosis. This and other more subtle tests[50] have been used to discover an example of this type of chain-thermal ignition in sensitized explosions of hydrogen and oxygen.[51,52] Under appropriate conditions the presence of a little nitric oxide will bring about ignition in the region above the second limit of the unsensitized reaction. The nitric oxide causes ϕ to become positive by reacting with an HO_2 radical

$$HO_2 + NO \rightarrow NO_2 + OH \qquad 6\text{–XXII}$$

thus converting an inactive HO_2 to an active OH radical. The effect of this is opposed mainly by the second-order reaction:

$$HO_2 + HO_2 \rightarrow H_2O_2 + O_2 \qquad \mathbf{10}$$

(when the surface is inactive). Hence ignition cannot take place isothermally but only when ϕ is increased sufficiently to cause thermal instability, as illustrated in Fig. 6–7.

The second and most prevalent kind of chain-thermal ignition is found with reactions which proceed by slow or 'degenerate' branching. An account of degenerate branching is given in the next section. Here we shall simply note for completeness that in reactions of this type the chains branch much more slowly than in the reactions hitherto considered. The result is that, although ϕ may be positive initially, the acceleration is not fast enough for the reaction to become explosive isothermally before the consumption of the reactants causes ϕ to become negative. Once more the graph of the rate against time is S-shaped, qualitatively similar to curve I of Fig. 6–7 but with a longer time scale. An explosion occurs only if there is a sufficient temperature rise during the period of acceleration (curve II); and again the critical condition is determined by both thermal and kinetic factors. Oxidations of organic compounds furnish the outstanding examples of this type of ignition (see Section 6–5).

The general facts just considered show that purely thermal and purely isothermal mechanisms of ignition are actually the limiting cases of a continuous spectrum of behaviour. In general, whether or not explosion is to occur in any particular circumstances is likely to be influenced in greater or less proportions by both thermal and mechanistic considerations. The mathematical treatment required to express these rigorously in unified form has been developed by Yang and Gray.[53] A more elementary account of chain-thermal explosions will be found in papers by Dainton and Norrish.[50,51]

6–4 Delayed branching and degenerate explosions

In the hydrogen–oxygen reaction and others of the same type the state of balance or otherwise between the branching and terminating of the chains normally depends directly upon fast radical–molecule or radical–radical reactions. The corresponding condition of slow reaction or explosion is therefore established very rapidly, usually within a few seconds or less. But it is also possible for chains to branch much more slowly by inter- or intra-*molecular* reactions. In these circumstances most of the familiar characteristics of branched chains remain, but the time scale is much longer and a different kinetic pattern tends to emerge. For reasons that will be clear shortly, branching of this type is known as 'degenerate' branching. The phenomenon is very widespread for it occurs in reactions of oxygen with hydrocarbons and organic compounds generally. Our examples will be taken from reactions of the kind but, in this section, they are to be regarded principally as illustrating general behaviour. The peculiarities of the oxidation of organic compounds are considered later.

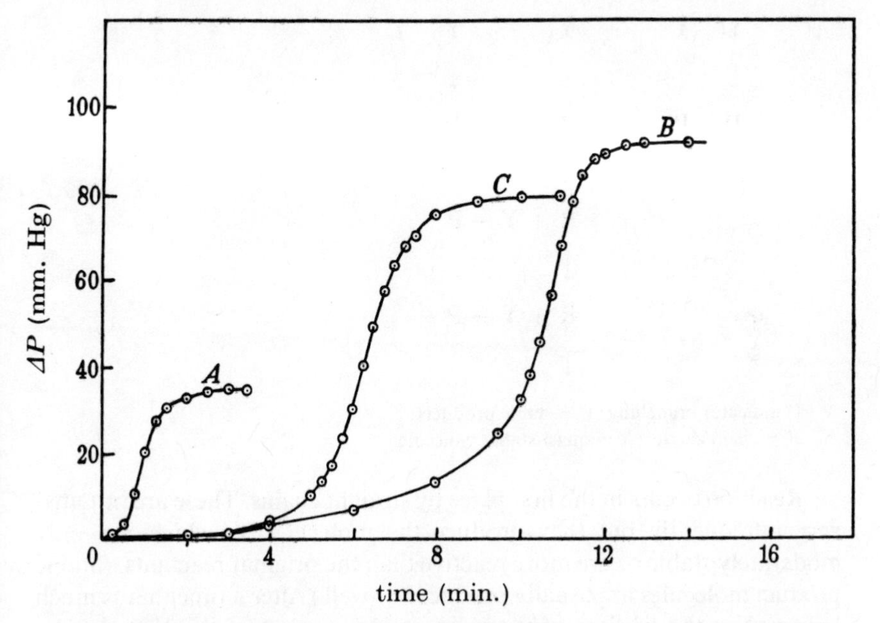

Fig. 6–8 Degenerate explosions. Progress of slow oxidations of hydrocarbons as shown by pressure–time curves. A, CH_4+O_2 at 530 °C; B, $C_2H_6+2O_2$ at 485 °C; C, $C_3H_8+3O_2$ at 430° C (Norrish and Reagh[54]). (Reproduced by permission of The Royal Society.)

The prime experimental characteristic of reactions with delayed branching is the occurrence of long induction periods. Such a reaction may remain imperceptible for many minutes or even hours after the reactants are brought together and then accelerate more or less rapidly to explosion. At lower temperatures or pressures, the induction period is followed not by ignition but by a non-explosive reaction which gradually accelerates to a maximum rate and then dies away. This is illustrated by the pressure–time curves for hydrocarbon oxidations shown in Fig. 6–8. If the time taken to reach the maximum rate is graphed against the total pressure or some other initial condition of the reactants, the graph is frequently found to be continuous with a similar graph of the induction period before explosion. This is an indication that the ignition is chain-thermal. The induction period, whether leading to slow reaction or ignition, is generally highly sensitive to the presence of inhibitors, but it is never reduced to zero, however much the reactants are purified. These two facts show that the reaction is at once a chain reaction and auto-catalytic. They are explained by Semenov's theory of delayed branching. The basic idea is shown below:

$$
\begin{array}{cccccccc}
 & & & & * & & & \\
R \rightarrow & R \rightarrow & R \rightarrow \cdots \rightarrow & & R \rightarrow \cdots \rightarrow & & R \rightarrow & R \rightarrow \\
 & + & + & & + & & + & + \\
 & Y & Y & & Y & & Y & Y \\
 & \downarrow & \downarrow & & \downarrow & & \downarrow & \downarrow \\
 & P & P & & R & & P & P \\
 & & & & \downarrow & & & \\
 & & & & R + Y \rightarrow P & & & \\
 & & & & \downarrow & & & \\
 & & & & R + Y \rightarrow P & & & \\
 & & & & \downarrow & & &
\end{array}
$$

(* indicates branching; P = stable products;
R = chain carrier; Y = quasi-stable molecule.)

Reaction occurs in the first place by straight chains. These are initiated very infrequently but they produce the molecules Y, which are only moderately stable or are more reactive than the original reactants. (Stable product molecules are usually produced as well.) After a time that is much longer than the lifetime of the primary chain, most of the Y molecules decompose or otherwise react to form stable products, but some undergo another reaction to produce chain carriers. Thus new chains are initiated. In this way branching occurs but not until long after the original chain has come to an end. On the other hand, Y is sufficiently reactive to produce the new carriers before an appreciable proportion of the reactants is consumed by the primary chains. This is ensured by the infrequency of the primary initiation. Hence, after a time, most of the chains are new chains initiated by Y, and, as Y accumulates, the reaction accelerates.

Eventually the acceleration is checked but we shall consider this aspect later. The net rate of formation of Y in the early stages of reaction is given by

$$\frac{d[\mathrm{Y}]}{dt} = P_{\mathrm{i}} + (F - G')[\mathrm{Y}] \qquad 6\text{–}38$$

$$\equiv P_{\mathrm{i}} + \phi[\mathrm{Y}] \qquad 6\text{–}39$$

where P_{i} is the rate of formation of Y in the primary chains and F is the specific rate of reproduction as the result of branching; G' is the rate constant for the destruction of Y to inert products (which is assumed to be first order) and ϕ is the net branching factor. This equation is formally identical with equation 6–17 given in Section 6–2–1 for the development of ordinary branching chains. Here, as there, the behaviour depends on the relative values of F and G'. If ϕ is negative, [Y] eventually reaches a steady low value; but if it is positive, [Y] at any time t is given by

$$[\mathrm{Y}] = \frac{P_\mathrm{i}}{\phi}\exp(\phi t) - 1 \qquad 6\text{–}40$$

$$\approx \frac{P_\mathrm{i}}{\phi}\exp(\phi t) \qquad 6\text{–}41$$

(cf. equations 6–18 to 6–20 and Fig. 6–4). Thus [Y] rises exponentially with time as shown by curve B in Fig. 6–4. Neglecting the small contribution from the primary initiation, the reaction rate is proportional to [Y] and therefore also increases in the same way. The quantity of product formed in the time t from the beginning of the reaction, obtained by integrating equation 6–41, is

$$[\text{Products}]_t = \frac{KP_\mathrm{i}}{\phi^2}\exp(\phi t) \qquad 6\text{–}42$$

where K is a stoichiometric constant. When the concentration of the products becomes measurable, for example by pressure change, as in Fig. 6–8, equation 6–42 can be tested by graphing $\log[\text{Products}]_t$ against t:

$$\log[\text{Products}]_t = \log(KP_\mathrm{i}/\phi^2) + \phi t/2.303 \qquad 6\text{–}43$$

Linear relations between $\log[\text{Products}]_t$ or $\log \Delta p_t$ and time have been found experimentally for the early stages of a number of reactions, including those illustrated in Fig. 6–8. The slope of the graph is frequently used to determine ϕ.

The induction period

This is a time when the chains are developing exponentially but often imperceptibly. If it is defined as the time τ for the reaction to generate a fixed concentration of products, equation 6–43 leads to the relation

$$\phi\tau = \text{const} - \log(KP_\mathrm{i}/\phi^2) \qquad 6\text{–}44$$

Since the logarithmic term varies very little with reaction conditions compared with ϕ, this becomes

$$1/\tau = \text{const} \times \phi \qquad 6\text{–}45$$

to a good approximation. The reciprocal induction periods therefore provide a measure of the variation of ϕ under different reaction conditions.

The identity of the basic equation 6–38 with equation 6–17 which applies to ordinary branching was noted previously, but it is instructive to compare the relative time scales of the corresponding phenomena. Consider a mixture of hydrogen and oxygen at 460 °C at a pressure just below the first ignition limit (4 torr in a vessel of 4 cm radius). The value of $f = g'$ in equation 6–17 can be obtained from equation 6–23. At 4 torr, D (for H atoms) $\approx 2000\ \text{cm}^2\ \text{s}^{-1}$ and hence $f = g' = 1200\ \text{s}^{-1}$. If now f is caused to be 1 per cent greater than g', thus exceeding the ignition limit, $\phi = f - g' = 12\ \text{s}^{-1}$. Hence the reaction rate would treble in about every tenth of a second, and the induction period before explosion would be perhaps a few tenths of a second, as in fact is shown by experiment. Returning to the case of delayed branching, suppose that an average of 1 in 200 of the molecules of Y formed decomposes to initiate two chains,

$$\text{Y} \rightarrow 2\text{R} \qquad \text{6–XXIII}$$

the remainder yielding inert products. Suppose, further, that each chain produces on the average 100 new molecules of Y, that is the chain length (υ) is 100 ($= \upsilon_0$). Thus, the concentration of Y is exactly replenished, $F = G'$ and $\phi = 0$. Now suppose that an adjustment is made so as to increase the chain length by 1 per cent all else remaining the same. Noting that

$$F = 2k_{23}(\upsilon - 1) \qquad \text{6–46}$$

we have

$$\phi = 2k_{23}(1.01\upsilon_0 - 1) - 2k_{23}(\upsilon_0 - 1) = 0.02\upsilon_0 k_{23} = 2k_{23}$$

The decomposition of organic peroxide molecules is responsible for delayed branching in certain oxidation reactions at about 200 °C. In this case, a typical value for k_{23} would be about $5 \times 10^{-4}\ \text{s}^{-1}$. If this were applicable to the present example, ϕ would be $10^{-3}\ \text{s}^{-1}$. Hence the rate would require about 20 min to treble itself, and the induction period would be expected to be several times longer than this.

The maximum rate

One may ask why reactions with delayed branching do not eventually accelerate to explosion but merely arrive at a finite maximum rate (unless a thermal explosion intervenes). This can come about in two distinct ways: one is due to the consumption of the reactants; the other to particular kinetics of the destruction of Y. Figure 6–9 is relevant to the first way. It is derived from equation 6–40 and shows the relative progress in time of the rates of two reactions which differ only in the value of ϕ. In each case the amount of reactants consumed at a given time is proportional to the area under the curve. To reach the same arbitrary rate shown by the horizontal dotted line the reaction with $\phi = 0.1$ consumes about five times the amount consumed by the reaction with $\phi = 1.0$. Naturally a greater difference in ϕ will produce a still greater disparity. Continued acceleration of the reaction depends on $\phi = F - G'$ remaining positive. Reference to equation 6–46 shows, however, that F depends on the chain length υ, which in turn depends on the concentrations of the reactants. When the initial value of ϕ is small, the consumption of the reactants eventually causes it to become negative before an explosive rate is attained, and from this point forward the reaction decelerates. A specific example will be found in Section 6–5–2. This is the origin of the term *degenerate explosion* for this type of behaviour,[55] though nowadays the term is applied whether explosion is prevented by reactant consumption or in the alternative manner now to be described.

The second way in which the reaction may be prevented from becoming explosive occurs when Y (to which we can now refer as the 'agent of degenerate branching') is destroyed by a reaction of higher order in [Y] than the branching reaction. To give a concrete example, in the oxidation of ethylene at 400 °C, Y is the molecule of formaldehyde.[56] This partly undergoes the reaction

$$HCHO + O_2 \rightarrow HO_2 + CHO \qquad \text{6–XXIV}$$

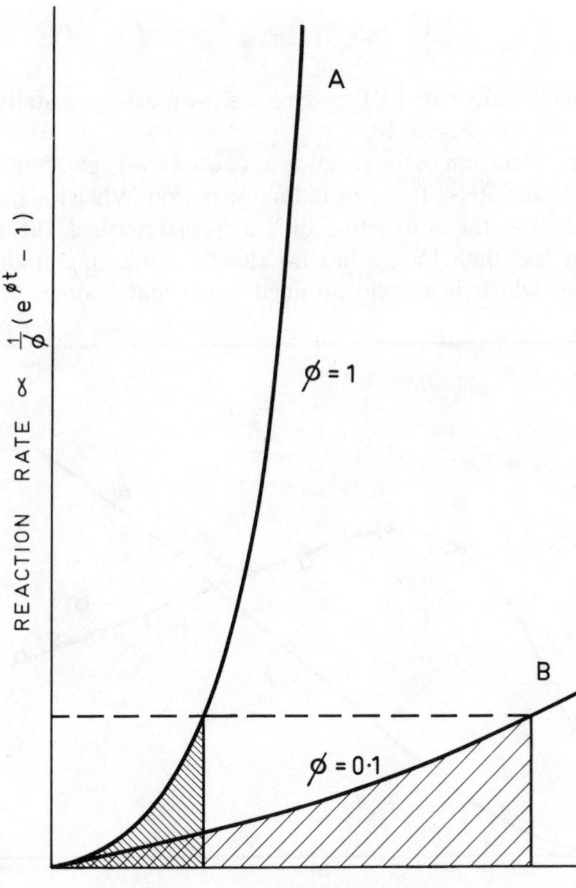

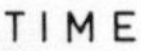

Fig. 6–9 The effect of the net branching factor ϕ on the consumption of the reactants during the development of branching chains. The shaded area shows the amount consumed by each reaction before reaching the same rate. Reaction B consumes about five times the amount consumed by reaction A.

to initiate the chains in which the ethylene is oxidized to more formaldehyde. In addition, it is itself oxidized to inert products ($CO+H_2O$) by a complex reaction the rate of which is proportional to $[HCHO]^2$. Thus

$$\frac{d[Y]}{dt} = P_i + \phi[Y] - H[Y]^2 \qquad 6\text{–}47$$

The circumstances are entirely analogous to the case of ordinary branching with quadratic termination of the chains, as described by equation 6–16 and curve C of Fig. 6–4. A maximum value of [Y], and therefore of the rate, is attained even in the absence of consumption of the reactants. The value of $[Y]_{max}$ is found by equating the rates of formation and destruction of Y. Assuming it is in fact reached before the reactants are appreciably depleted and neglecting P_i, the maximum rate is given by

$$\left(\frac{d[\text{Products}]}{dt}\right)_{max} = \phi[Y]_{max} = \frac{\phi^2}{H} \qquad 6\text{–}48$$

It occurs coincidentally with $[Y]_{max}$. This is shown experimentally for the ethylene oxidation by Fig. 6–10.

Since the acceleration of the reaction is caused by Y, prior addition of a little Y to the reactants reduces the induction period. When the branching is checked solely by the destruction of Y as just described, the addition of an amount less than $[Y]_{max}$ has no effect on the magnitude of the maximum rate which is merely attained somewhat sooner. When an

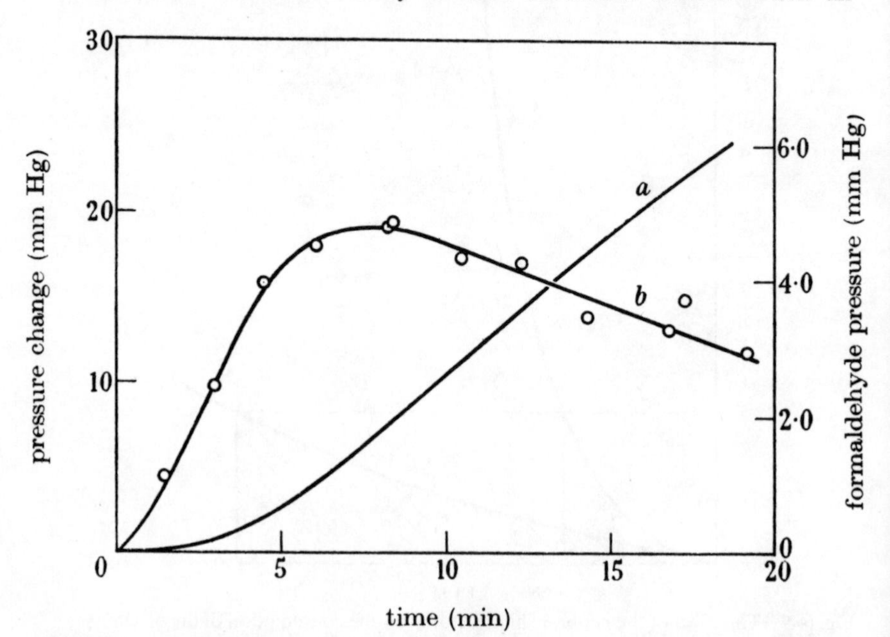

Fig. 6–10 Degenerate branching in the oxidation of ethylene. Curve *a* shows the change in total pressure; curve *b* the partial pressure of HCHO, the agent of degenerate branching (Harding and Norrish[56]). (Reproduced by permission of The Royal Society.)

amount equivalent to $[Y]_{max}$ is added, the induction period is eliminated and the reaction begins immediately at the maximum rate. A somewhat different effect is to be expected when the maximum rate is caused by the consumption of the reactants. In this case an initial addition of Y, however small, has the effect of increasing the maximum rate as well as shortening the induction period. Tests of this kind are also used to determine the identity of Y, which is not always clear from chemical analysis of the reacting gas. In oxidation reactions the concentrations of several species are commonly found to pass through maxima more or less coincidentally with the maximum rate, but some of these species may be produced merely by side reactions. Their relative importance for chain-branching is easily assessed from their effects on the induction period. A variation of this procedure was used by Shtern[57] to identify the branching agent in the oxidation of propane at 310 °C. The rival candidates were acetaldehyde and an unidentified peroxidic compound (or compounds) which reached their maximum concentrations together at the maximum rate. The reaction mixture was withdrawn from the reaction vessel just before the end of the induction period and stored over mercury until the peroxides were completely destroyed. It was then transferred to a second reaction vessel at 310 °C, whereupon the reaction was observed to begin almost without an induction period. This showed that the peroxides had little or no branching effect. The 'victory' of acetaldehyde was confirmed by the observation of a powerful accelerating effect when it was added to the reaction mixture.

6–5 Slow oxidation and ignition of saturated hydrocarbons and other simple organic compounds

Few chemical facts are better known than that an organic vapour will react explosively with oxygen. Though less obvious, it is equally true that under different circumstances a non-explosive reaction will occur. The explosion normally produces a more or less equilibrated mixture of the oxides of the elements present, but the products of the slow reaction usually contain oxygenated organic compounds of lower molecular weight than the original substrate in addition to carbon oxides and water. When the reaction is conducted at temperatures about 250°, the composition of the products from a molecule containing more than three or four carbon atoms may be very complex indeed, particularly if the reaction is interrupted before completion. As already indicated, kinetically the slow reaction almost always has the characteristics of a degenerate explosion and ignition, when it occurs, is chain-thermal. But within these generalities a number of extraordinary phenomena are found that reflect the presence of complicated reaction mechanisms. The great variety of organic compounds might suggest a similar multiplicity of mechanisms, but, although this is the case to some extent, a common if rather variegated pattern of behaviour is clearly visible. On the other hand, it cannot be said that the mechanism of oxidation of even the simplest individual compound, such as methane or formaldehyde,[58] is understood sufficiently well to cover all the observed phenomena. Substantial agreement exists on some aspects, at least in a qualitative sense, but there is much that is still conjectural or in dispute. It is therefore important to separate the experimental facts from the mechanistic details currently proposed to explain them; so we shall begin by summarizing the main facts and afterwards discuss some examples of what can be concluded or partly concluded about reaction mechanisms.

6–5–1 Basic experimental facts

The range of reactivity among different compounds is very great. Under comparable conditions, acetaldehyde is oxidized at an appreciable rate at 80 °C, whereas methane scarcely begins to react below 400 °C. Large differences in reactivity are not always to be attributed to substituents in the hydrocarbon molecule. The rates of oxidation of the *n*-alkanes, or of the members of almost any homologous series, increase in a remarkable way with increasing length of the hydrocarbon chain; *n*-decane, for example, in roughly 10000 times as reactive as propane at about 250 °C.[59] (The maximum rate (Section 6–4) is usually taken as the parameter for comparative measurements.) The opposite influence is exerted by branching in the hydrocarbon skeleton: *n*-butane is over 10 times more reactive than isobutane.§

The effect of temperature is unusual. The rate generally passes through a maximum with increasing temperature and shows a negative temperature coefficient over a range of about 50° before increasing again. Under the conditions to which Fig. 6–11[61] refers, the rate of oxidation of propane at 340 °C is only half what it is at 320°. Under different conditions the rate can decrease by as much as an order of magnitude over the temperature interval. Obviously, this behaviour marks a change in the reaction mechanism, and the temperature range over which it occurs is commonly regarded as separating the *high-temperature* from the *low-temperature* regime of reaction. Thus in the experiments recorded in Fig. 6–11, *n*-butane reacted by a low-temperature mechanism, ethane by a high-temperature mechanism and propane by both. At different pressures or with different mixture compositions, however, butane and ethane and indeed most organic compounds also exhibit a negative temperature coefficient and may react by one or other type of mechanism in the appropriate temperature range.

The character of the reaction products also changes with reaction temperature. In the low-temperature regime carbon oxides, alcohols, aldehydes, ketones, and the like predominate, but as the reaction tem-

§ These effects are of great practical importance for the internal combustion engine. Premature oxidation of the hydrocarbon vapour before the advent of the spark leads to the undesirable effect of 'knock'; and for this reason branched-chain are much to be preferred to straight-chain hydrocarbons as fuel.[60]

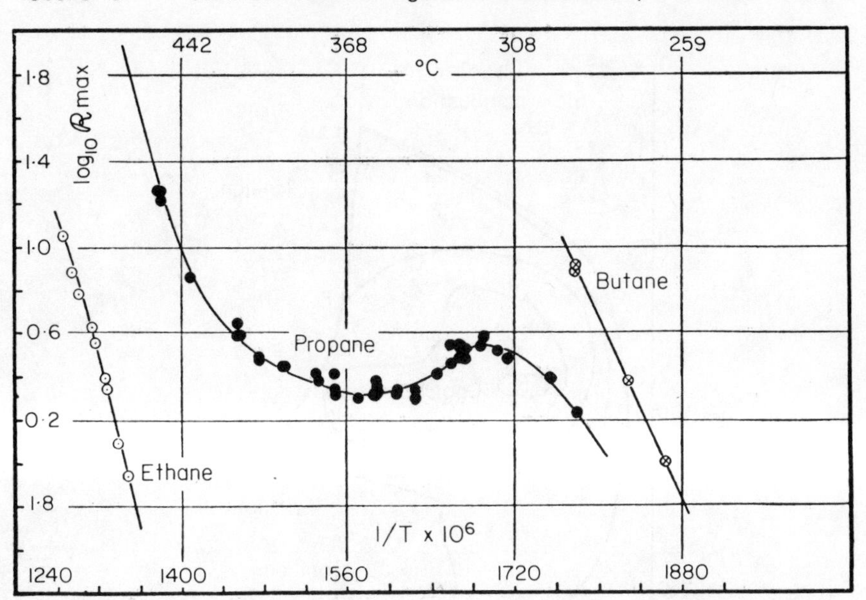

Fig. 6–11 Slow oxidation of hydrocarbons. Variation of the reaction rate with temperature. Propane shows a negative temperature coefficient between 'low-temperature' and 'high-temperature' regimes of reaction.

perature is raised, these gradually give way to unsaturated hydrocarbons, together with saturated hydrocarbons of lower molecular weight. Thus Shtern[57] found that a 2:1 mixture of propane and oxygen produced oxygenated and unsaturated compounds in the ratio 4:1 at 300 °C, but at 465° the ratio was 1:2.4. At the higher temperatures, the oxidation amounts to an oxygen-sensitized 'cracking' of the substrate molecule. At the same time, substantial amounts of hydrogen peroxide are formed.[62] Indeed the high-temperature oxidation of hydrocarbons has been the subject of several patents for the industrial manufacture of hydrogen peroxide.

Particularly interesting phenomena are associated with ignition. In the first place, the conditions for ignition to take place are profoundly affected by the negative temperature coefficient. As a chain-thermal process, ignition depends on the slow reaction reaching a critical rate, and in ordinary circumstances the $p^* - T$ diagram would show a uniform increase in ignition temperature with decreasing initial pressure of the gas mixture. The actual behaviour is shown typically by Fig. 6–12.[63] Below about 250 °C and above about 300 °C the ignition is as just described. This is shown by the sections a–c and d–h of the ignition boundary. In between, however, the ignition temperature actually decreases with decreasing pressure. This is because a rise in temperature in the range c–d causes a decrease in reaction rate. The rather extraordinary consequence follows that, as the diagram shows, a gas mixture held initially at say 200 torr and 290 °C explodes when its temperature is *lowered* by about 10 °C.

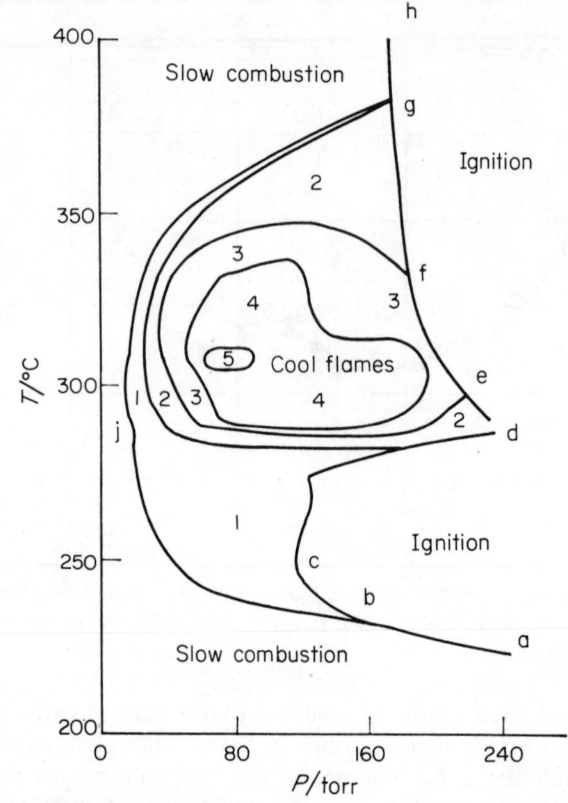

Fig. 6–12 Ignition diagram for *n*-heptane–oxygen mixtures (1:2). Note change of axes from Fig. 6–5 (Cullis, Fish, and Gibson[63]).

Still more remarkable is the occurrence of *cool flames*. Under the conditions indicated by the boundary a–d–g in Fig. 6–12, ignition is preceded by one or more sharp, though feeble, pressure pulses lasting not longer than one or two seconds. Each of these is produced by a temporary self-generated rise in the temperature of the gas. It is accompanied by an intensification of the extremely weak bluish luminescence that is generally emitted during the normal course of the slow reaction.§ The intensity of the light is still quite feeble, however, and not at all comparable with that emitted by the final explosion. Similarly, the maximum temperature rise in the pulse seldom exceeds 250 °C, whereas the explosion leads to temperatures of 2000–3000 °C, hence the term 'cool flame'. A typical temperature record taken with a fine thermocouple inside the reaction vessel is given in Fig. 6–13. Here ignition is preceded by three cool flames. The temperature pulses are superimposed on a much slower temperature rise in the reacting gas and ignition finally occurs in gas that is heated by the third cool flame. The last event is known as *two-stage ignition*, though

§ The light is emitted from a minute concentration of electronically excited formaldehyde molecules.

in these particular circumstances it follows the two earlier stages of temporary thermal instability. In Fig. 6–12 the conjugate values of p^* and T required for two-stage ignition are represented by the section a–g of the ignition boundary. The latter is subdivided according to the total number of cool flames observed. For example the section e–f indicates the conditions for which ignition follows three cool flames (and could therefore be described as 'four stage'). At the section of the boundary marked g–h there is no cool flame and thus *single-stage* ignition takes place.

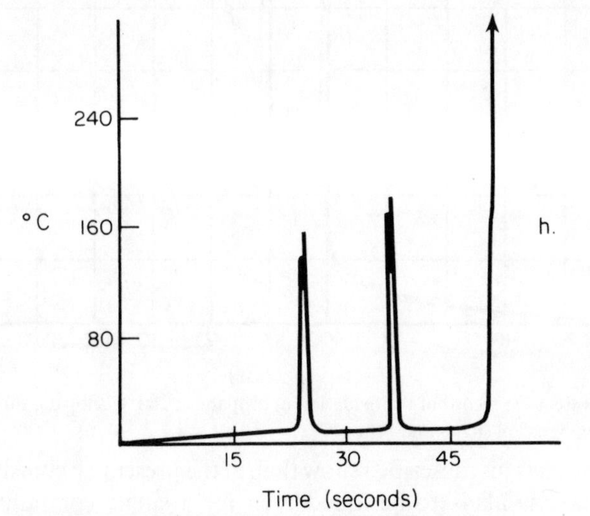

Fig. 6–13 Temperature changes during the oxidation of propane showing two cool flames followed by two-stage ignition. Note discontinuity at *h* (Griffiths, B. F. Gray, and P. Gray[64]).

Cool flames may occur without subsequent ignition. This happens at $p^* - T$ values within the area b–j–g–d–b of the diagram, and b–j–g is the sharp boundary for a single cool flame to appear. Likewise, the areas shown within represent the conditions for multiple cool flames (in this case up to five). A pressure–time record of a reaction giving rise to three cool flames is shown in Fig. 6–14.[65] This reveals another curiosity of the cool flame, namely that, in spite of its quasi-explosive character, it may involve surprisingly little consumption of the reactants. In such cases, the rates of formation of some products are scarcely perturbed by the events associated with the cool flames. It is significant, however, that the concentration of hydroperoxides which builds up more or less exponentially during the previous reaction periods usually disappears abruptly during the cool flames.[66]

6–5–2 Reaction mechanisms

In approaching the mechanism of a branched-chain reaction the key question naturally relates to the identity of the species responsible for

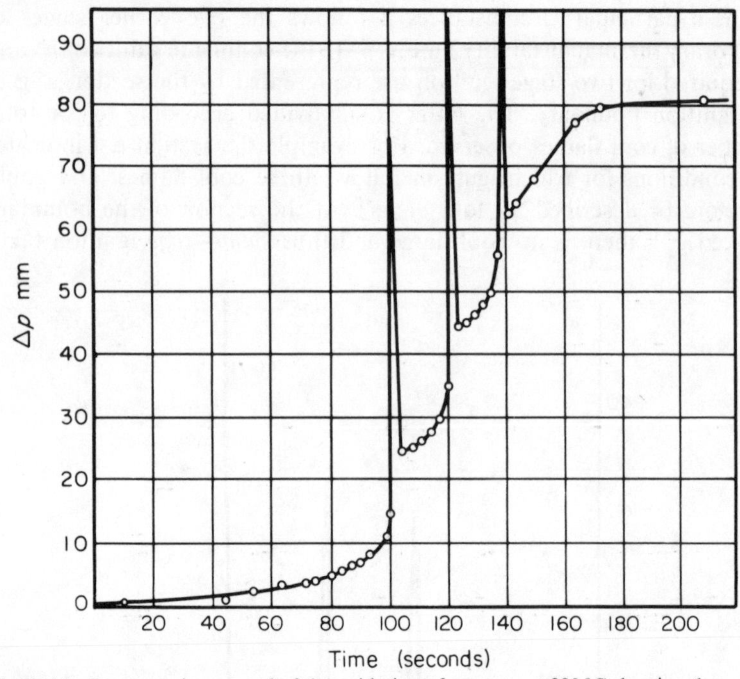

Fig. 6–14 Pressure–time record of the oxidation of propane at 280 °C showing three cool flames but no ignition (Repa and Shtern[65]).

branching. The facts just described show that, in the present circumstances, no single answer is likely to be found even for a single compound. In particular, the region of negative temperature coefficient shows that the nature of the branching process, or at least its efficiency, changes with temperature. Hence for any particular system there are really three questions: 'How does branching occur in the low-temperature region?' 'How does it occur in the high-temperature region?' and 'What happens in between?' The answers have proven more difficult to obtain than may at first appear and, particularly for the low-temperature region, have not yet been answered very definitely for more than a few compounds. To these questions we must add the intriguing problems of the cool flame and its relation to true ignition. Here again, although the broad facts are reasonably well understood as thermokinetic phenomena, the underlying chemistry is known at best very imperfectly. In short, as indicated earlier, a complete mechanistic picture of organic oxidation is not yet available. Nevertheless the following discussion will provide a glimpse of some of the areas in which progress has been made.

Low-temperature mechanisms

It was mentioned in Chapter 3 that photochemical oxidation of an organic vapour at room temperature produces almost exclusively the hydroperoxide,

$$RH + O_2 \rightarrow ROOH \qquad 6\text{–XXV}$$

The reaction occurs by the straight-chain mechanism given in Section 3–5–4. Higher temperatures are required to initiate reaction thermally and, in general, the organic molecule becomes increasingly degraded to products of lower molecular weight as the reaction temperature is increased. However, organic peroxides can usually be detected in the products up to about 300 °C. Hydroperoxides are known to decompose in the relevant temperature range by fission of the weak O—O bond

$$ROOH \rightarrow RO + OH \qquad \text{6–XXVI}$$

Thus delayed branching becomes possible, and it is natural to suggest that the formation and decomposition of the hydroperoxide are responsible for the autocatalytic character of the kinetics. This is supported by the well-established fact that the oxidations of various compounds in the liquid phase occur in this way.[67] Nevertheless, we shall see shortly that under most conditions gas-phase oxidations are more complex. On the other hand, when the molecule contains an especially reactive R—H bond, oxidation can be initiated thermally at relatively low temperatures; and in these circumstances substantial amounts of ROOH are found in the products. The outstanding compounds with which this is the case are acetaldehyde and higher aldehydes. At temperatures below 150 °C, oxidation occurs exclusively at the aldehydic C—H bond, and hydroperoxide, that is the peracid, is practically the sole product:

$$RCHO + O_2 \rightarrow RC(O)OOH \qquad \text{6–XXVII}$$

The *low-temperature oxidation of aldehydes*[68] therefore offers the opportunity to examine a purely 'hydroperoxidic' mechanism. In fact this turns out to be basically the same as that of the photochemical oxidation at lower temperatures except for the means of initiation:

$$\text{Initiation} \rightarrow R \text{ or } RO_2 \qquad \textbf{(i)}$$

$$R + O_2 \rightarrow RO_2 \qquad \text{6–XXVIII}$$

$$RO_2 + RH \rightarrow ROOH + R \qquad \text{6–XXIX}$$

$$RO_2 + RO_2 \rightarrow \text{Termination} \qquad \text{6–XXX}$$

where $R = R'CO$ and the ratio $[O_2]/[RH]$ is not too far below unity (cf. Section 3–5–4). Soon after the beginning of the reaction the hydroperoxide becomes the main source of initiation and the reaction accelerates as it builds up. The hydroperoxide accumulates during the whole course of the reaction, but consumption of the aldehyde eventually causes the rate of formation to pass through a maximum. (Unlike the reactions illustrated in Fig. 6–8 the reaction, of course, produces a decrease in pressure.) Thus the reaction has the characteristics of a degenerate explosion, but it differs from the schematic mechanism in Section 6–4 in that the agent of degenerate branching is itself virtually the only product. This means that it reacts only to produce radicals and in doing so it initiates *long* chains. The simple stoichiometry enables a clear picture to be formed of the progress of the reaction. Following the discussion in Section 3–5–4, the rate at any time is given by

$$\frac{d[\mathrm{ROOH}]}{dt} = k_{29}\left(\frac{\rho_i}{2k_{30}}\right)^{1/2}[\mathrm{RH}] \tag{6–49}$$

If it is assumed that the rate of primary initiation by the reactants is negligible and that

$$\rho_i = 2k_{26}[\mathrm{ROOH}] \tag{6–50}$$

we have, at any time during the reaction,

$$= -\frac{d[\mathrm{O_2}]}{dt} = -\frac{d[\mathrm{RH}]}{dt} = \frac{d[\mathrm{ROOH}]}{dt}$$

$$= k_{29}\left(\frac{k_{26}}{k_{30}}[\mathrm{ROOH}]\right)^{1/2}\{[\mathrm{RH}]_0 - [\mathrm{ROOH}]\} \tag{6–51}$$

where $[\mathrm{RH}]_0$ is the initial concentration of the aldehyde. Solution of the equation for the concentration of hydroperoxide at any time gives

$$[\mathrm{ROOH}] = [\mathrm{RH}]_0 \tanh^2(bt) \tag{6–52}$$

where $b = k_{29}(k_{26}[\mathrm{RH}]_0/4k_{30})^{1/2}$. This accounts for the 'sigmoidal' progress of the reaction. The expression for the maximum rate is

$$\mathscr{R}_{max} = 0.38k_{29}(k_{26}/k_{30})^{1/2}[\mathrm{RH}]_0^{1.5} \tag{6–53}$$

This relation correctly describes the experimental fact that $\mathscr{R}_{max}$ is independent of the oxygen concentration (over a certain range); but the observed dependence on the initial aldehyde concentration is closer to 2 than 1.5.[68] The cause of this appears to be peculiar to aldehyde oxidation. Peracids react reversibly with aldehydes to produce less stable peroxides and, at low temperatures, it is the small concentration present of the latter that principally initiates the chains:

$$\mathrm{R'C(O)OOH + R'CHO \rightleftharpoons R'C(O)OOCH(OH)R' \rightarrow radicals} \tag{6–XXXI §}$$

Substitution of this reaction for reaction 6–XXVI leads to an expression for the maximum rate in better accord with the facts:

$$\mathscr{R}_{max} = 0.32k_{29}(k_{31}/k_{30})^{1/2}[\mathrm{RH}]_0^2$$

(k_{31} is a composite rate constant). With increasing temperature, however, the equilibrium in reaction 6–XXXI moves to the left, with the result that branching probably occurs mostly by direct fission of the peracid above 200 °C. At these temperatures, however, the character of the reaction also changes in other ways, as is apparent from the fact that substantial amounts of degradation products (CO_2, CO, H_2O, R′OH, etc.) are formed.

§ The structure of the intermediate peroxide is conjectural.

The relatively high activation energy of the branching reaction compared with that of the overall reaction causes the chains to shorten at higher temperatures, thus creating a greater proportion of products resulting directly from peroxide fission. In addition, decomposition of the acyl radicals (R) competes with their addition to oxygen in reaction 6–XXVIII:

$$R'CO \rightarrow R' + CO \qquad \text{6–XXXII}$$

Unlike most oxidations, that of an aldehyde begins at a measurable rate, and kinetic analysis[68,69] shows that this is due to the effect of the primary initiation reaction

$$R'CHO + O_2 \rightarrow R'CO + HO_2 \qquad \text{6–XXXIII}$$

We shall see presently that the exceptional reactivity of the relatively weak aldehydic C—H bond enables aldehydes to act as agents of degenerate branching when they are produced as intermediates in the oxidation of less reactive compounds.

With *hydrocarbons* and other compounds less reactive than aldehydes the importance of the corresponding hydroperoxides as branching agents is less clear. Addition of the hydroperoxide usually promotes reaction, but it is not easy to establish that the necessary concentration is actually generated by the reaction. The chief difficulty is one of chemical analysis. Normally, a mixture of peroxides is produced. Some of these may be derived from aldehydes or other degradation products or from secondary reactions of hydrogen peroxide, and it is often difficult to distinguish the various species quantitatively.

Nevertheless when oxidation can be initiated below about 250 °C, that is, with fairly reactive compounds, the nature of the products shows that some of these at least are produced by the fission of ROOH molecules. Thus oxidation of 2-methylpentane at 240 °C produces substantial amounts of acetone and pentan-2-one evidently by the reactions:[70]

$$C_3H_7(CH_3)_2COOH \xrightarrow{\text{branching}} C_3H_7(CH_3)_2CO + OH \qquad \text{6–XXXIV}$$

$$C_3H_7(CH_3)_2CO \rightarrow C_3H_7COCH_3 + CH_3 \qquad \text{6–XXXV}$$

$$C_3H_7(CH_3)_2CO \rightarrow CH_3COCH_3 + C_3H_7 \qquad \text{6–XXXVI}$$

There is evidence, however, that between about 250° and 300° branching by hydroperoxide (ROOH) decomposition becomes progressively less important (at sub-atmospheric pressure). The RO_2 (peroxy) radicals formed in the manner of reaction 6–XXVIII above become increasingly liable to decompose before they can react to produce hydroperoxide by reactions such as 6–XXIX. The decomposition is preceded by an internal rearrangement (cf. p. 102). In the oxidation of propane, for example, it seems that the following reactions occur:

$$\begin{array}{l} CH_3-CH_2-CH_2 \\ \quad\vdots \qquad\quad | \\ \quad O\text{——}O \end{array} \rightarrow CH_3CH_2O+HCHO \qquad \text{6–XXXVII}$$

and

$$\begin{array}{l} CH_3-CH-CH_3 \\ \qquad | \qquad\quad \vdots \\ \qquad O\text{——}O \end{array} \rightarrow CH_3CHO+CH_3O \qquad \text{6–XXXVIII}$$

Very little hydroperoxide is found in the products and the acetaldehyde formed in reaction 6–XXXVIII takes over the function of degenerate branching via reaction 6–XXXIII.§ (At these temperatures the formaldehyde formed in reaction 6–XXXVII is unreactive and does not participate in branching.)

With longer-chain hydrocarbons another type of decomposition of the peroxy radicals predominates. This is preceded by internal transfer of a hydrogen atom to the terminal oxygen. Thus, with *n*-pentane[71]:

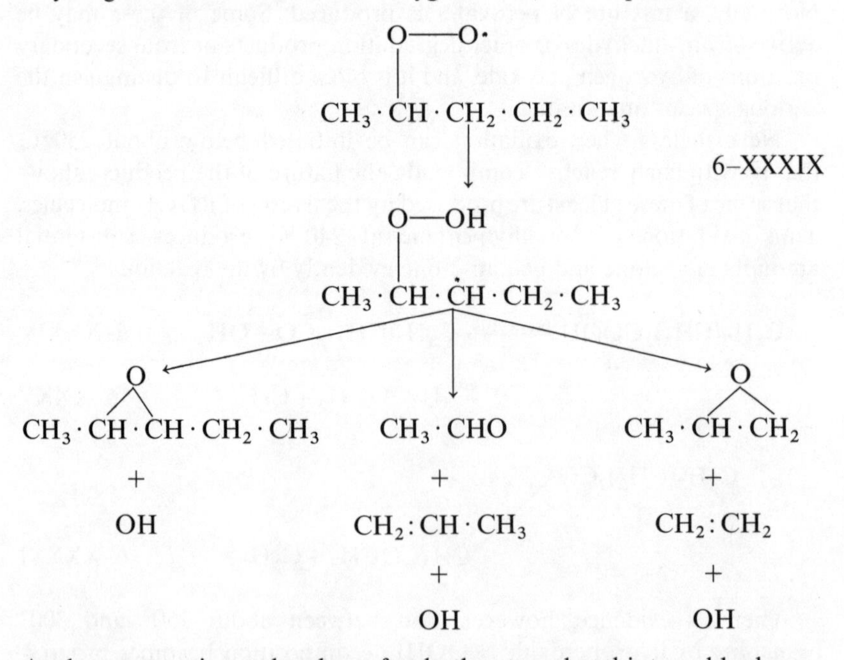

Analogous reactions take place after both external and internal hydrogen

§It is not certain, however, that branching is due *simply* to the provision of radicals by reaction 6–XXXIII. It may be that this reaction initiates the more facile oxidation of the aldehyde and so leads to further chain-branching by producing a small concentration of the peracid.

abstraction from different carbon atoms. In this way, a great variety of oxygenated and olefinic products are formed, among which 0-heterocyclic compounds are prominent. These reactions are most prevalent at temperatures corresponding to the cool-flame region of the $p^* - T$ diagram.§

The negative temperature coefficient and high-temperature mechanisms
The effect of temperature on the kinetics (Fig. 6–11) shows first that the branching mechanism fails in some way in the region of the negative temperature coefficient, and second that branching resumes at higher temperatures. The failure must occur in the sequence:

$$R + O_2 \xrightarrow{A} RO_2 \xrightarrow{B} \text{Branching} \qquad \text{6–XL}$$

and the initial effect of increasing the temperature is evidently to introduce alternative reactions to A or B, which either do not lead to branching or substitute a slower branching mechanism.

Decomposition of the RO_2 radicals to aldehydes and other relatively inactive oxygenated molecules, as illustrated previously, provides a slower alternative to branching via formation and decomposition of hydroperoxide. But, though the position is not completely clear, it appears that the decrease in the rate of branching frequently results largely from alternatives to part A of the sequence. Two factors, one thermodynamic and the other kinetic, conspire to bring this about. Sufficient is known or can be estimated about the thermodynamic properties of R and RO_2 radicals to enable equilibrium constants for reaction A to be calculated; and from this it emerges that reactions of this type become reversible at relevant temperatures. For example, Benson(73) estimates that the equilibrium concentrations of R and RO_2 in the reaction

$$(CH_3)_2CH + O_2 \rightleftharpoons CH_3C(H)O_2 \qquad \text{6–XLI}$$

are equal at 500 °C in the presence of 76 torr O_2; and with 7.6 torr O_2 they are equal at 410 °C. This effect alone, however, does not explain the decrease in branching rate, which sets in at lower temperatures; but, by increasing the stationary concentration of the R radicals, it enables another bimolecular reaction with an appreciable activation energy to compete for them. The second reaction, in the case of propane oxidation, is

$$(CH_3)_2CH + O_2 \rightarrow CH_3CH{:}CH_2 + HO_2 \qquad \text{6–XLII}$$

and analogous reactions take place in other systems. A reaction of this type requires an appreciable activation energy and, without the reversibility effect, could not compete effectively with the formation of RO_2, which does not. Thus, with increasing temperature, the reaction

$$R + O_2 \rightleftharpoons RO_2 \rightarrow \begin{cases} \text{branching via ROOH or other} \\ \text{oxygenated organic compounds} \end{cases} \qquad \text{6–XLIII}$$

§ Barat, Cullis, and Pollard(72) detected 86 products from the oxidation of 3-methylpentane in the cool flame region.

is replaced by

$$R+O_2 \rightarrow \text{unsaturated products}+HO_2 \qquad \text{6–XLIV}$$

Since neither the unsaturated compounds nor the HO_2 radical lead to branching at these temperatures, the reaction rate decreases with increasing temperature.

The mechanism also accounts for the progressive appearance of unsaturated compounds at higher temperatures. As the increase in temperature is continued, the HO_2 radical becomes more active and chains are propagated by reactions of the type

$$RH+HO_2 \rightarrow HOOH+R \qquad \text{6–XLV}$$

Unlike the RO_2 radicals, HO_2 is thermally stable at all relevant temperatures. Hydrogen peroxide is likewise much more stable than organic hydroperoxides though, like them, it is highly susceptible to decomposition at the surface. However, its decomposition in the gas phase becomes apparent at about 400 °C and branching is eventually resumed via the reaction

$$HOOH+M \rightarrow 2OH+M \qquad 7$$

Thus, the rate once more assumes a positive temperature coefficient. In essence the high-temperature mechanism for the oxidation—or 'oxidative pyrolysis'—of a hydrocarbon such as propane is:

$$\text{Branching}\begin{cases} HOOH+M \rightarrow 2OH+M & 7 \\ OH+C_3H_8 \rightarrow C_3H_7+H_2O & \text{6–XLVI} \end{cases}$$

$$\text{Propagation}\begin{cases} C_3H_7+O_2 \rightarrow C_3H_6+HO_2 & \text{6–XLVII} \\ HO_2+C_3H_8 \rightarrow C_3H_7+HOOH & \text{6–XLVIII} \end{cases}$$

$$\text{Termination}\begin{cases} HO_2 \rightarrow \\ OH \rightarrow \end{cases}\left.\begin{matrix} \text{destruction at} \\ \text{wall and in gas phase} \end{matrix}\right\} \qquad \text{6–XLIX}$$

The transition from the low-temperature mechanisms is, of course, gradual.

A significant variation in the pattern of high-temperature oxidation is found when the reaction mechanism produces methyl radicals. Reactions akin to reaction 6–XLVII are replaced by the formation of formaldehyde. This apparently occurs by the reactions:[74]

$$CH_3+O_2+M \rightarrow CH_3O_2+M \rightarrow HCHO+OH+M \qquad \text{6–L}$$

which compete with redissociation of the peroxy radical to CH_3 and O_2. The reactions

$$HCHO+O_2 \rightarrow HCO+HO_2 \qquad \text{6–LI}$$

$$HCO+O_2 \rightarrow CO+HO_2 \qquad \text{6–LII}$$

$$HO_2+RH \rightarrow HOOH+R \qquad \text{6–LIII}$$

enable formaldehyde to initiate branching at temperatures at which hydrogen peroxide is stable. The special influence of methyl radicals naturally is particularly evident in the oxidation of methane,[58] which rather paradoxically we come to last in our discussion of isothermal oxidation. Because of its exceptionally strong C—H bonds, methane is the least reactive hydrocarbon towards oxidation, but above about 420 °C methyl radicals are produced by hydrogen abstraction

$$R' + CH_4 \rightarrow CH_3 + R'H \qquad \text{6–LIV}$$

and are oxidized to formaldehyde and hydroxyl radicals as above. Degenerate branching takes place via formaldehyde. The oxidation normally takes place, however, at temperatures where the hydrogen peroxide formed in reaction 6–LIII also gives rise to branching. The presence of two branching reactions, together with the variable propensity for hydrogen peroxide to be destroyed heterogeneously, make the kinetics more complicated than might be expected. Nevertheless, it appears they can be understood in terms of formation and destruction of formaldehyde as just described, together with parallel oxidation of the formaldehyde by reactions such as

$$HCHO + OH \rightarrow HCO + H_2O \qquad \text{6–LV}$$

$$HCHO + HO_2 \rightarrow HCO + HOOH \qquad \text{6–LVI}$$

In general outline, the overall kinetics are similar to those described on p. 249 for ethylene.

To return finally to the questions posed at the beginning of our discussion concerning mechanisms of degenerate branching, the position can be summarized very roughly as follows. At low temperatures, up to say 250 °C, branching occurs by fission of organic hydroperoxide formed by the sequence $R \rightarrow RO_2 \rightarrow ROOH$. Above about 450 °C it occurs by fission of hydrogen peroxide formed by $R \rightarrow HO_2 \rightarrow HOOH$. At intermediate temperatures the first mechanism fails before branching by the second becomes effective; but branching may take place via the oxidation of aldehydes formed by the decomposition of RO_2 or oxidation of methyl radicals.

Cool flames and ignition

The essential phenomena were described in Section 6–5–1. When ignition takes place outside the two-stage region, that is to the right of g–h in Fig. 6–12, ignition is a relatively uncomplicated chain-thermal event with no preliminaries other than the normal generation of heat by slowly branching chains. Leaving aside two-stage ignition for the moment, it is reasonably certain that the conditions for the appearance of a single cool flame (i.e. inside the boundary b–j–g) are also chain-thermal in nature;[75] but naturally the peculiar properties of self-quenching and multiplicity of cool flames need special explanations.

It is significant that the cool-flame region of the p^*-T diagram invariably adjoins the ignition boundary in the temperature range which includes the negative temperature coefficient (c–d). Indeed the accepted view of the mechanism of the self-quenching is that it is due basically to the same processes that produce the negative temperature coefficient. If hydroperoxide is assumed to be responsible for chain-branching, the basic picture in accord with our previous discussion is as follows:

$$(RH, O_2) \longrightarrow R \begin{cases} \xrightleftharpoons[\Delta H<0]{O_2} RO_2 \xrightarrow{RH} ROOH \xrightarrow[E\gg 0]{\text{branching}} 2R \begin{cases} \nearrow \\ \searrow \end{cases} \\ \xrightarrow[E>0]{O_2} R' \xrightarrow[\text{chains}]{\text{straight}} \text{Inert Products} \end{cases}$$

The upper reaction path leads to branching and the lower, which comes into prominence with increasing temperature, does not. During the stage before the appearance of a cool flame, the reaction follows the branching path and accelerates as the hydroperoxide accumulates. The rate eventually becomes very fast, resulting in a correspondingly large and rapid rise in the temperature of the gas. This has the effect of suddenly 'switching' the reaction to the non-branching path and the rate immediately falls. At the same time, the accumulated hydroperoxide decomposes rapidly at the higher temperature, but, because of the prevalence of the non-branching path, it is unable to regenerate itself. Hence there is a sharp fall in the concentration of the hydroperoxide. This may be virtually to zero. The decrease in reaction rate entails a correspondingly rapid decrease in the rate of heat generation; consequently the temperature also falls as the result of heat loss to the walls. With hydroperoxide absent, chain initiation can occur only by a 'primary' reaction involving the reactants, and ordinarily this is much slower than initiation by the branching process. The result is that the reaction may be brought almost to a halt, and, in this case, the temperature returns practically to its initial value. In short, the mechanism produces a rapid rise and fall in temperature—and, because of this, in pressure—together with a very sharp fall in the concentration of the branching agent, namely hydroperoxide.

These events, of course, are as characteristically observed. Moreover they can occur after only a little of the reaction mixture has been consumed. Hence, with the destruction of the branching agent and the return of the temperature to its original level after the cool flame, the reaction is ready to begin again. The hydroperoxide gradually accumulates once more and the whole cycle may repeat itself. This continues until insufficient of the reactants remains to generate the temperature rise required to produce the 'thermal switch'. Thus a succession of cool flames is accounted for.

So far in this context we have made no reference to the existence of high-temperature mechanisms of oxidation. But we have seen previously that if the mixture is heated to a temperature above the range of the negative temperature coefficient, the rate increases again as another branching agent—hydrogen peroxide or formaldehyde or both—begins to come into play. Under suitable conditions, the temperature in the cool flame may rise sufficiently to take the reaction into this region. The high-temperature mechanism then takes over and the reaction, unchecked by any further mechanistic impediment, accelerates to a chain-thermal explosion. Thus we have two-stage ignition.

This qualitative thermokinetic picture of cool flames has been put on a quantitative basis by Halstead, Prothero, and Quinn[76] by means of a computer-simulation study of the oxidation of acetaldehyde, the simplest compound to exhibit the phenomena in a well-developed way. They were able to show that the evolution of heat which accompanies the mechanism of oxidation via peracid branching acting in competition with the non-branching reactions 6–XXXII and 6–L, can account for most of the observed facts. These include multiple cool flames, two-stage ignition, and the general features of the ignition diagram. Several substantial details of cool-flame behaviour, however, have yet to be filled in, particularly in relation to the oxidation of hydrocarbons. One of these is the occurrence of multi-stage ignition, for, with a single branching mechanism, it is difficult to see how any but the first cool flame should be followed by ignition.[64] It seems that two concurrent branching reactions are required, the one as it were riding upon the other; as perhaps is suggested by Figs. 6–13 and 6–14.

Bibliography

(A) P. Gray and P. R. Lee, *Thermal Explosion Theory* in *Oxidation and Combustion Reviews*, Vol. 2, C. F. H. Tipper (ed.), p. 1, Elsevier, 1967.

(B) N. N. Semenov, *Some Problems in Chemical Kinetics and Reactivity*, Vol. 2, Princeton U.P., 1959 and Pergamon Press, 1958.

(C) F. S. Dainton, *Chain Reactions*, 2nd edn, Methuen, 1966.

(D) G. J. Minkoff and C. F. H. Tipper, *Chemistry of Combustion Reactions*, Butterworths, 1962.

(E) R. R. Baldwin and R. W. Walker, 'Branching-chain reactions: the hydrogen–oxygen reaction', in *Essays in Chemistry*, Vol. 3, Bradley, Gillard, and Hudson (eds), Academic Press, 1972.

(F) S. W. Benson, Effects of resonance and structure on the thermochemistry of organic peroxy radicals and the kinetics of combustion reactions, *J. Am. Chem. Soc.*, **87**, 972, 1965.

References

1 B. Lewis and G. von Elbe, *Combustion, Flames and Explosions of Gases*, 2nd edn, pp. 323 ff, Academic Press, 1961.

2 N. N. Semenov, *Some Problems in Chemical Kinetics and Reactivity*, Vol. 2, Chap. 8, Princeton U.P., 1959 and Pergamon Press, 1958.

3 D. A. Frank-Kamenetskii, *Diffusion and Heat Transfer in Chemical Kinetics*, 2nd edn, Chap. 7, Plenum Press, 1969.

4 H. S. Carslaw and J. C. Jaeger, *Conduction of Heat in Solids*, pp. 6 ff, Oxford, 1947.

5 P. Gray and P. R. Lee, *Thermal Explosion Theory* in *Oxidation and Combustion Reviews*, Vol. 2, pp. 39 ff, C. F. H. Tipper (ed.), Elsevier, 1967.

6 P. Gray and P. R. Lee, *Thermal Explosion Theory* in *Oxidation and Combustion Reviews*, Vol. 2, p. 21, C. F. H. Tipper (ed.), Elsevier, 1967.

7 P. Gray and E. P. O'Neill, *J. Chem. Soc., Faraday Trans. I*, **68,** 564, 1972.

8 A. O. Appin, J. Chariton, and O. M. Todes, *Acta Physicochim. URSS*, **5,** 655, 1936; P. Gray and P. R. Lee, *11th Symposium (Internat.) on Combustion*, p. 1123, The Combustion Institute, 1967.

9 A. O. Allen and O. K. Rice, *J. Am. Chem. Soc.*, **57,** 310, 1935.

10 E. J. Harris, *Proc. Roy. Soc.*, **A175,** 254, 1940.

11 D. H. Fine, P. Gray, and R. Mackinven, *12th Symposium (Internat.) on Combustion*, p. 545, The Combustion Institute, 1969.

12 P. L. Hanst and J. G. Calvert, *J. Phys. Chem.*, **63,** 104, 1959.

13 P. G. Ashmore, B. J. Tyler, and T. A. B. Wesley, *11th Symposium (Internat.) on Combustion*, p. 1133, The Combustion Institute, 1967.

14 F. Kaufman and N. J. Gerri, *8th Symposium (Internat.) on Combustion*, p. 619, Williams and Wilkins, Baltimore, 1962.

15 P. Gray and P. R. Lee, *Thermal Explosion Theory* in *Oxidation and Combustion Reviews*, Vol. 2, C. F. H. Tipper (ed.), p. 1, Elsevier, 1967.

16 D. R. Warren, *Proc. Roy. Soc.*, **A211,** 86, 96, 1952.

17 N. Semenova quoted by N. N. Semenov, *Some Problems in Chemical Kinetics and Reactivity*, Vol. 2, Chap. 10, Princeton U.P., 1959 and Pergamon Press, 1958.

18 R. N. Pease, *Equilibrium and Kinetics of Gas Reactions*, p. 196, Princeton U.P., 1942.

19 V. R. Bursian and V. S. Sorokin, *Z. phys. Chem.*, **B12,** 247, 1931; N. Semenoff, *Chemical Kinetics and Chain Reactions*, Chap. 3, Oxford, 1935.

20 F. S. Dainton and H. M. Kimberley, *Trans. Faraday Soc.*, **46,** 624, 1950.

21 H. W. Melville and E. B. Ludlam, *Proc. Roy. Soc.*, **A132,** 108, 1931.

22 C. N. Hinshelwood and E. A. Moelwyn-Hughes, *Proc. Roy. Soc.*, **A138,** 311, 1932.

23 Quoted by N. N. Semenov, *Some Problems in Chemical Kinetics and Reactivity*, Vol. 2, pp. 159–161, Princeton U.P., 1959 and Pergamon Press, 1958.

24 A. H. Willbourn and C. N. Hinshelwood, *Proc. Roy. Soc.*, **A185,** 353, 1946.

25 B. Lewis and G. von Elbe, *Combustion, Flames and Explosion of Gases*, 2nd edn, p. 24, Academic Press, 1961.

26 N. N. Semenov, *Some Problems in Chemical Kinetics and Reactivity*, Vol. 2, Princeton U.P., 1959 and Pergamon Press, 1958.

27 B. Yakovlev and P. Shantarovich, quoted by C. F. Cullis and M. F. R. Mulcahy, *Combustion and Flame*, **18,** 225, 1972.

28 N. Semenoff, *Zeit. Physik.*, **46,** 109, 1927; for an English translation see M. H. Back and K. J. Laidler (eds), *Selected Readings in Chemical Kinetics*, Pergamon Press, 1967.

29 C. N. Hinshelwood, *Proc. Roy. Soc.*, **A188,** 1, 1946.

30 R. R. Baldwin *et al.*, *Trans. Faraday Soc.*, **63,** 1665, 1676, 1967.

31 R. R. Baldwin *et al.*, *Trans. Faraday Soc.*, **56,** 80, 93, 103, 1960.

32 D. E. Hoare, J. B. Protheroe, and A. D. Walsh, *Trans. Faraday Soc.*, **55,** 548, 1959.

33 C. J. Jachimowski and W. M. Houghton, *Combustion and Flame*, **17,** 25, 1971.
34 R. R. Baldwin, D. E. Hopkins, and R. W. Walker, *Trans. Faraday Soc.*, **66,** 189, 1970.
35 D. E. Hoare and A. D. Walsh, *Trans. Faraday Soc.*, **50,** 37, 1954.
36 E. J. Buckler and R. G. W. Norrish, *Proc. Roy. Soc.*, **A167,** 292, 318, 1938.
37 G. Dixon-Lewis, J. W. Linnett, and D. F. Heath, *Trans. Faraday Soc.*, **49,** 756, 766, 1953.
38 F. Gaillard-Cusin and H. James, *J. Chem. Phys.*, **63,** 379, 1966.
39 K. G. Sulzmann, B. F. Myers, and E. R. Bartle, *J. Chem. Phys.*, **42,** 3969, 1965.
40 A. M. Dean and G. B. Kistiakowsky, *J. Chem. Phys.*, **53,** 830, 1970.
41 R. S. Brokaw, *11th Symposium (Internat.) on Combustion*, pp. 1063, 1072, The Combustion Institute, 1967.
42 A. S. Gordon and R. H. Knipe, *J. Phys. Chem.*, **59,** 1160, 1955.
43 P. G. Dickens, J. E. Dove, and J. W. Linnett, *Trans. Faraday Soc.*, **60,** 539, 1964.
44 F. S. Dainton, *Chain Reactions*, 2nd edn, Methuen, 1966.
45 J. B. Levy and B. W. K. Copeland, *J. Phys. Chem.*, **72,** 3168, 1968.
46 G. A. Kapralova, E. M. Margolina, and A. M. Chaikin, *Kinetics and Catalysis*, **10,** 23, 245, 1969; *Combustion and Flame*, **13,** 557, 1969.
47 V. I. Vedeneev, V. I. Propoi, and O. M. Sarkisov, *Kinetics and Catalysis*, **11,** 26, 1970.
48 N. N. Semenov, *Photochemistry and Reaction Kinetics*, Ashmore, Dainton, and Sugden (eds), p. 243, C.U.P., 1967.
49 O. D. Krogh and G. C. Pimental, *J. Chem. Phys.*, **56,** 969, 1972.
50 F. S. Dainton, *Trans. Faraday Soc.*, **38,** 227, 1942.
51 F. S. Dainton and R. G. W. Norrish, *Proc. Roy. Soc.*, **A177,** 393, 421, 1941.
52 P. G. Ashmore, *Photochemistry and Reaction Kinetics*, Chap. 11, Ashmore, Dainton, and Sugden (eds), C.U.P., 1967.
53 C. H. Yang and B. F. Gray, *11th Symposium (Internat.) on Combustion*, p. 1099, The Combustion Institute, 1967.
54 R. G. W. Norrish and J. D. Reagh, *Proc. Roy. Soc.*, **A176,** 429, 1940.
55 N. Semenoff, *Chemical Kinetics and Chain Reactions*, p. 84, Oxford, 1935.
56 A. J. Harding and R. G. W. Norrish, *Proc. Roy. Soc.*, **A212,** 291, 1952.
57 (a) V. Ya. Shtern, *The Gas-Phase Oxidation of Hydrocarbons*, Pergamon Press, 1964; (b) see also J. J. Batten and M. J. Ridge, *Aust. J. Chem.*, **8,** 370, 1955.
58 D. E. Hoare, The combustion of methane, *Low Temperature Oxidation*, Chap. 6, W. Jost (ed.), Gordon and Breach, 1965; D. E. Hoare and G. S. Milne, *Trans. Faraday Soc.*, **63,** 101, 1967.
59 C. F. Cullis, Sir Cyril Hinshelwood, and M. F. R. Mulcahy, *Proc. Roy. Soc.*, **A196,** 160, 1949.
60 A. D. Walsh, 'Knock' in spark-ignition engines, *Low Temperature Oxidation*, Chap. 9, see ref. 58.
61 M. F. R. Mulcahy, *Disc. Faraday Soc.*, **2,** 128, 1949.
62 T. Kunugi *et al.*, *Adv. Chem. Series*, **76,** 326, 1968; C. N. Satterfield *et al.*, *Ind. Eng. Chem.*, **46,** 1001, 1954.
63 C. F. Cullis, A. Fish, and J. F. Gibson, *Proc. Roy. Soc.*, **A311,** 253, 1969.
64 J. F. Griffiths, B. F. Gray, and P. Gray, *13th Symposium (Internat.) on Combustion*, p. 239, The Combustion Institute, 1971.
65 L. A. Repa and V. Ya. Shtern, *Zhur. Fiz. Khim.*, **28,** 414, 1954.
66 J. Bardwell and Sir Cyril Hinshelwood, *Proc. Roy. Soc.*, **A205,** 375, 1951.
67 L. Bateman, *Q. Rev. Chem. Soc.*, **8,** 147, 1954.

68 A. Combe, M. Niclause, and M. Letort, *Rev. Inst. franc., Pétrole*, **10,** 786, 1955; J. F. Griffiths and G. Skirrow, *Oxidation and Combustion Rev.*, **3,** 47, 1968.
69 R. R. Baldwin, R. W. Walker, and D. H. Langford, *Trans. Faraday Soc.*, **65,** 792, 806, 1969.
70 A. Fish, *Proc. Roy. Soc.*, **A298,** 204, 1967.
71 C. F. Cullis, M. Saed, and D. L. Trimm, *Proc. Roy. Soc.*, **A300,** 455, 1967.
72 P. Barat, C. F. Cullis, and R. T. Pollard, 13*th Symposium* (*Internat.*) *on Combustion*, p. 179, The Combustion Institute, 1971.
73 S. W. Benson, *J. Am. Chem. Soc.*, **87,** 972, 1965.
74 J. A. Barnard and A. Cohen, *Trans. Faraday Soc.*, **64,** 396, 1968.
75 F. E. Malherbe and A. D. Walsh, *Trans. Faraday Soc.*, **46,** 824, 1960; R. Ben-Aïm, *J. Chem. Phys.*, **57,** 683, 1960.
76 M. P. Halstead, A. Prothero, and C. P. Quinn, *Proc. Roy. Soc.*, **A322,** 377, 1971.

Problems

6–1 The first order decomposition of azomethane $(CH_3)_2N_2$ is 43 kcal mole^{-1} exothermic and has Arrhenius parameters $A = 2 \times 10^{16}\,s^{-1}$ and $E_a = 51.3$ kcal mole^{-1}. Assuming the heat is lost from the decomposing vapour entirely by conduction ($\lambda = 1 \times 10^{-4}$ cal cm^{-1} s^{-1}K^{-1}), estimate to the nearest 5 K the maximum initial temperature T_0 at which the vapour can fill a 200 cm^3 flask at 200 torr pressure without exploding.

Dilution of the vapour with an equal volume of helium increases λ to about 2×10^{-4}. Estimate T_0 for the mixture at the same total pressure as the above.

6–2 (a) Draw up a table showing the qualitative effects (if any) of (i) temperature, (ii) addition of inert gas and (iii) size of vessel on the critical partial pressure p^* of reactants required for a purely thermal ignition and for ignitions at the first and second limits of an isothermal linearly branched chain system. Where appropriate, note constitutive factors which affect the influence of factor (ii).
(b) Sketch a p^*–T diagram for a reaction which manifests all three types of ignition under appropriate conditions and indicate temperatures at which only one ignition limit would be observed. Using the same axes, draw diagrams showing the behaviour observed when (i) the reactants are diluted by a heavy inert gas (say SF_6), and (ii) the size of the reaction vessel is reduced.

6–3 (a) Show that, on the basis of the accepted reaction mechanism, the pressure p_1 torr at the first ignition limit of a H_2:O_2 mixture in a long cylindrical vessel with 'active' walls is given by the relation: $p_1 d = 852(D_0 T/xk_2)^{1/2}$, where d cm is the vessel diameter, D_0 cm^2 s^{-1} is the diffusion coefficient for H atoms in the particular mixture at 1 torr, x is the mole fraction of O_2, and k_2 cm^3 mole s^{-1} refers to reaction 2 in Table 6–2. Does this relation apply when inert gas is present?
(b) The following data were obtained by Kurzius and Boudart (*Comb. & Flame*, **12,** 477, 1968) with $2H_2 + O_2$ mixtures under the above conditions.

Temp. K	985.7	921.4	854.7	794.2
p_1 torr	2.32	2.83	3.45	4.56

D_0 for $(2H_2 + O_2)$ is given by: $D_0 = 0.0440T^{1.8}$. By means of an appropriate Arrhenius-type plot (cf. Fig. 6–2) determine the Arrhenius constants of reaction 2.
(c) Combine your value of E_2 with that obtained for the reverse reaction in Problem 4–6 to obtain ΔH_2. Hence check the attribution of chain-branching to reaction 2 by comparing this value with the value obtained by consulting a table of thermochemical data.

6–4 (a) Mixtures of hydrogen and oxygen with the compositions shown below were successively prepared in a vessel at 560 °C (ref. 37). Each mixture was gradually withdrawn from the vessel until the remaining gas exploded. This occurred when the pressure had fallen to p_c.

$\% O_2$	20	33	50	60	67
p_c torr	122	135	155	169	182

Determine (graphically) the efficiency (a_{O_2}) of O_2 relative to that of H_2 as third body in the reaction $H+O_2+M \rightarrow HO_2+M$.

(b) Similar experiments to the above conducted with O_2—H_2—N_2 mixtures (at 550 °C) gave the following results:

$[O_2]:[H_2]:[N_2]$	(1:1:0)	(5:4:1)	(7:5:2)	(5:3:2)	(5:2:3)
p_c torr	130	142	148	153	171

Extend the high pressure form of equation 6–32 to take account of the presence of an inert gas and derive the value of a_{N_2}.

6–5 (a) The slow oxidation of 2:1 methane–oxygen mixtures initially at 150 torr was followed by the pressure change (Hoare and Walsh, *Proc. Roy. Soc.*, **A215,** 454, 1952). Estimate the overall activation energy E_ϕ for the degenerate branching process from the following measurements made in the early stages of the reaction.

500 °C	Time s	0	300	350	400	450	500
	Δp torr	0	0.30	0.45	0.70	1.00	1.50
525 °C	Time s	0	150	175	200	225	250
	Δp torr	0	0.35	0.60	0.90	1.30	1.90

(b) If the induction period is defined as the time τ taken for Δp to reach 1 torr, how well is the relation $\phi\tau$ = const. obeyed in the above circumstances?

6–6 Peroxy radicals (RO_2) and hydroperoxides (ROOH) are destroyed by mildly alkaline surfaces, the former with high collision efficiency, the latter requiring a small activation energy. Supposing that the hypothetical oxidation

$$RH+O_2 \rightarrow ROOH$$
$$ROOH \xrightarrow{\text{surface}} \text{Products}$$

is carried out in a vessel with walls of this type and that termination of the (long) chains is exclusively heterogeneous, derive an expression relating the initial net branching factor to the radius of the vessel and the total pressure. (Assume constant composition with O_2 in excess.) Show that the reaction rate would be virtually zero below a certain radius or pressure.

6–7 (a) Describe, with appropriate diagrams, the main phenomena associated with the spontaneous ignition of mixtures of hydrocarbons with air or oxygen. Include reference to single- and two-stage ignition and cool flames, and indicate points of difference and similarity to the pattern of ignition behaviour shown typically by the hydrogen–oxygen reaction.

(b) Give a qualitative explanation for the negative temperature coefficient of the rate of oxidation of hydrocarbons. Show how the transient appearance of cool flames can be explained along similar lines. How are multiple cool flames possible?

7 Kinetics of combustion

7–1 What is combustion?

What is combustion? The shortest answer is that it is self-sustained chemical reaction. A chemical system capable of undergoing combustion has an inbuilt feedback mechanism: the energy released by the reaction of some molecules of the reactants supplies the activation energy required to initiate reaction of more molecules and so on. Consider two examples. When, under appropriate conditions, a source of heat is applied to a mixture of propane and air issuing continuously from a tube, a *flame* appears; that is, a zone of rapid exothermic reaction is established and remains in existence after the original source of heat is withdrawn. Again, application of a spark or a flash of light, in suitable circumstances, to a mixture of hydrogen and chlorine in a closed vessel produces an immediate explosion. A ciné-camera would show that this is caused by a flame spreading out rapidly from the region of the spark or light beam to all parts of the vessel. These examples reveal the existence of a process whereby reaction, once initiated, sustains itself until the reactants are consumed. In both cases some kind of motion is involved: in the first, the reaction zone is stationary and the reactants flow into it; and in the second the zone propagates itself through the stationary reactive medium. The situations are, of course, interchangeable: hydrogen and chlorine can be burned as a stationary flame by supplying the mixture to the end of a tube at the same speed as the reaction zone propagates into it, and a flammable mixture of propane and air can be exploded readily enough in a closed vessel.

Such commonplace phenomena show that the rate of a reaction in the state of combustion is likely to be determined not only by the factors considered in previous chapters as applicable to reactions occurring under isothermal and static conditions but also by thermal and kinematical considerations. The last two factors are those which, for the most part, ordinary kinetic studies are at pains to exclude. Nevertheless the occurrence of thermal explosions discussed in Chapter 6 has already provided an example of the mutual influence of reaction rate and heat of reaction; and simple instances of the interaction of flow, diffusion, and reaction were considered in Chapters 2 and 4. Since chemical reaction is invariably accompanied by a heat change and consequently some disturbance of the thermal balance between the system and its surroundings, it is obvious that the occurrence of reaction in an absolutely isothermal and motionless gaseous medium is an abstraction not actually realized in Nature. With combustion, the interaction between reaction, heat, and motion is perhaps at its most complex. Nevertheless, our quantitative understanding of combustion phenomena has advanced considerably in recent years, though much yet remains to be discovered. The following sections, which necessarily are selective, are intended to provide a brief introduction to this challenging and practically important area of reaction kinetics.

7–2 Ignition and extinction of combustion

If self-sustenance is the chief characteristic of combustion, the phenomenon of ignition by a temporary stimulus comes a close second, as is evident from the phenomena recalled above. Everyday experience with fire shows, however, that whether or not ignition can be brought about in any particular circumstances depends on a rather delicate balance of physical and chemical factors. The same applies to the extinction of combustion after it is fully initiated. Furthermore, it is apparent that limiting conditions for extinction are frequently more than marginally different from those applicable to ignition: the draught which prevents a flame from being lit will by no means necessarily put it out when it is burning brightly.

It is a remarkable fact that the essentially discontinuous phenomena of ignition and extinction follow naturally from standard principles of reaction kinetics. To demonstrate this it will be convenient to consider the conditions for ignition and extinction to occur in a gas which is capable of exothermic reaction and flows continuously through a thermally insulated chamber. The practically minded reader might consider the system to be an idealization of a rocket engine, or better, a ram-jet engine.

7–2–1 Exothermic reaction in flowing gas

The system just described is illustrated in Fig. 7–1. Whether or not the gas ignites on entering the vessel naturally depends both on constitutive factors and on the initial physical conditions, for example the temperature

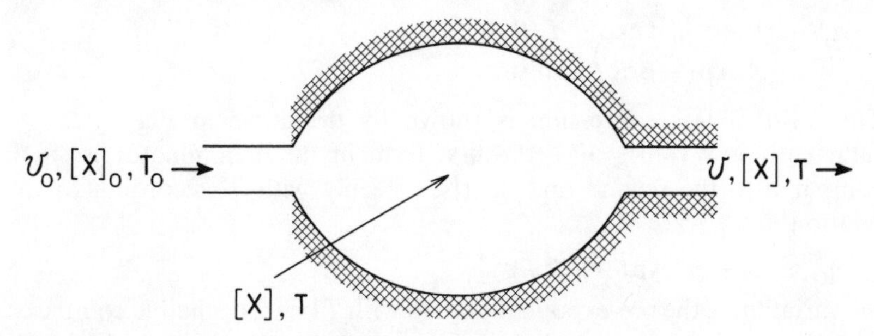

Fig. 7–1 Perfectly stirred and thermally insulated reaction vessel.

and flow velocity. An ether–air mixture, for example, is 'more flammable' than one of propane and air in similar proportions. And we may expect that, all else being constant, ignition will be brought about by heating the inflowing reactants; and combustion, once established, will be extinguished by chilling them. Thus, there will be critical values of the initial temperature T_0 for ignition and extinction and it will be shown presently that these, in general, are not the same.

We shall assume that a condition of *adiabatic perfectly stirred flow* prevails in the reactor. Firstly, turbulence is sufficiently vigorous to make the mixture of reactants and products uniform in composition and temperature throughout the vessel and secondly, no heat is lost to or through the walls. In addition, we assume there is no pressure drop; that the specific heat of the effluent gas is independent of its composition and temperature; and finally that the reaction rate is controlled by a first-order reaction of one of the reactants (X), that is,

$$-\frac{d[\mathrm{X}]}{dt} = k[\mathrm{X}] \qquad 7\text{–}1$$

Under steady operation, the number of X molecules entering the reaction vessel in unit time is equal to the sum of those consumed and those which pass through unreacted. Thus

$$v_0[\mathrm{X}]_0 - Vk[\mathrm{X}] - v[\mathrm{X}] = 0 \qquad 7\text{–}2$$

where v_0, v are the volumetric flow rates of gas into and out of the vessel measured at the reaction temperature T; $[\mathrm{X}]_0$, $[\mathrm{X}]$ are the corresponding concentrations and V is the volume of the vessel. The rate of generation of heat is given by

$$\dot{q}_{\mathrm{G}} = QVk[\mathrm{X}] \qquad 7\text{–}3$$

where Q is the heat of reaction. Eliminating $[\mathrm{X}]$ from equations 7–2 and 7–3 yields

$$\dot{q}_{\mathrm{G}} = \frac{Qkv_0[\mathrm{X}]_0}{k+(v/V)} \qquad 7\text{–}4$$

Substituting the Arrhenius relation for k gives the dependence of $\dot{q}_{\mathrm{G}}$ on temperature:

$$\dot{q}_{\mathrm{G}} = \frac{Qv_0[\mathrm{X}]_0 A\exp(-E/RT)}{A\exp(-E/RT)+(v/V)} \qquad 7\text{–}5$$

The form of this expression is shown by the curve in Fig. 7–2a. At sufficiently low values of T the first term in the denominator is small compared to the second and $\dot{q}_{\mathrm{G}}$ rises steeply with T according to the relation

$$\dot{q}_{\mathrm{G}} \approx \mathrm{const} \times \exp(-E/RT) \qquad 7\text{–}6$$

the variation of the pre-exponential term with T being negligible compared with that of the exponential. At sufficiently high temperatures, however, the first term in the denominator supervenes and the relation becomes

$$\dot{q}_G \approx Q[X]_0 v_0 \qquad 7\text{–}7$$

This simply expresses the fact that, with increasing T, the rate of heat release eventually becomes limited by the rate at which the reactants are supplied to the vessel. This, of course, is independent of T.

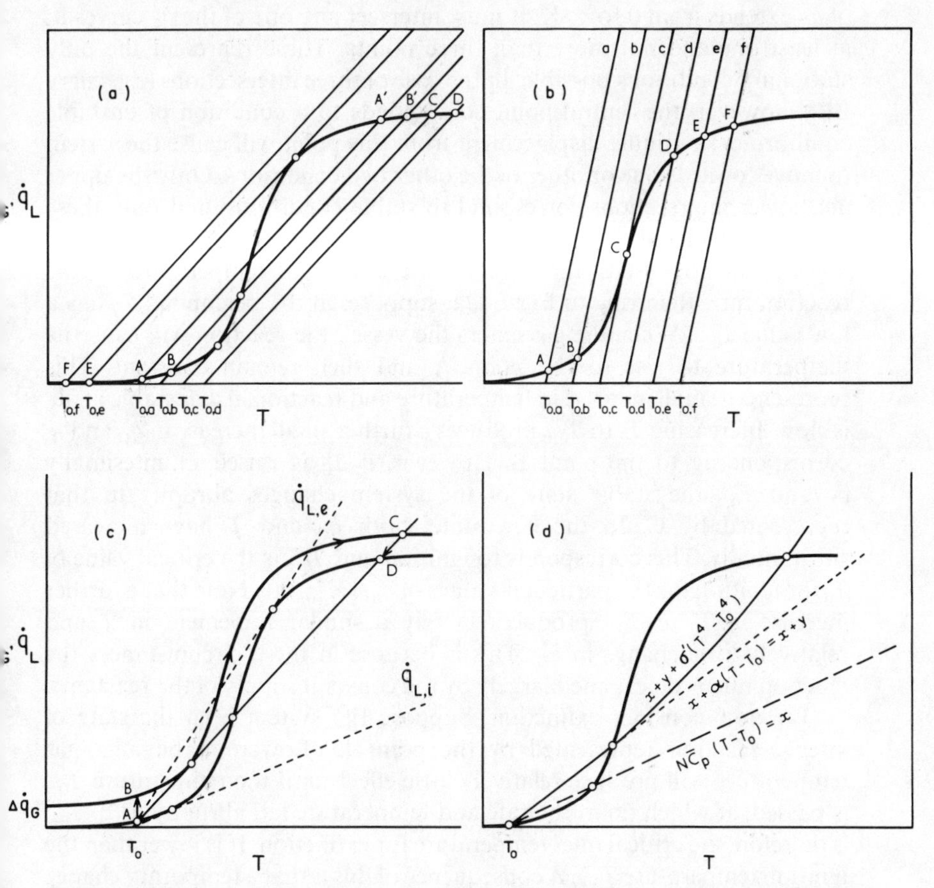

Fig. 7–2 Rates of heat generation $\dot{q}_G$ and heat loss $\dot{q}_L$ versus absolute temperature for an exothermic reaction in a perfectly stirred reactor.

The diagrams are schematic. (a), (b), and (c) refer to the reaction conducted under adiabatic conditions. In (d), $\dot{q}_L$ includes heat lost to the walls. In (a), (c), and (d) critical conditions for ignition and extinction are shown. In (b) no critical conditions occur.

Since the only heat lost is that carried out of the vessel by the gas, the rate of heat loss is given by

$$\dot{q}_L = N\bar{C}_p(T - T_0) \equiv K(T - T_0) \qquad 7\text{–}8$$

where N is the total molar flow and $\bar{C}_p$ is the mean heat capacity of the effluent gas. On the basis of our assumptions, K is a constant for constant

N. Graphs of equation 7–8 for several values of T_0 are shown by the straight lines in Fig. 7–2a.

So far no restriction has been placed on the values of T. In fact, however, the values of T assumed by the reacting gases are determined by the steady-state relation $\dot{q}_G = \dot{q}_L$. They are given by the points of intersection of the $\dot{q}_G$ and $\dot{q}_L$ curves in Fig. 7–2a. Since, in principle, the curve of $\dot{q}_G$ extends from 0 to ∞ K, it must intersect any one of the $\dot{q}_L$ curves in at least one and not more than three points. These represent the only stationary conditions possible. In the case of three intersections (e.g. curve BB′), however, the central point corresponds to a condition of unstable equilibrium: a minute displacement from this point will cause the system to move towards one or other of the other two conditions. Only the upper and lower intersections correspond to stable conditions, and only these need be considered.

We can now study the effect of the initial gas temperature on the reaction rate. Referring to Fig. 7–2a, suppose, in the beginning, T_0 has a low value $T_{0,a}$. When the gas enters the vessel, the reaction will cause its temperature to rise to the point A and then remain constant. This represents a small increase in temperature and reaction, if detectable at all, is slow. Increasing T_0 to $T_{0,b}$ produces a further small increase in $\dot{q}_G$ and T corresponding to the point B. However, if T_0 is raised infinitesimally beyond $T_{0,c}$ the stable state of the system changes abruptly to that represented by C. In the new state, both $\dot{q}_G$ and T have increased substantially. This corresponds to ignition and $T_{0,c}$ is the critical value of T_0 for ignition for the particular values of v_0, $[X]_0$, etc. Note that a further increase in T_0 to $T_{0,d}$ produces merely a similar increment in T and relatively little change in $\dot{q}_G$. This is because in these circumstances the reaction rate is determined largely by the constant supply of the reactants.

We now consider extinction. Suppose the system is in the state of intense reaction represented by the point D. Lowering the inlet gas temperature will produce relatively little effect until the temperature $T_{0,e}$ is passed, at which point the rate and temperature fall abruptly to E. $T_{0,e}$ is therefore the critical inlet temperature for extinction. It is lower than the ignition temperature $T_{0,c}$. A consequence of this is that a temporary change in the temperature (or other condition) of the inlet gas can produce a permanent change in the reaction rate. Suppose the initial temperature, originally at $T_{0,a}$, is raised briefly to $T_{0,d}$ and then returned to $T_{0,a}$; the condition of the inlet gas is the same before and after the adjustment but the system will have passed from quiescence (A) to intense reaction (A′). Similarly, when the system is in the state of intense reaction represented by B′, chilling the inlet gas from $T_{0,b}$ to $T_{0,f}$ would permanently extinguish the reaction notwithstanding a subsequent return of T_0 to $T_{0,b}$.

Returning to ignition, another way by which this can be brought about is to generate extra heat temporarily in the reaction vessel, for example by replacing temporarily some of the intake gas by a more flammable reactant. The extra heat generation is represented by $\Delta\dot{q}_G$ in Fig. 7–2c.

The system, initially in the sub-critical state A, now follows the path ABC to the state of intense reaction C and remains in a similar state D after the extraneous source of heat is cut off.

Thus simple theory explains not only the critical nature of ignition and extinction but also the fact, well attested by experience, that in appropriate circumstances either event may be brought about by a transitory change in conditions. A steady-state treatment naturally has nothing to say about the duration of the perturbation required to produce these effects, but we may expect it to be of similar magnitude to the residence time of the gas in the vessel (V/v).

These considerations show how ignition and extinction occur, but they do not show that such critical, that is discontinuous, phenomena are inevitable. Suppose, without otherwise altering the conditions, the value of K in equation 7–8 is increased so as to be represented by the straight lines a, b etc. in Fig. 7–2b. The $\dot{q}_L$ line for any value of T_0 now intersects the $\dot{q}_G$ curve at one point only. Progressively increasing or decreasing the inlet temperature therefore simply causes a smooth alteration in $\dot{q}_G$ and T and the critical effects we have identified with ignition and extinction are absent. The occurrence of the critical phenomena obviously depends on the values of the variables in equations 7–5 and 7–8; for example, the change represented by the transition between Figs. 7–2a and 7–2b could be effected by replacing a diluent gas by one of greater heat capacity.

It is reasonable to enquire to what extent the ideal conditions of perfect thermal insulation and mixing are essential for critical ignition and extinction phenomena to occur. For simplicity, suppose the reaction vessel, instead of being thermally insulated, is provided with a cooling jacket that maintains the temperature of the walls at T_0. In place of equation 7–8 we now have

$$\dot{q}_L = N\bar{C}_p(T-T_0)+\alpha(T-T_0) \qquad 7\text{–}9$$

In general, α, the coefficient of heat transfer from gas to wall, increases in a sublinear fashion with increasing N. It will be seen from Fig. 2d that, qualitatively, this introduces no new element into the situation. At high reaction temperatures, heat loss by radiation must also be considered; thus

$$\dot{q}_L = N\bar{C}_p(T-T_0)+\alpha(T-T_0)+\sigma'(T^4-T_0^4) \qquad 7\text{–}10$$

where σ' is a constitutive constant. Again, Fig. 7–2d shows that qualitatively the situation is much the same as previously. By reasoning analogous to that used in deriving equation 7–5 it is easily shown that, provided the internal temperature remains uniform, plug-flow through a tubular vessel gives an S-shaped $\dot{q}_G$–T curve and consequently the possibility of critical phenomena. On the other hand critical effects cannot occur if each successive layer of gas retains its heat of reaction; that is, if, instead of complete heat exchange as just assumed, there is no internal heat exchange. Thus, although in principle mass mixing is not required, critical phenomena are only possible when there is a feedback of heat from products to reactants. As noted earlier, this is the essential characteristic of combustion.

7–3 Kinetics of reactions in flames

Much research on flames is carried out with different objects in view from those of chemical kinetics. In the past, this has tended to produce a vague impression that the reactions occurring in flames are somehow different in kind from those more familiar as the subjects of 'conventional' laboratory studies. To the extent that the impression remains, it is a historical accident; and it is hoped that the previous section will have shown that combustion phenomena do no more (and no less) than add an interesting dimension to kinetics studies; a dimension which, perhaps it is not irrelevant to add, extends into areas of much practical consequence.

The most familiar combustion system is, of course, the stationary flame. In principle it is also the simplest to set up for kinetic studies though, in practice, because of the sensitivity of a flame to its environment and the great speed of the reactions involved, very careful experimentation is needed to obtain meaningful results. The remainder of this chapter is intended to provide an introduction to kinetic studies conducted on stationary flames supported by previously mixed reactants. Such investigations fulfil the reciprocal purposes of providing insight into combustion phenomena and contributing to knowledge of the kinetics of elementary radical reactions at high temperatures.

7–3–1 Macroscopic structure of flames

A flame invariably possesses structure. This is apparent not only with a diffusion flame, such as that of a candle, where the 'fuel' and oxygen molecules must diffuse together from opposite directions to be able to react, but also with flames of the Bunsen type where the reactants are pre-mixed. With a hydrocarbon–oxygen flame of the latter type, the brightly luminous inner cone—a zone of intense reaction about 0.1 mm thick at atmospheric pressure—is surrounded by a sheath of less luminous but still reacting gas. Reactions take place beyond the inner cone even when the flame is isolated from surrounding air. This is most simply revealed by the fact that the maximum temperature under such conditions usually occurs at some distance downstream. The inner cone constitutes the *primary reaction zone*, and this is followed by a more diffuse region in which carbon monoxide and hydrogen are oxidized much more slowly than they are produced by the primary reactions. Thus the visible structure depends upon kinetic considerations. It is not confined to hydrocarbon–oxygen flames; indeed in some flames, such as those of cyanogen–oxygen and hydrocarbon–nitrogen dioxide flames, still more complex structure can be clearly seen in certain circumstances. It might be supposed that the pre-mixed hydrogen–oxygen flame, since it can produce only one stable product, would not possess macroscopic structure. Nevertheless, though less sharply defined and less evident to the eye than in flames of organic compounds, such structure is present. It will be seen later that it is the consequence of a well-recognized kinetic effect, namely the relative slowness of three-body radical recombinations.

Kinetic studies have been conducted both on the gas downstream of the primary reaction zone (known synonymously but inaccurately as the *burned* or *post-flame gas*) and, with rather more difficulty, on the primary reaction zone itself. Examples of both types of investigation will be given presently. Since the reactions in the burned gas are relatively slow, it is not too difficult to arrange for the temperature to remain reasonably constant for one or two centimetres above the primary zone; and analysis of measured changes in concentration with reaction time, that is with height, can be performed by much the same methods as are applied to ordinary fast-flow systems (see Chapter 4). In the primary reaction zone, on the other hand, reactions are much faster and investigations of this part of the flame must deal with steep temperature and concentration gradients. The experimental difficulties can be alleviated by working at low pressure or with highly dilute reactants so as to expand the reaction zone for greater spatial resolution. Even so, rather elaborate procedures are required to extract the purely chemical factors, such as rate constants or Arrhenius parameters, from the experimental data. In the present state of knowledge, however, this is not generally the immediate purpose of such studies. More frequently, information about reaction rates derived from other sources is used in analysing the data with the object of understanding the reaction mechanism and the way in which the mechanism interacts with the diffusional and thermal processes it creates. It is this interaction which gives rise to the characteristic flame phenomena.

7–3–2 Determination of free-radical concentrations in flames

It will be no surprise to the reader acquainted with the existence of branching chain reactions (Section 6–2) to learn that flames contain high concentrations of free atoms and radicals. Indeed the presence of the hydroxyl radical in hydrogen–air flames was recognized by its emission spectrum over 50 years ago; and the spectrum of the familiar blue-green emission from the inner cone of a Bunsen-type hydrocarbon flame has long been assigned to the labile species CH and C_2. Again, the greenish-yellow light often to be seen at the tip of a welder's oxy-hydrogen flame is indicative of the presence of oxygen atoms. It is due to their chemiluminescent reaction with nitric oxide formed from atmospheric nitrogen and, as in the discharge-flow-type experiments discussed in Chapter 4, it can be used to determine relative concentrations of oxygen atoms in flames.

The products of a hydrogen–oxygen flame burning at atmospheric pressure with the 'stoichiometric' ratio of the reactants ($2H_2/O_2$) contain a total partial pressure of more than 0.2 atm free O, H, and OH radicals at thermodynamic equilibrium at the 'adiabatic flame temperature'. The latter is the final temperature attained by the products on complete reaction when there is no loss of the heat of reaction to the surroundings. When the flame is cooler, by reason of the presence of an excess of one of the reactants or of an inert diluent, the concentration of radicals at final equilibrium is lower. The concentrations of radicals in the reaction zone, on the other hand, are determined by kinetic factors and, in cooler flames, they can be several thousand times greater than the equilibrium values. In the case of OH radicals, this has been demonstrated by measuring their ultraviolet absorption. Likewise, high concentrations of other radicals, such as CH_3, C_2, CH, S_2, NH, have been detected in various flames and, in some cases, they have been measured by absorption photometry.[1]

The concentration of hydrogen atoms in the burned gas is most conveniently measured indirectly by adding a trace of lithium salt to the reactants (as a fine mist of aqueous solution from an atomizer).[2] The lithium exists in the flame both as free atoms and lithium hydroxide and these species come very rapidly to equilibrium with the hydrogen atoms present by the balanced reaction:

$$\mathrm{Li} + \mathrm{H_2O} \rightleftharpoons \mathrm{LiOH} + \mathrm{H} \qquad \text{7–I}$$

The required concentration is given by

$$[\mathrm{H}] = \frac{K_1[\mathrm{H_2O}]}{([\mathrm{Li}]_0/[\mathrm{Li}]) - 1} \qquad \text{7–11}$$

where $[\mathrm{Li}]_0$ represents the total lithium concentration added to the flame. Since the values of K_1 are known for the temperature range of interest, measurement of the ratio $[\mathrm{Li}]_0/[\mathrm{Li}]$ at any point in the flame yields the local concentration of hydrogen atoms. The value of [Li] can be determined conveniently by atomic absorption photometry.[3] In practice neither $[\mathrm{Li}]_0$ nor [Li] is determined absolutely, but the experimental system is calibrated by carrying out measurements on the upper regions of hot flames where the gases are known to be at thermodynamic equilibrium.[4] Under these conditions, the concentration of hydrogen atoms can be calculated from thermodynamic data, and equation 7–11 is then used in reverse to relate measured quantities to $[\mathrm{Li}]_0/[\mathrm{Li}]$. (The calibration procedure, however, does not eliminate the need for knowledge of K_1.)

Several methods of measuring radical concentrations based on emission photometry are also in use. For example, relative hydrogen atom concentrations can be determined from the intensity of emission from CuH radicals which are formed in an electronically excited state when a copper salt is added to the flame.[5] Similarly, concentrations of oxygen atoms can be compared by adding a little nitric oxide, as already mentioned; alternatively, an organic iodine compound may be added, whereupon the emission from the radical IO is proportional to the atomic oxygen concentration.[6]

Mass spectrometry combined with molecular-beam sampling may also be used to determine the concentrations of radicals such as O, OH, and CH_3 as well as more complex species[7] (see Section 4–2–3). The presence of the HO_2 radical in hydrogen–oxygen flames was established by this means. As we have seen in Chapter 6, its existence was first postulated on purely kinetic grounds to explain the properties of the second explosion limit. More recently, radical concentrations have been determined by ESR spectrometry. Either the flame (burning at low pressure) is situated in the cavity of the spectrometer or, with more precision, it is sampled continuously into the cavity by expanding the gases rapidly through a fine probe.[8, 9] Figure 7–3 shows radical concentrations in the post-flame gases of a hydrogen sulphide–oxygen flame determined by ESR, together with the concentrations of stable species determined by other means. The variation of the concentrations of SO

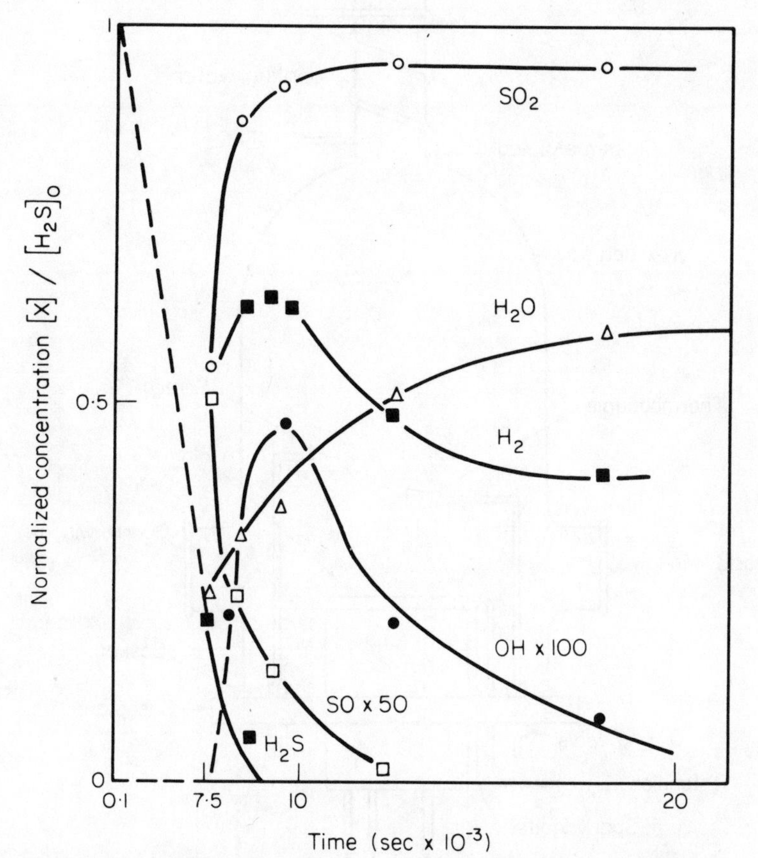

Fig. 7–3 Relative concentrations of OH and SO radicals (determined by ESR) and molecular species in a hydrogen sulphide–oxygen flame at 2.5 torr pressure (Balakhnin, Gershenson, and Nalbandyan[8]).

and OH with reaction time is seen clearly and gives direct insight into the reaction mechanism. At higher pressures, O and H atoms have also been determined in this and other flames by ESR.

7–3–3 Flat flame apparatus

As with other kinetic investigations conducted in linear-flow systems, the object in flame studies is to produce a unidirectional time scale for the reacting medium. The standard apparatus for this is the flat-flame burner, which basically is a more elaborate version of the laboratory Meker burner. Constructional details of several versions of the apparatus and of the ancillary instruments, probes, and other equipment required for this type of work will be found in reference (D) of the Bibliography. Only the essentials will be mentioned here.

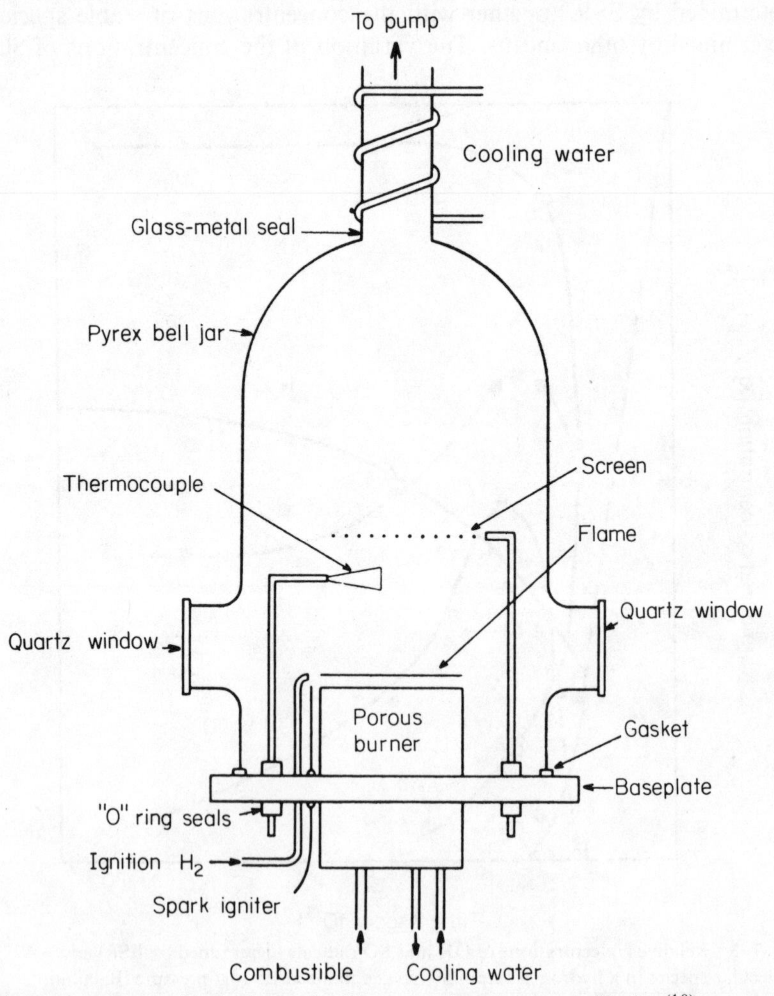

Fig. 7–4 Apparatus for producing flat flames at reduced pressures (Kaskan[10]).

Typical apparatus suitable for use at reduced pressure is shown in Fig. 7–4. A plug-flow velocity profile is produced by passing the pre-mixed gases through a porous plate.§ After ignition, a horizontal flat flame floats above the plate at the distance where its linear burning velocity downwards is exactly balanced by the upwards flow of the gaseous medium. The stability is produced in the following way. If, at constant flow velocity, the flame would move back to its original position. Similarly, if the flow is to the plate. This would lower its burning velocity and consequently the flame would move back to its original position. Likewise, if the flow is increased, the flame moves away from the plate until its burning velocity is sufficiently high for the balance to be restored. Thus, by varying the initial flow velocity or by water-cooling the plate or both, a range of reaction temperatures can be achieved. The apparatus shown in Fig. 7–4 was designed for absorption spectrophotometry and measurements at different heights in the flame were made by moving the whole apparatus vertically relative to a fixed observation point. When a thermocouple or analytical probe is to be inserted into the flame it may be held fixed and the height of the burner adjusted through an O-ring seal or vice versa. The flame is protected from surrounding air by a flowing concentric sheath of inert gas. When photometric measurements are made on a substance added to the flame, for example the lithium atoms mentioned previously, the flame is surrounded by a second flame of similar composition but without the additive. This helps to ensure that the temperature and concentration of the additive is uniform over the cross-section of the inner flame.

Temperature measurements naturally are most important. Of the several methods available,[11] the thermocouple and line-reversal methods are the most commonly used in kinetics work. In the former, noble metal thermocouples made from 10 μm wire coated with a film of silica to reduce catalytic effects are inserted in the flame; they can yield temperatures reliable to ± 15 K and spatially resolved to within about 10 μm. The line-reversal method is carried out by observing the emission from a trace of sodium or other metal added as a salt to the flame. The D-lines, in the case of sodium, are brought to a focus on the slit of a spectroscope together with light passing through the flame from the filament of a standard tungsten lamp. The lines appear dark when the flame is cooler, and bright when it is hotter than the filament, and they disappear when the temperatures are equal. It is assumed that the metal atoms are in thermodynamic equilibrium with their environment. The method, therefore, is generally suitable only for the post-flame region where the radical concentrations are insufficient to give rise to 'abnormal' excitation of the metal atoms. With modern instrumentation[12] differences of 1 K can be detected and resolved spatially to within about 0.5 mm.

§ Other forms of the burner use a series of fine screens or a roll of crimped metal tape or a cluster of hypodermic tubing to achieve the same effect. Burners in current use range in diameter from about 4 to 30 cm.

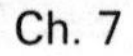

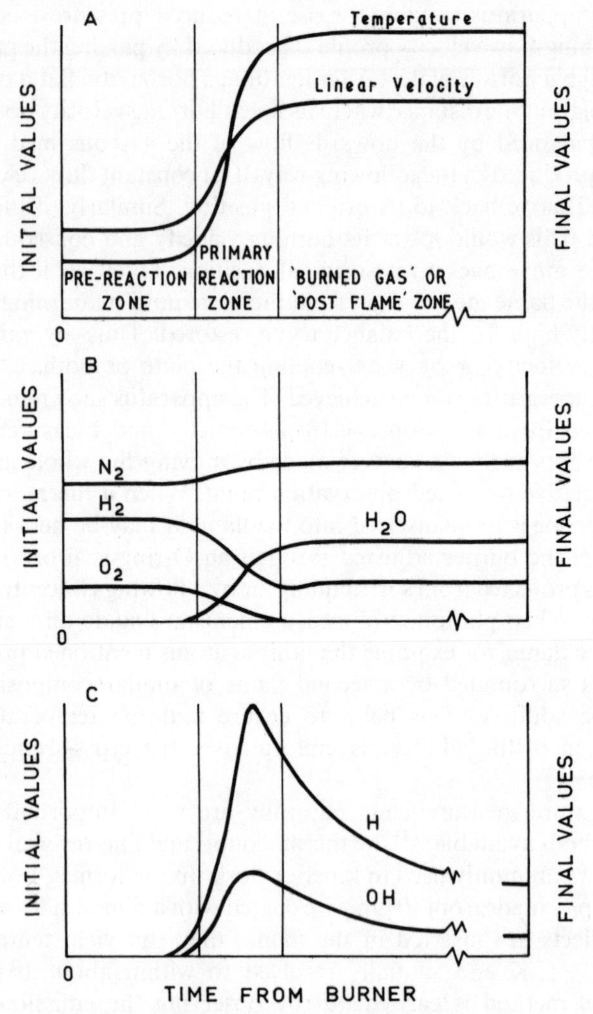

Fig. 7–5 Diagrammatic representation of profiles of a unidimensional, pre-mixed hydrogen–oxygen–nitrogen flame. Original proportions of reactants 3:1:4. A. Temperature and linear flow velocity. B. Concentrations of molecular species. C. Concentrations of radicals (after Jenkins and Sugden (ref. F of Bibliography)).

Anticipating a little the following two sections, Fig. 7–5 shows the type of unidimensional concentration, temperature, and gas velocity profiles that occur in a flat-flame system. The acceleration of flow in the flame zone is due to the expansion caused by the rise in temperature; this more than counteracts the stoichiometric contraction that occurs in the particular case of the hydrogen–oxygen flame.

7–3–4 Reactions in the 'post-flame' gas

The high concentrations of radicals that emerge from the primary reaction zone normally decline towards their equilibrium values over times of the order of milliseconds. Studies of the kinetics of the decay in flat flames have cast much light on the nature of the reactions occurring not only in the post-flame region but also, by inference, in the primary reaction zone itself. Kaskan[10] used absorption spectrometry to follow the disappearance of OH radicals downstream from the reaction zone of hydrogen-rich hydrogen–air flames and showed that this occurred by a second-order process. From this it might be supposed that the OH concentration decays by reactions of the type:

$$OH + OH + M \rightarrow H_2O_2 + M \qquad 7\text{–II}$$

$$OH + OH \rightarrow H_2 + O_2 \qquad 7\text{–III}$$

$$OH + OH \rightarrow H_2O + O \qquad 7\text{–IV}$$

This, however, is not the case. At the temperatures (1300–2000 K) and concentrations involved, the equilibrium for reaction 7–II is well over to the left and the lifetime of any hydrogen peroxide molecules formed would be only about a microsecond. Likewise, reactions 7–III and 7–IV do not account for the experimental findings that the second-order rate constant depends on the composition of the flame gases and is approximately proportional to the total pressure. The decay of H atoms is also second order and behaves in a very similar way to that of the OH radicals. This was shown by Bulewicz and Sugden[13] who followed the H concentration by the lithium–lithium hydroxide and copper hydride methods. They also found, very significantly, that the ratio of the H and OH concentrations in any one flame remains substantially constant during the period of decline.

For the explanation of these facts we must return to the mechanism of chain-branched hydrogen–oxygen explosions in closed vessels. It was shown in Chapter 6 that the fast exponential increase in radical concentrations that occurs during an explosion takes place by the sequence of (forward) reactions:

$$H + O_2 \rightleftharpoons OH + O \qquad 7\text{–V}$$

$$O + H_2 \rightleftharpoons OH + H \qquad 7\text{–VI}$$

$$OH + H_2 \rightleftharpoons H_2O + H \qquad 7\text{–VII}$$

Since the reactants are present initially in a fixed amount, and disappear at a rate proportional to the concentration of radicals, the explosive reaction is 'over' in a very short time. It is clear, however, that, even if the reactants consumed during the explosion were to be continuously replaced —as, in fact, they are in a stationary flame—the radical concentrations cannot increase indefinitely. At high radical concentrations two kinds of reaction must inevitably become important which were neglected in considering simply the conditions for the build-up of radicals to occur.

These are first, the reverse reactions of 7–V, 7–VI, and 7–VII and, second, radical recombinations represented by the reactions

$$H+H+M \rightarrow H_2+M \qquad \text{7–VIII}$$

$$H+OH+M \rightarrow H_2O+M \qquad \text{7–IX}$$

Further consideration shows that the reactions of the first type are all bimolecular with small activation energies and therefore are likely to be faster than the termolecular recombinations at the temperatures involved. That this is at least partly the case can be seen very simply by comparing the rate of reaction 7–−VI with that of reaction 7–IX at, say, 1 atm pressure and 2000 K. The rate constants for both reactions can be estimated from independent experimental data. Reaction 7–VI was investigated in a flow tube by Clyne and Thrush.(14) Combining their extrapolated value of k_6 with the equilibrium constant for the reaction gives the value $10^{11.9}$ cm^3 mole^{-1} s^{-1} for k_{-6}. Similarly, the value for k_{-9} derived from shock-tube studies of the decomposition of water vapour in argon (Section 2–4–1) leads to the value $10^{15.0}$ cm^6 mole^{-2} s^{-1} for k_9. The corresponding value for the pseudo second-order constant ($k_9[M]$) at 1 atm and 2000 K is $10^{9.8}$ cm^3 mole^{-1} s^{-1}. This is probably somewhat low because of the greater third-body efficiency of the water molecules in the flame gas, but it is clear that $k_9[M]$ is considerably smaller than k_{-6}. In other words, reaction 7–IX, which causes the final disappearance of radical species, lags behind reaction 7–−VI, which does not. This result can be considered representative of the relative rates of the two types of reaction under flame conditions.

A second important point is that under these conditions reactions 7–−V, 7–−VI, and 7–−VII are fast enough to come to equilibrium with their forward reactions in times which are short compared with the milliseconds over which the decay of radicals is observed. Thus we have an interesting situation of partial equilibrium whereby reactions 7–V to 7–VII are effectively balanced at all times after the period of the primary reaction zone. In this way, the concentrations of the three radicals O, H, and OH are coupled together as they gradually decay towards complete equilibrium as the result of the slower termolecular recombinations 7–VIII and 7–IX. (Except when oxygen is in large excess, the reaction

$$O+O+M \rightarrow O_2+M \qquad \text{7–X}$$

can be neglected as being intrinsically two orders of magnitude slower than reactions 7–VIII and 7–IX.)

On this basis it is a simple matter to derive an expression for the rate of decay of H or OH. In hydrogen-rich flames the concentration of O is negligible and the total rate of radical decay is given by

$$-\frac{d([H]+[OH])}{dt} = 2\{k_8[M][H]^2+k_9[M][H][OH]\} \qquad \text{7–12}$$

Since reaction 7–VII is balanced:

$$[H]/[OH] = K_7[H_2]/[H_2O] \qquad 7\text{–}13$$

where, with hydrogen in excess, the concentrations of hydrogen and water are effectively constant. Equation 7–13 enables either [OH] or [H] to be eliminated from equation 7–12. Thus, if the approximation is made that the efficiency of M is independent of the composition of the gases, the following expression for the decay of the H atoms results:

$$-\frac{d[H]}{dt} = 2\left[\frac{(k_8K_7+k_9[H_2O]/[H_2])[M]}{K_7+[H_2O]/[H_2]}\right][H]^2 \qquad 7\text{–}14$$

The term in the square brackets corresponds to the experimental pressure- and concentration-dependent second-order decay constant to which reference was made at the beginning of the discussion. By determining the decay constants for a series of isothermal flames with different values of $[H_2O]/[H_2]$ Bulewicz and Sugden[13] were able to show that equation 7–14 is in fact obeyed for hydrogen-rich flames; and Halstead and Jenkins,[15] using a more refined but essentially similar treatment of similar experiments, obtained values of k_8 and k_9 which agree well with those obtained by shock-tube technique.

With oxygen-rich flames the kinetics of radical decay are less clear largely because of uncertainty about the role played by the radical HO_2. This is formed by the reaction

$$H+O_2+M \rightarrow HO_2+M \qquad 7\text{–XI}$$

and exists in higher concentrations in such flames. Nevertheless McEwan and Phillips[16] have shown that, whatever is the precise mechanism of radical decay, it is too slow to disturb the equilibration of reactions 7–V, 7–VI, and 7–VII. A similar result has been obtained with hydrocarbon–air flames. Here, the reaction

$$OH+CO \rightleftharpoons CO_2+H \qquad 7\text{–XII}$$

is also equilibrated and the intrinsically slow reaction

$$O+CO+M \rightarrow CO_2+M \qquad 7\text{–XIII}\S$$

contributes to the radical decay.[17]

Summarizing, the details of the picture of pre-mixed hydrogen–oxygen flames illustrated by Fig. 7–5 can now be filled in a little further. Following upon initiation of reaction in the gas issuing from the burner, a rapid build-up of radicals occurs by fast bimolecular reactions which are soon brought to equilibrium with their bimolecular reverse counterparts. These events give rise to the primary reaction zone and result in an 'overshoot'[18] of the concentrations of the radicals beyond those corresponding to complete thermodynamic equilibrium. In the post-flame region the excess concentrations, while remaining 'balanced', are more slowly brought to

§ Cf. Section 6.2.5.

true equilibrium by termolecular recombination. Since these reactions are exothermic and, further, their result is to allow the composition of the system to move towards that of the final products, the production of an appreciable proportion of the heat of reaction in the post-flame region is explained.

With hydrocarbon flames the reactions taking place in the primary zone are not able to be so precisely specified at present as for hydrogen flames. They evidently result in less overshoot of the radical concentrations since the radical concentrations in the post-flame gas of these flames are generally closer to equilibrium than in hydrogen flames. Nevertheless when sufficient oxygen is present in the original mixture to consume the hydrocarbon completely, a basically similar picture still applies. In fact, in these circumstances, the reactions in the burned gas are similar to those in a hydrogen flame containing carbon monoxide. Post-flame oxidation to carbon dioxide occurs mainly by the gradual displacement of the balance of reaction 7–XII as the hydrogen atoms are removed by recombination.

7–3–5 The burned gas as a source of radicals

The understanding achieved of the reactions occurring in the burned gas makes it possible to use a flame as a source of radicals for 'wall-less' studies of elementary reactions at high temperatures. A considerable number of such investigations has been carried out,[19] of which the following is a typical example.

Fenimore and Jones[20] studied the kinetics of the reaction

$$O + N_2O \rightarrow 2NO \qquad \text{7–XIV}$$

in the temperature range 1450–1730 K by adding a small amount of nitrous oxide to the burned gas of oxygen-rich hydrogen–air flames. The rates of formation of nitric oxide and disappearance of nitrous oxide were measured by abstracting samples for mass-spectrometric analysis at various points downstream. The amount of nitrous oxide added was too small to disturb the equilibria in the flame gas and it was shown that the nitric oxide formed did not undergo further reactions in the time available. Since reactions 7–V and 7–VII are equilibrated, it follows by addition that the reaction

$$O + H_2O \rightleftharpoons H_2 + O_2 \qquad \text{7–XV}$$

is also equilibrated. Hence by determining the (trace) concentration of residual hydrogen in the gas, the concentration of O could be calculated from the known equilibrium constant K_{15} and the effectively constant concentrations of oxygen and water vapour. The concentration of O varied by only $\pm 15\%$ in the sampling region and it was regarded as constant in any one flame. The total rate of disappearance of nitrous oxide by *all* the reactions it undergoes in the flame gases was found to obey first-order kinetics (with a composite rate constant k_a). Hence the rate of reaction 7–XIV was given by

$$\frac{d[\mathrm{NO}]}{dt} = 2k_{14}[\mathrm{O}][\mathrm{N_2O}]_0 \exp(-k_a t) \qquad 7\text{–}15$$

which, on integration, becomes

$$[\mathrm{NO}] = (2k_{14}/k_a)[\mathrm{O}][\mathrm{N_2O}]_0\{1-\exp(-k_a t)\} \qquad 7\text{–}16$$

A plot of [NO] against $\{1-\exp(-k_a t)\}$ gave a straight line from the slope of which the value of k_{14} was derived. The temperature of the burned gas was varied by partly replacing the reactant air or hydrogen with nitrogen or carbon monoxide or by burning acetylene; and, in this way, the Arrhenius parameters of the reaction were determined:

$$k_{14} = 10^{14.3} \exp[-(32 \pm 4)/RT] \text{ cm}^3 \text{ mole s}^{-1} \qquad 7\text{–}17§$$

It is worth mentioning that from this result Fenimore and Jones were able to come to an interesting conclusion concerning the mechanism of the thermal decomposition of nitric oxide to nitrogen and oxygen. This is known to be second order at temperatures near those of the flame studies. Conceivably the decomposition could proceed either directly by the four-centre reaction

$$2\mathrm{NO} \rightarrow \mathrm{N_2} + \mathrm{O_2} \qquad 7\text{–XVI}$$

or by the reverse of reaction 7–XIV followed by decomposition of the nitrous oxide and recombination of the O atoms. Since the values of k_{-14} calculated from the measured values of k_{14} and the equilibrium constant agreed with the experimentally determined rate constant for the decomposition, it was reasonable to conclude that the rate-determining step of the latter reaction is the reverse of reaction 7–XIV.

7–3–6 The primary reaction zone

The information required to elucidate the processes occurring in the primary flame zone is necessarily similar in kind to that required from any other experimental kinetic study: measurements must be made of the reaction pressure and temperature and the changes with time of the concentrations of the species present including, as far as possible, those of intermediate radicals. The concentration changes are normally expressed as concentration profiles such as are shown in Figs. 7–5 and 7–6. Apart from the fine spatial resolution required, perhaps the chief difference from more conventional flow studies is the existence of a temperature profile in the reacting gas. This cannot be avoided and indeed is an essential feature of the phenomena. In ordinary flames at atmospheric

§ Subsequent work by the same authors using another method of determining [O] gave a slightly different relation, $k_{14} = 10^{14.0} \exp[-(28 \pm 3)/RT]$, which they consider to be the more accurate.[21]

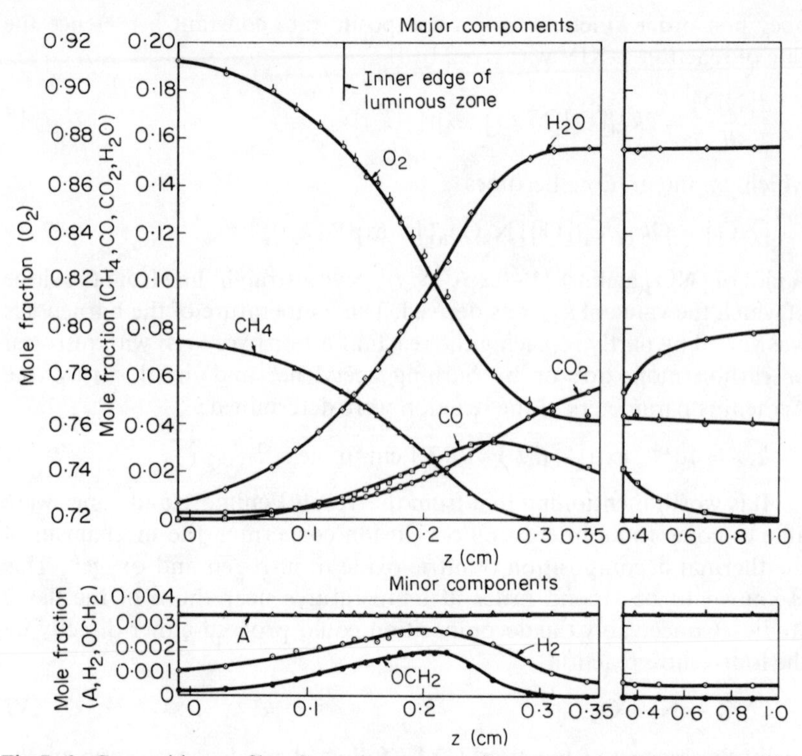

Fig. 7–6 Composition profiles of a methane–oxygen flame at 0.1 atm. Initial composition of unburned gas 7.8 CH_4 + 91.5 O_2 with traces of Ar, N_2, and CO_2 (Fristrom, Grunfelder, and Favin[24]).

pressure, the greater parts of the changes in temperature and concentration are accomplished in distances of the order of tenths of a millimetre. Even when the velocity of the reactions is reduced by operating at lower pressures or diluting the reactants by inert gas, it is seldom possible to extend the length of the reaction zone to more than a few millimetres.

The reactions, therefore, occur under conditions whereby the temperature and the linear-flow velocity as well as the concentrations of the products increase steeply with time, that is, with distance. The steep concentration gradients cause the reactant molecules to diffuse rapidly forwards and the product molecules backwards within the flowing medium. The direction of diffusion of a radical or other intermediate species naturally depends on whether its concentration profile slopes downwards or upwards in the downstream direction at the point under consideration; because of the overshoot phenomenon described in Section 7–3–4, the direction of diffusion usually changes at some point. Thus, the concentration of a species observed at any point is not simply the consequence of the reactions by which the species is formed and consumed and the flow time: it is also determined to a very significant degree by diffusion. Indeed in the early part of the flame the concentrations of the reactants may be considerably below their initial values before any reaction has occurred at all. This is because the molecules diffuse rapidly forward from this region towards the region of lower concentration where reaction is taking place.

The influence of diffusion

The theoretical relation between the concentration of any species and the distance from the burner can be found by basically the same procedure as was used in Section 2–2–2 to deal with the reaction of a single molecular species in a flowing diffusive medium; that is, by setting up and solving the steady-state continuity equation. For the unidimensional flow occurring in a flat flame, this is analogous to equation 2–29 but needs to be expressed in a more general form:

$$\frac{d}{dz}\{u[\mathrm{A_i}]-D_\mathrm{i}d[\mathrm{A_i}]/dz\} = \mathcal{R}_\mathrm{i} \qquad 7\text{–}18$$

where z is the distance from the surface of the burner, $\mathcal{R}_\mathrm{i}$ is the net rate of *formation* (mole cm^{-3} s^{-1}) of the species $\mathrm{A_i}$ by chemical reaction and is a function of $[\mathrm{A_i}]$, T, and usually of the concentration of at least one other species $[\mathrm{A_j}]$. The linear velocity (u) of the total gas and the diffusion coefficient (D_i) of $\mathrm{A_i}$ are now functions of z. $\mathrm{A_i}$, $\mathrm{A_j}$ may be molecules or radicals and $\mathcal{R}_\mathrm{i}$ is negative when there is a net consumption of $\mathrm{A_i}$ by reaction at the point z under consideration.

The term $-D_\mathrm{i}(d[\mathrm{A_i}]/dz)$ expresses the number of moles of $\mathrm{A_i}$ which diffuse across unit area in unit time. This can be converted to a linear *diffusion velocity* U_i cm s^{-1} by the relation

$$U_\mathrm{i} = -D_\mathrm{i}(d[\mathrm{A_i}]/dz)/[\mathrm{A_i}] \qquad 7\text{–}19$$

Equation 7–18 then becomes

$$\frac{d}{dz}\{[\mathrm{A_i}](u+U_\mathrm{i})\} = \mathcal{R}_\mathrm{i} \qquad 7\text{–}20$$

where U_i can be positive or negative. It is usual to express equation 7–20 in terms of a quantity G_i, the ratio of the mass flow carried by $\mathrm{A_i}$ to the total average mass flow of the gas. Thus,

$$G_\mathrm{i} = M_\mathrm{i}[\mathrm{A_i}](u+U_\mathrm{i})/\rho u \qquad 7\text{–}21$$

where M_i is the molecular weight of $\mathrm{A_i}$ and ρ is the density of the gas. Since the flow is unidimensional, mass conservation requires that

$$\rho u = \rho_0 u_0 = \text{const} \qquad 7\text{–}22$$

The suffix 0 refers to the unburned gas. Equation 7–18 finally becomes

$$\frac{\rho_0 u_0}{M_\mathrm{i}}\frac{dG_\mathrm{i}}{dz} = \mathcal{R}_\mathrm{i} \qquad 7\text{–}23$$

Since the advance of the flame is exactly balanced by the flow of the unburned gas, u_0 is the cold gas *burning velocity*. Returning to equation

7–21, since $M_i[A_i]/\rho$ is equal to the mass fraction of A_i in the total gas (f_i), G_i is conveniently found from the relation

$$G_i = f_i(u+U_i)/u \qquad 7\text{–}24$$

Equation 7–23 enables the effect of chemical reaction ($\mathscr{R}_i$) on the concentration of a given species to be separated from the effect of diffusion at any point on the concentration profile. The procedure is first to calculate a profile of values of G_i by means of equation 7–24. The values of U_i are computed via equation 7–19 by taking the slopes of the concentration profile and using the values of D_i appropriate to the temperatures at successive values of z. The values of u are usually calculated from a single measurement at a convenient point using equation 7–22 and the equation of state

$$\rho = \overline{M}P/RT \qquad 7\text{–}25$$

($\overline{M}$ is the mean molecular weight and P is the total pressure). Finally the slopes of the G_i profile are determined at successive points to obtain corresponding values of $\mathscr{R}_i$ via equation 7–23. It will be observed that the determination of $\mathscr{R}_i$ requires double differentiation of the experimental data, and therefore the values obtained are considerably less precise than the original measurements of concentration, gas velocity, and so forth.

It should be mentioned that the above treatment neglects two effects relating to diffusion. First, the value of D_i depends in a complex way on the composition of the gas which, of course, changes during the reaction. In practice the difficulty is avoided by working with flames containing a large excess of one component, usually oxygen or an inert diluent. Each of the remaining species can then be regarded as a trace component of a binary mixture and D_i is taken to be the corresponding binary diffusion coefficient. The second effect neglected is that of thermal diffusion, which is brought about by the temperature gradient. This may act in the same or the opposite direction from that of ordinary diffusion, depending on the nature of the species and other circumstances. Its contribution to the total diffusional flux is often but not always small compared to that of ordinary diffusion.[22,23]

The methane–oxygen flame

The primary reaction zone of the simplest hydrocarbon flame has been investigated very carefully by Fristrom and Westenberg and collaborators by means of the method just outlined.[23,24] The concentration profiles they obtained by probing a flame burning with excess oxygen at 0.1 atm pressure are shown in Fig. 7–6. The corresponding profiles of the net reaction rates ($\mathscr{R}_i$) are given in Fig. 7–7. In both figures the profiles extend into the post-flame region. Similar results were obtained with a flame at 0.05 atm. Figure 7–7 shows that the rate of consumption of methane and the rates of generation of the products pass through maxima which occur at different positions in the flame. The early production of carbon monoxide and hydrogen and the subsequent long-drawn-out oxidation of carbon monoxide are clearly seen. Figure 7–6 shows that about half the carbon dioxide finally produced is formed in the burned gas after the methane has been totally consumed. The influence of diffusion can be appreciated by comparing Fig. 7–6 with Fig. 7–7.

It appears from Fig. 7–7 that chemical reaction begins at a distance of about 0.2 cm from the burner. Figure 7–6 shows, however, that at this point the concentrations of both reactants have already fallen more than

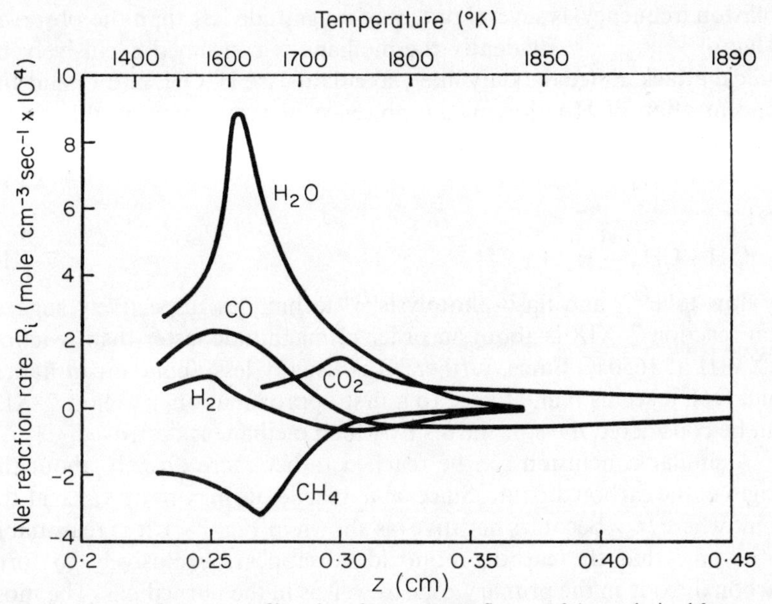

Fig. 7–7 Net reaction rate profiles of methane–oxygen flame at 0.1 atm derived from composition profiles shown in Fig. 7–6 (Westenberg and Fristrom[23]).

halfway towards their final values as the result of diffusion. Likewise the maximum rate of *reaction* of methane occurs at about 0.27 cm, whereas its maximum rate of *disappearance* by combined physical and chemical processes, as given by the maximum slope of the concentration profile, is at about 0.22 cm. An additional point of interest is that the maximum reaction rate takes place late in the flame when only a few per cent of methane remains. This is because the rise in temperature has a greater influence on the rate than the decrease in concentration.

When the true chemical reaction rates have been extracted from the experimental data, the question will be asked to what degree they cast light on the underlying chemical mechanism. The general answer is that this depends on the amount of guidance to be obtained from other sources as to the rates of various elementary reactions that are likely to be involved. Unlike the reactions in hydrogen flames, for which a substantial body of kinetic information is available from more conventional studies, present knowledge of the rates of elementary reactions applicable to methane flames, or indeed to any hydrocarbon flame, is limited. Nevertheless some progress towards interpreting the rate profiles has been made. For example, it seems certain that the endothermic reaction

$$O_2+CH_4 \rightarrow HO_2+CH_3 \qquad \text{7–XVII}$$

makes a negligible contribution to the consumption of methane. Its maximum possible rate, calculated at 1650 K by equating its activation energy to the enthalpy change and its pre-exponential factor to the collision frequency, is several orders of magnitude less than the observed value of $-\mathscr{R}_{CH_4}$.[23] Evidently the methane is consumed exclusively by radical attack. Since the only likely assailants are H, OH, and O and the concentration of H is low in the presence of excess oxygen, this leaves OH and O. Studies of the reactions

$$O+CH_4 \rightarrow OH+CH_3 \qquad \text{7–XVIII}$$

and

$$OH+CH_4 \rightarrow H_2O+CH_3 \qquad \text{7–XIX}$$

by flow-tube[25] and flash-photolysis[26] techniques respectively suggest that reaction 7–XIX is about an order of magnitude faster than reaction 7–XVIII at 1650 K. Since, further, O is usually less abundant in flames than OH, it seems that, at least to a first approximation, reaction 7–XIX can be considered the sole means by which methane is destroyed.

A similar conclusion can be reached rather more directly about the origin of the carbon dioxide. Since $\mathscr{R}_{CO_2}$ reaches its maximum value at the point where $\mathscr{R}_{CO}$ becomes negative (as shown in Fig. 7–7), it is reasonable to assume that the carbon monoxide disappears exclusively to form carbon dioxide in the primary zone as well as in the burned gas. The most likely reaction is

$$OH+CO \rightarrow CO_2+H \qquad \text{7–XX}$$

If this and the previous conclusion about the occurrence of reaction 7–XIX are correct, the relation

$$(-\mathscr{R}_{CH_4}/-\mathscr{R}_{CO}) = k_{19}/k_{20} \qquad \text{7–26}$$

should hold throughout the primary reaction zone. Values of k_{19}/k_{20} derived from the rate profiles are shown on an Arrhenius diagram in Fig. 7–8, where it will be seen that they are reasonably compatible with values derived from other sources. Apart from the value at 300 K, all the values plotted are based on mechanistic assumptions of a similar degree of probability to those just discussed. It is therefore difficult to judge the extent to which the discrepancies revealed in the diagram are due to failure of the assumptions or to inaccuracy of the experiments. The situation is not atypical of flame kinetics where firm conclusions about reaction mechanisms can seldom be reached by a single investigation, but rather are approached by the gradual confluence of evidence.

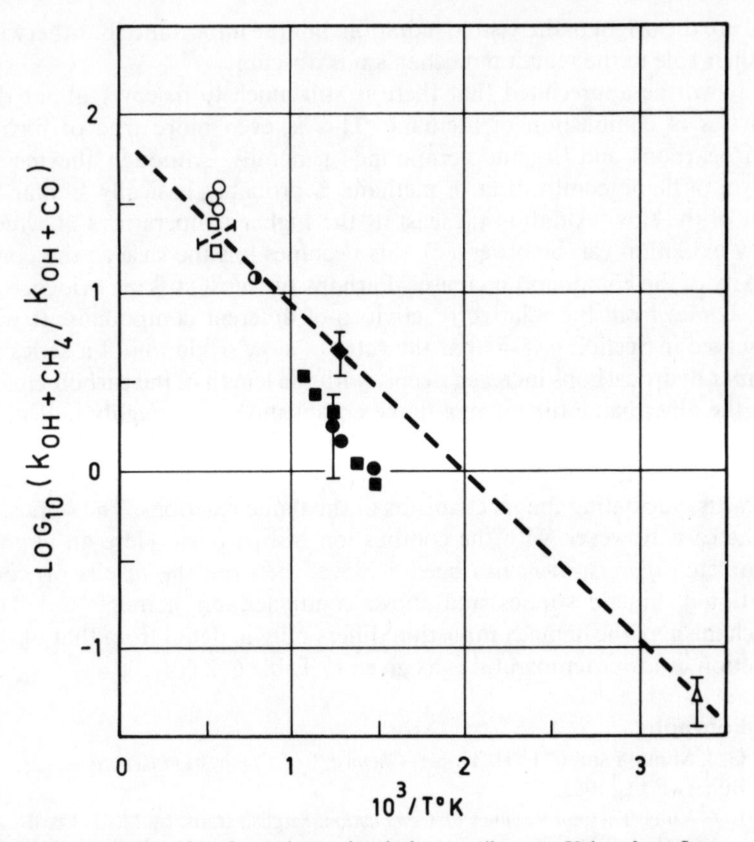

Fig. 7–8 Arrhenius plot of experimental ratio k_{OH+CH_4}/k_{OH+CO}. Values from flame studies: ○, □, Westenberg and Fristrom,[23] |---|, Fenimore and Jones (1961), ◑, Dixon-Lewis and Williams (1967); from explosion limit: ◆, Baldwin and Walker;[27] from pyrolysis of $H_2O_2+CO+CH_4$: ■, Hoare (1962); from slow oxidation of methane: ●, Blundell *et al.*, (1965); from OH + CH_4 and OH + CO in flow tube: △, Wilson and Westenberg.[27] (after Wilson and Westenberg.[27])

The reaction path from the methyl radicals generated in reaction 7–XIX to the formation of carbon monoxide is still less certain. At present the balance of evidence favours a sequence similar to that by which the oxidation of methane is believed to occur at lower temperatures (p. 263):

$$CH_3+O_2 \rightarrow CH_2O+OH \qquad 7\text{–XXI}$$

$$CH_2O+OH \rightarrow H_2O+CHO \qquad 7\text{–XXII}$$

$$HCO+OH \rightarrow H_2O+CO \qquad 7\text{–XXIII}$$

or possibly

$$HCO+M \rightarrow CO+H+M \qquad 7\text{–XXIV}$$

On this view, the formaldehyde present in the flame in trace amount (Fig. 7–6) is an essential intermediate. The species CH and C_2 are also present in the primary reaction zone in both ground and excited states,[28] and are the origin of the visible radiation, but the importance or otherwise of their role in the reaction mechanism is obscure.

It will be appreciated that there is still much to discover about the kinetics of combustion of methane. This is even more true of higher hydrocarbons and organic compounds generally. Although the mechanism of flame combustion of methane is probably basically similar to that of the slow oxidation (at least at the higher temperatures at which slow oxidation can be observed), this becomes less the case as the complexity of the compound increases. Perhaps the most striking evidence of this comes from the relative reactivities of different compounds. It was observed in Section 6–5–1 that the rates of slow oxidation of a series of similar hydrocarbons increase steeply with the length of the carbon chain. On the other hand, the rates of flame combustion—as roughly indicated by the relative burning velocities—are little different. Studies of the slow oxidation of hydrocarbons, therefore, do not contribute a great deal towards elucidating the mechanisms of the flame reactions. The opposite is the case, however, with the combustion of hydrogen. Here an almost complete *rapprochement* has been achieved between the results of 'conventional' kinetic studies and those conducted on flames.[19,22] The mechanism of the flame combustion differs only in detail from that of the reaction at lower temperatures as given in Table 6–2.

Bibliography

(A) G. J. Minkoff and C. F. H. Tipper, *Chemistry of Combustion Reactions*, Butterworths, 1962.

(B) L. A. Vulis, *Thermal Regimes of Combustion*, English trans. by M. D. Friedman and G. C. Williams, McGraw-Hill, 1961.

(C) A. G. Gaydon and H. G. Wolfhard, *Flames Their Structure, Radiation and Temperature*, 3rd edn, revised, 1970.

(D) R. M. Fristom and A. A. Westenberg, *Flame Structure*, McGraw-Hill, 1965.

(E) C. P. Fenimore, *Chemistry in Premixed Flames*, Pergamon Press, 1964.

(F) D. R. Jenkins and T. M. Sugden, Radicals and molecules in flame gases, Chap. 5 of *Flame Emission and Atomic Absorption Spectrometry*, Vol. I, J. A. Dean and T. C. Rains (eds), Marcel Dekker, 1969.

References

1 A. G. Gaydon and H. G. Wolfhard, *Flames Their Structure, Radiation and Temperature*, p. 197, 3rd edn, revised, 1970.

2 E. M. Bulewicz, C. G. James, and T. M. Sugden, *Proc. Roy. Soc.*, **A235,** 89, 1956.

3 D. H. Cotton and D. R. Jenkins, *Trans. Faraday Soc.*, **64,** 2988, 1968.

4 C. J. Halstead and D. R. Jenkins, *Combustion and Flame*, **11,** 362, 1967.

5 E. M. Bulewicz and T. M. Sugden, *Trans. Faraday Soc.*, **52,** 1475, 1956.

6 L. F. Phillips and T. M. Sugden, *Trans. Faraday Soc.*, **57,** 914, 1961.

7 S. N. Foner and R. L. Hudson, *J. Chem. Phys.*, **21,** 1374, 1953; K. H. Homann, M. Mochizuki, and H. Gg. Wagner, *Z. phys. Chem.*, N.F., **37,** 299, 1963.

8 (a) V. P. Balakhnin, Yu. M. Gershenson, and A. B. Nalbandyan, *Doklady Phys. Chem.*, **172,** 27, 1967; (b) G. A. Sachyan, Yu. Gershenzon, and A. B. Nalbandyan, *Dokl. Akad. Nauk SSSR*, **175,** 1328, 1967.

9 A. A. Westenberg and R. M. Fristrom, *10th Symposium (Internat.) on Combustion*, p. 473, The Combustion Institute, 1965.
10 W. E. Kaskan, *Combustion and Flame*, **2,** 229, 286, 1958.
11 R. M. Fristrom and A. A. Westenberg, *Flame Structure*, McGraw-Hill, 1965.
12 J. H. Gibson, W. E. L. Grossman, and W. D. Cooke, *Anal. Chem.*, **35,** 266, 1963.
13 E. M. Bulewicz and T. M. Sugden, *Trans. Faraday Soc.*, **54,** 1855, 1958.
14 M. A. A. Clyne and B. A. Thrush, *Proc. Roy. Soc.*, **A275,** 544, 1963.
15 C. J. Halstead and D. R. Jenkins, *12th Symposium (Internat.) on Combustion*, p. 978, The Combustion Institute, 1969.
16 M. J. McEwan and L. F. Phillips, *Combustion and Flame*, **11,** 63, 1967.
17 P. J. Th. Zeegers and C. Th. J. Alkemade, *Combustion and Flame*, **9,** 247, 1965.
18 G. Schott, *J. Chem. Phys.*, **32,** 710, 1960.
19 C. P. Fenimore, *Chemistry in Premixed Flames*, Pergamon Press, 1964.
20 C. P. Fenimore and G. W. Jones, *J. Phys. Chem.*, **62,** 178, 1958.
21 C. P. Fenimore and G. W. Jones, *8th Symposium (Internat.) on Combustion*, p. 127, Williams and Wilkins, 1962.
22 G. Dixon-Lewis, *Proc. Roy. Soc.*, **A317,** 235, 1970.
23 A. A. Westenberg and R. M. Fristrom, *J. Phys. Chem.*, **64,** 1393, 1960; **65,** 591, 1961.
24 R. M. Fristrom, C. Grunfelder, and S. Favin, *J. Phys. Chem.*, **64,** 1386, 1960; **65,** 587, 1961.
25 J. M. Brown and B. A. Thrush, *Trans. Faraday Soc.*, **63,** 630, 1967.
26 D. G. Horne and R. G. W. Norrish, *Nature*, **215,** 1373, 1967.
27 W. E. Wilson and A. A. Westenberg, *11th Symposium (Internat.) on Combustion*, p. 1143, The Combustion Institute, 1967.
28 R. P. Porter, A. H. Clark, W. E. Kaskan, and W. E. Browne, *11th Symposium (Internat.) on Combustion*, p. 907, The Combustion Institute, 1967.

Problems

7–1 (a) Show that the fraction of the total heat available which is actually generated by an exothermic first-order reaction taking place in a perfectly stirred adiabatic flow reactor of volume V is given by

$$h_G = \exp(-1/\theta)/[\exp(-1/\theta)+1/\tau]$$

where $\theta = (R/E)T$ and $\tau = AV/v$ (i.e. dimensionless temperature and residence time respectively). T is the reaction temperature, E and A are the Arrhenius parameters, and v cm^3 s^{-1} is the exit velocity of gas from the reactor. If T_a is the temperature attained by the products when the reaction proceeds to completion, the fraction of the available heat passing out of the reactor is

$$h_L = (\theta-\theta_0)/(\theta_a-\theta_0)$$

Show that, under steady conditions,

$$\tau = [(\theta-\theta_0)/(\theta_a-\theta)]\exp(1/\theta)$$

and use an analogous procedure to that given in Section 6–1 to derive the following expression for the temperatures at ignition and extinction:

$$\theta_{cr} = [1+2u\theta_0 \mp (1-4\theta_0-4u\theta_0^2)^{1/2}]/2(1+u)$$

where $u = 1/(\theta_a-\theta_0)$.

(b) Hydrazine hydrate vapour at 423 K is fed to a 100 cm^3 reactor of the above type where it undergoes vigorous exothermic decomposition. The input velocity is increased until the reaction is extinguished. Calculate the reaction temperature, and hence the per cent reaction, and the exit velocity (v_e) at the point of extinction. ($A = 10^{9.3}\,s^{-1}$; $E = 36.2$ kcal (151 kJ) mole^{-1}; T_a (the 'adiabatic flame temperature') = 1500 K.) (v_e is in fact the maximum value of v at which combustion can be sustained; the corresponding rate of input of the reactant (s) is known as the 'blow-out' velocity; see Longwell and Weiss, *Ind. Eng. Chem.*, **47**, 1634, 1955.)

7–2 The 'water-gas shift' reaction

$$CO + H_2O \rightarrow CO_2 + H_2 \qquad \text{(I)}$$

which comes to equilibrium in the 'post-flame' gas of hydrocarbon flames, taken place *via* the reactions

$$OH + CO \rightleftharpoons CO_2 + H \qquad \text{(II, −II)}$$

With the object of determining k_{-2}, the approach of reaction II to equilibrium from the reverse direction was studied by adding a small concentration of CO_2 to the post-flame gas of a H_2-rich $H_2/O_2/N_2$ flame, the amount added ($[CO_2]_0$) being too small to affect other equilibria in the flame. (Dixon-Lewis *et al.*, *Trans. Faraday Soc.*, **61**, 255, 1965.) Show that k_{-2} may be derived from measurements of [H], $[CO_2]$, etc. at various points in the flame by means of the relation:

$$-\frac{d[CO_2]}{dt} = k_{-2}[H][CO_2]\left\{1 - K_1\left(\frac{[CO_2]_0}{[CO_2]} - 1\right)\frac{[H_2O]}{[H_2]}\right\}$$

7–3 Emission of traces of nitric oxide formed by oxidation of atmospheric nitrogen in combustion devices is an important factor in the occurrence of photochemical smog. The oxidation is much slower than the primary combustion of the fuel and, when *excess air* is present, it occurs in the post-flame gas by the 'Zeldovich' chain cycle

$$O + N_2 \rightarrow NO + N \qquad \text{(I)}$$

$$N + O_2 \rightarrow NO + O \qquad \text{(II)}$$

reaction I being rate-determining. Assuming the reverse reactions can be neglected ([NO] and [N] very small), show that the rate of formation of NO at any point in the gas under these conditions can be derived approximately from the concentrations of stable species by means of the expression:

$$d[NO]/dt = 2k_1 K_W K_O [H_2]([N_2][O_2]/[H_2O]) \qquad (1)$$

where K_W and K_O refer to the reactions $H_2 + \frac{1}{2}O_2 \rightleftharpoons H_2O$ and $\frac{1}{2}O_2 \rightleftharpoons O$ respectively. (*Note.* When O_2 is in excess, $[H_2]$ is small and varies with the reaction time.) Does the expression account for the fact that $d[NO]/dt$ is highly temperature sensitive? (Consult a table of thermodynamic data; cf. also Table 4–1.)

7–4 The reactions in and beyond the primary reaction zone of a flat flame of an acetylene–oxygen mixture ($3C_2H_2 + 97O_2$) burning at 0.1 atm pressure were investigated by Fristrom *et al.* (*7th Comb. Symposium*, 1959, p. 304). Some of their measurements made on the axis of the flame at various positions z cm from the burner are tabulated below. T refers to the temperature, u to the gas velocity and X_{CO} to the mole fraction of CO in the reacting gas at z.

z cm	0.20	0.24	0.28	0.32	0.36	0.40	0.60	0.80
T K	729	913	1047	1134	1182	1213	1288	1341
u cm s^{-1}	152	188	213	228	235	238	239	236
$10^2 X_{CO}$	1.0	1.8	2.3	2.54	2.4	2.1	1.2	0.7
D_{CO,O_2} cm^2 s^{-1}	10	15	18	21	22	23	28	29

From these measurements and the values of D_{CO,O_2}, the diffusion coefficient for CO—O_2 at 0.1 atm at T K, determine graphically, as precisely as the data will allow, relative values of the net rate of appearance (or disappearance) of CO by chemical reaction as a function of z. At what positions are the net rates of appearance and disappearance greatest and where are they exactly balanced? (If possible, use a desk calculator for this problem.)

Physical constants and conversion factors

Avogadro constant	N	6.023×10^{23} mole^{-1}
Boltzmann constant	$\bar{k}$	1.381×10^{-23} J K^{-1}
Gas constant	R	1.987 cal K^{-1} mole^{-1}
		8.314 J K^{-1} mole^{-1}
		82.05 cm^3 atm K^{-1} mole^{-1}
Planck constant	h	6.626×10^{-34} J s
Speed of light	c	2.998×10^8 m s^{-1}
1 Ångstrom	1 Å	$= 10^{-10}$ m
1 atmosphere (pressure)	1 atm	$= 1.013 \times 10^5$ N m^{-2}
1 kilocalorie (thermochem.)	1 kcal	= 4.184 kJ
1 kilojoule	1 kJ	= 0.2390 kcal
1 torr (= 1 mm Hg)	1 torr	$= 1.333 \times 10^2$ N m^{-2}
1 electron volt	1 eV	= 23.06 kcal mole^{-1}
		= 96.49 kJ mole^{-1}
1 wavenumber	1 cm^{-1}	$= 2.859 \times 10^{-3}$ kcal mole^{-1}
		$= 1.196 \times 10^{-2}$ kJ mole^{-1}
1 torr ideal gas at 298.15 K		$= 5.379 \times 10^{-8}$ mole cm^{-3}
	2.3026 R	$= 4.576 \times 10^{-3}$ kcal K^{-1} mole^{-1}
		$= 1.9145 \times 10^{-2}$ kJ K^{-1} mole^{-1}
	$h\nu/\bar{k}$	$= 1.439\omega$ K (ω in cm^{-1})
	π	= 3.1416
	e	= 2.7183

Answers to problems

Chapter 1

1–1 approx. 546, 1093, 2185, 47 700 K; 800 K.

1–2 27.8 kcal (116 kJ) mole^{-1}; 2.0×10^{-12}; (a) 2.3×10^{-3}, (b) 1.3×10^{-3}; P would be 1/15 the correct value.

1–3 (b) $p_{0.5} \simeq 5 \times 10^{10}$ torr; see p. 18.

1–4 10; 0.56.

1–5 1.5×10^{13} year; 2nd estimate is smaller by about 25%.

1–6 (a) $$kr = \frac{h^2 N}{(8\pi^3 kT)^{1/2}} \left(\frac{1}{m^*_{\mathrm{H_2Br}}}\right)^{3/2} \times \frac{I^\ddagger_{\mathrm{HHBr}}}{I_{\mathrm{H_2}}} \frac{(1-\exp[-h\upsilon_{\mathrm{H_2}}/kT])\exp[-E_0/RT]}{(1-\exp[-h\upsilon^\ddagger_{\mathrm{s}}/kT])(1-\exp[-h\upsilon^\ddagger_{\mathrm{b}}/kT])^2} \mathrm{cm^3\ mole^{-1}\ s^{-1}}$$

where $m^*_{\mathrm{H_2Br}} = m_{\mathrm{H_2}} m_{\mathrm{Br}}/(m_{\mathrm{H_2}} + m_{\mathrm{Br}})$ and $\upsilon^\ddagger_{\mathrm{b}}$ and $\upsilon^\ddagger_{\mathrm{s}}$ are respectively the bending and sym. stretching frequencies of the activated complex.

(b) $+5/2$, $-1/2$.

1–7 1.6, 4.0; 1.1, 2.0; 1.1, 1.3.

1–8 $E_{\mathrm{a}}^{1000} - E_{\mathrm{a}}^{200} = 2.1$ kcal (8.9 kJ) mole^{-1} [According to Clyne and Stedman (*Trans. Faraday Soc.*, **62,** 2164, 1966) the experimental value is 2.6 kcal (11.1 kJ) mole^{-1}, but this is not certain (Westenberg, *loc. cit.*]; $\Delta E_z = -0.27$ kcal (-1.14 kJ) mole^{-1}.

1–9 $+11.3$, -11.3, $+4.0$, -3.3, $+5.4$, -10.2 cal mole^{-1} K^{-1} ($+47$, -47, $+17$, -14, $+23$, -43 J mole^{-1} K^{-1}); 13.6, 14.0.

1–11 $\Delta S^\ddagger = -17.55R$; $\log(A\ \mathrm{cm^3\ mole^{-1}\ s^{-1}}) = 10.4_3$ (assuming $g_\ddagger = g_{\mathrm{OH}}$). Cf. Dryer *et al.*, *Combustion & Flame*, **17,** 270, 1971.

Chapter 2

2–1 $t_{0.5} = \ln 2/k[B]_0$.

2–2 (a) 1.5; (b) 11.5 cm^3 mole^{-1} s^{-1}; (c) 1.54×10^{-4} s^{-1}; $\delta = 0.04$ (k would then be 1.50×10^{-4} s^{-1}).

2–3 Equation 2–20 reproduces approximately the observed effect (see also ref. 30); about -1%; $< |-1|\%$; about -5%.

2–4 Values of $(1-f)$ are similar at very small but diverge at large values of kt_{c}; $(t_{\mathrm{c}})_{0.5} = 0.69/k$, $1/k$.

2–5 (b) 3.39×10^{-4}; 2.05×10^{-3} s^{-1}.

2–6 880 K; 0.14 (0.137) s^{-1}; $\log(A\ \mathrm{s^{-1}}) = 13.1$, $E = 56.1$ kcal (235 kJ) mole^{-1}.

2–7 See Armitage and Cullis (*Combustion & Flame*, **16,** 125, 1971) for an application of the static method.

Chapter 3

3–1 See (a) Maccoll *et al.*, *J. Chem. Soc.*, **1955,** 973; **1957,** 5033; (b) ref. 10; (c) Poppleton and Mulcahy, *Aust. J. Chem.*, **19,** 65, 1966.

3–2 $\phi = 520$.

3–3 See Laidler and McKenney, Chap. 4 of *The Chemistry of the Ether Linkage*, S. Patai (ed.), Interscience, 1967 (especially p. 170).

3–4 See (a) James and Steacie, *Proc. Roy. Soc.*, **A244**, 289, 1958; (b) Loucks and Laidler, *Can. J. Chem.*, **45,** 2795, 1967.

3–5 $k = 1.6 \times 10^7 \text{ cm}^3 \text{ mole}^{-1} \text{s}^{-1}$.

3–6 (a) See Foon and Tait, *loc. cit.*, (b) 87.4, 18.1 kcal mole^{-1}.

Chapter 4

4–2 (a) $n = 2$; $k = 3.1 \times 10^{11} \text{ cm}^3 \text{ mole}^{-1} \text{s}^{-1}$. (b) $10^{-15}k = 1.3$, 4.3, and 700 cm^6 mole^{-2} s^{-1} for M = He, Kr, and I_2 respectively.

4–3 (a) 14%; 0.028 cm^3 min^{-1}. (b) $\gamma = 5.0 \times 10^{-5}$; 22 s.

4–4 (a) $\log_{10}(A_a \text{ cm}^3 \text{ mole}^{-1} \text{s}^{-1}) = 9.1$; $E_a = 0.62$ kcal (2.6 kJ) mole^{-1}. $\log_{10}(A_b \text{ cm}^6 \text{ mole}^{-2} \text{s}^{-1}) = 14.4$; $E_b = -0.96$ (4.0). (b) cf. ref. 76.

4–5 (b) Equation 1 becomes equation 4–6; hence there is no extra error introduced. [A useful treatment is given by Morgan and Schiff (*J. Chem. Phys.*, **38,** 2631, 1963) who, however, neglect the effect of the probe at $z = 0$.

4–6 $A \approx 2 \times 10^{13} \text{ cm}^3 \text{ mole s}^{-1}$; $E \approx 0$.

Chapter 5

5–1 (a) The higher I and lower ω_b of the LEP complex contribute about equally to making C_{LEP} 1.35 times greater than C_{LEPS}. (b) $\Gamma^*_{LEPS}/\Gamma^*_{LEP} = 1.6$.

5–2 $E_c = 7.7$ kcal mole^{-1}; $r^{\ddagger}_{H-H} = 1.11$ Å; $r^{\ddagger}_{H-Cl} = 1.34$ Å.

5–3 (b) See Timmons *et al.*, *J. Am. Chem. Soc.*, **90,** 5996, 1968.

5–4 $k_{HF}/k_{DF} = 1.2$.

5–5 (a) 50×10^{-16} cm^2. (b) Boltzmann equilibrium; $S_r(\varepsilon)$ independent of vibrational and rotational states.

5–6 $C = \pi\sigma^2_{AB} = S_r$ ('hard sphere'); it would contain the additional factor $(1+\varepsilon_0/kT)$. (For an experimental determination of $S_r(\varepsilon)$ see the paper cited in problem 5–5(a).)

5–7 About 50%. (The more accurate figure obtained by Morokuma *et al.* (*loc. cit.*) is about 25%.)

5–8 Not at all; $k/k_\infty = 0.91, 0.5, 0.09$; $9 \times 10^7 \text{ s}^{-1}$; 10^5.

5–9 $P = [1-(m/n)]^{s-1}$.

5–11 $s \simeq 14$ (The experimental data are from Casas *et al.*, *J. Chem. Soc.*, **1964,** p. 3655).

5–12 (a) See Golden *et al.*, *J. Phys. Chem.*, **75,** 1333, 1971; (b) see Grotewald *et al.*, *J. Chem. Soc.*, **A1968,** p. 375; **1970,** p. 1001. The azomethane ratio behaves in a similar (but less marked) way, because the higher vibration frequencies of $C_2H_6^*$ are less excited than those of the deuterated ethanes: see also Kobrinsky and Pritchard, *J. Phys. Chem.*, **76,** 2196, 1972; (c) see Kerr *et al.*, *Chem. Comm.*, **1967,** p. 365.

Chapter 6

6–1 605 K; 625 K.

6–3 (a) Yes, provided due regard is paid to the influence of the inert gas on D_0; (b) $\log(A_2 \text{ cm}^3 \text{ mole}^{-1} \text{s}^{-1}) = 14.2$, $E_2 = 15.9$ kcal (66.5 kJ) mole^{-1}; (c) ΔH_2 (kinetic) = 16 kcal (67 kJ), ΔH_2 (thermochem.) = 16.6 kcal (69 kJ).

6–4 (a) $a_{O_2} = 0.38$; (b) $a_{N_2} = 0.45$ (obtained from a graph of $(P_{H_2}+0.38\,p_{O_2})$ *vs* p_{N_2}).

6–5 (a) $E_\phi \approx 37$ kcal (155 kJ) mole^{-1}; (b) to within a few per cent.

6–6 $\phi = A\mathbf{R}^2P^2 - B/\mathbf{R}$ approximately, where A and B are constants for the particular vessel, mixture composition, etc. Reaction rate = 0 when $\mathbf{R}^3P^2 \leqslant B/A$ (apart from the rate due to primary initiation).

Chapter 7

7–1 (b) $T_e = 1380$ K; 89%; $v_e = 5 \times 10^4 \text{ cm}^3 \text{ s}^{-1}$.

7–3 Since $E_{-1} \approx 0$, $E_{overall} \approx 75-58+59 = 76$ kcal (320 kJ) mole^{-1}. For an experimental test of equation 1 see Thompson *et al.*, *Comb. and Flame*, **19,** 69, 1972.

7–4 $(\mathscr{R}_{CO})_{max}$ at $z = 0.28$ cm; $(-\mathscr{R}_{CO})_{max}$ at $z = 0.40$ cm (approx.); $\mathscr{R}_{CO} = 0$ at $z = 0.36$ cm.

Index

Chemical species are indexed by their chemical formulae. Page numbers in *italic* refer to Problems.